Augustus Jay DuBois

Elements of graphical statics and their application to framed structures

Second Edition

Augustus Jay DuBois

Elements of graphical statics and their application to framed structures
Second Edition

ISBN/EAN: 9783337278137

Printed in Europe, USA, Canada, Australia, Japan

Cover: Foto ©berggeist007 / pixelio.de

More available books at **www.hansebooks.com**

THE ELEMENTS

OF

GRAPHICAL STATICS

AND THEIR APPLICATION TO

FRAMED STRUCTURES,

WITH NUMEROUS PRACTICAL EXAMPLES OF

CRANES—BRIDGE, ROOF AND SUSPENSION TRUSSES—BRACED
AND STONE ARCHES—PIVOT AND DRAW SPANS—
CONTINUOUS GIRDERS, &c.,

TOGETHER WITH THE BEST

METHODS OF CALCULATION,

AND CONTAINING ALSO

NEW AND PRACTICAL FORMULÆ

FOR THE

Pivot or Draw Span—Braced Arch—Continuous Girder, Etc.

BY

A. JAY DU BOIS, C.E., Ph.D.

PROFESSOR OF CIVIL AND MECHANICAL ENGINEERING, LEHIGH UNIVERSITY, PENNA.

WITH AN ATLAS OF THIRTY-TWO PLATES.

Second Edition, Revised and Corrected.

NEW YORK:
JOHN WILEY & SONS, PUBLISHERS,
15 ASTOR PLACE.
1877.

PREFACE TO FIRST EDITION.

It is now ten years since the appearance of the Graphical Statics of Culmann,* during which time the method has been greatly extended in its applications, and has met with such acceptance that there is now scarcely a *Polytechnikum* in Germany where it is not a prominent feature in the regular course of instruction.

This rapid spread of a new discipline is the more remarkable when we consider the obstacles which it encountered. Culmann, with a boldness which we might almost term rash, based his development upon the modern geometry of von Staudt, and assumed in his readers a familiarity with this very terse presentation of a subject then, as indeed now, but little known, and which, therefore, but few possessed. To practical engineers, therefore, to whom his methods specially recommended themselves, his presentation of those methods was almost unintelligible.

At a time when the students of the Zürich Polytechnic were already overburdened, the new discipline was introduced; while, owing to want of familiarity with the fundamental principles premised, they were unable to understand his lectures or read his work. Yet such was the intrinsic value of the new method that, notwithstanding these obstacles, even in spite of them, it made rapid headway; found friends everywhere; crept into other departments of the Polytechnic; and finally the aim of Culmann was completely attained when the *modern geometry* was itself introduced, and a special lecturer in that branch appointed. Thus, as a direct result of the Graphical Statics of Culmann, appeared the first and, till now, only complete text-book upon the modern geometry, viz., Reye's "*Geometrie der Lage*," Hannover, 1868. Since then, hand in hand and with remarkable rapidity, these two studies have made their way,

---

* *Die Graphische Statik.* Culmann. Zürich, 1866. Second Edition, 1st vol., 1875.

until, as already remarked, they now form a notable feature in the course of every technical institution in the land.

The acceptance which the method has found in France, and the attention which it has there excited, is sufficiently indicated by the work of *Lévy* (*La Statique Graphique et ses Applications*, Paris, 1874), which contains a very clear and elegant presentation of the principles, though the applications are of the simplest character, while, as was perhaps not unnatural in the author, the German origin of the system is very imperfectly indicated, and the special methods of Culmann but little more than hinted at.

In Italy also the method has found an ardent expounder in the distinguished mathematician *Cremona* (*Le figure reciproche nelle statica grafica*, Milan, 1872), and to his efforts and labors its introduction and acceptance is due.

In England, Prof. Clerk Maxwell, in the *Trans. of the Royal Society of Edinburgh*, 1869-70, has contributed a paper upon "Reciprocal Figures, Frames and Diagrams of Forces," and, among others, Jenkin, Ranken, Bow, and Unwin have contributed to the popularity and spread of "Maxwell's Method." Maxwell and his followers give, however, only the very simplest applications, based upon the resolution and composition of forces, such as will be found in our first chapter. The entire system developed by Culmann, the properties of the "equilibrium polygon," upon which the fruitfulness and value of the graphical statics wholly depend, are unnoticed both by our English and French authors.

The author feels, therefore, that no apologies are needed for the present work. Whatever its shortcomings and defects, he claims at least the honor of making the first attempt to introduce among American Colleges and American Engineers a knowledge of a subject of approved interest and practical value to both, whether regarded as a geometrical discipline or as a most efficient aid in investigations of stability. Nor is he without hope that the next ten years may find the method as universally accepted at home as now abroad.

The same difficulties certainly have not here to be encountered. The subject as here presented requires only a knowledge of the elements of geometry as universally taught, and can thus be readily introduced into our schools as well as read by those practical engineers for whose benefit the method seems so especially designed. A subject of such importance, which has already endured successfully so severe a test, and made headway against such obstacles, we cannot certainly afford any longer to ignore, and it is hoped that the present work may serve to excite a more general interest in the method.

For the practical engineer, the importance of graphical methods needs, indeed, to-day no demonstration. Such methods are everywhere in use. But a simple and general *system* which shall include all special solutions—the development of the few principles upon which all such solutions are based, and from which they all flow—is at least in this country unknown. Even in English literature there is to be found little more than the very elementary deductions of our first chapter, so that it may justly be said that the entire method owes its existence and development to the labors of German scholars and the enlightened appreciation of German engineers. How thorough have been these labors, how widespread this appreciation, and how various are the applications of the method itself, the reader may gather from the Introduction to this work, and from the appended *list of literature* upon the subject. A glance at this list will also show that the selection of what was of most value, and the omission of those applications of minor importance, necessary to bring the present work within reasonable limits, and at the same time preserve the logical unity and completeness of the whole, was not the least difficult portion of our task. It would, indeed, have been easy to have given the work twice its present dimensions, though without a corresponding increase in value sufficient to justify the additional cost. As it is, no application of real and practical value to the engineer strictly deducible from the graphical statics has been overlooked, and discrimination has been chiefly exercised in those departments where graphical and analytical processes are still of necessity combined. Here we have selected only those cases where such union shows itself most advantageous, and the graphical constructions most simplify, illustrate, or interpret the purely analytical process, and where such cases, moreover, presented a *useful*, practical, and not merely theoretical value.

As to the plan of the work, a word of explanation is necessary. We have endeavored to keep always in view the requirements of both students and practitioners, of technical schools and practical engineers, and thus to combine a text-book for school instruction, and a book of reference and manual for practice as well. The attempt is a difficult, if not a dangerous one, and one which, in other departments, has met with more failure than success. If we venture to indulge a hope that in this case at least partial success has been attained, and that the attempt to occupy the two stools at once has not been disastrous, our belief is due to the nature of the subject itself, and not to any overweening estimate of our own abilities to succeed where so many have failed. The subject seems, indeed, especially suited to such a method of treatment. In fact, no other would appear at this period to properly meet the necessities of the case. Its geometrical principles are simple, its applications eminently practical. To present the principles alone would be to deprive the study of its chief interest and attraction. To rest content with a few practical applications would be to sacrifice, in a great measure, system and clearness of presentation. In the accomplishment of our double task we are fortunate to have had at our disposal such works as those of *Bauschinger* in the one, and *Culmann* in the other direction. Our obligations to both authors are great, and are fully indicated in the text. The same acknowledgment is due, in greater or less degree, to *Mohr* and *Winkler*, *Ritter* and *Reuleaux*. In every case where such assistance has been received, due acknowledgment has been made.

For the historical and critical Introduction, we are indebted, with few alterations, to the pen of *Weyrauch*.* It will, we are sure, prove of value to the student, and serve to awaken an interest in those highly important developments which geometry has within the last decade undergone.

Thus collecting in a connected form the scattered results and researches of various authors, it has been a pleasurable duty to recognize the labors of those men who have chiefly contributed to this new branch of geometrical statics, and to whom our own obligations are so great. While thus crediting fully that which others have done, we have felt the more justified in calling attention to any deviations of our own. We have especially sought to extend the application of the *method by resolution of forces* (known best, perhaps, as *Maxwell's Method*)—a method

---

* *Ueber die graphische Statik—zur Orientirung.* Von Dr. phil. Jacob J. Weyrauch, Privat docent an der polytechnischen schule zu Stuttgart. Leipzig, 1874.

which bids fair to obtain widespread recognition, in directions in which it has hitherto been supposed of little service. This often, indeed, by the aid of analytical results, we have been enabled to do, and not, as we conceive, without a degree of success. The formulæ used are always simple and of ready application, and this union of analytical results and graphical processes the practical engineer will, we think, find of value. Thus, in the *braced arch* (Chap. XIV.) and *continuous girder* (Chap. XII.) new constructions will be found, and both these important and difficult cases may thus be solved with an ease, completeness and accuracy far superior to that of the pure graphical method itself. Those acquainted with the analytical investigation of the "*braced arch,*" as contained in *Capt. Eads' Report to the Ill. and St. Louis Bridge Co.*, May, 1868 (App.), will not, we feel sure, be slow to recognize the advantages of the present method. The subject in its present state is thus fairly brought within the reach of the *practical* Engineer and Constructor.

To simple girders, contrary to usually received opinions, by the means of *apex loads*, the above method applies directly, and without the aid of analytical results—a fact which has been too generally passed over without sufficient notice by writers upon the subject.

We have devoted considerable space to the subject of the *continuous girder*, but not, we feel sure, more than its importance demands. The subject deserves more attention at the hands of the practical engineer and constructor than it has hitherto received. That the present indifference upon the subject is due chiefly to lack of information can hardly be doubted, when the opinion is current, and is even endorsed by those who are considered as authorities, that the complete solution of the problem is "probably impossible by reason of its complexity," and "too complex for mathematical investigation." * Opinions like these are best met by the complete solutions of particular examples, and in Chapter XII. will be found the complete calculation and tabulation of the strains in every piece due to every apex load, for the central span of seven continuous successive spans, and, as far as any inherent difficulties are concerned, we might as well have taken 50 or 100 spans.

---

* *Graphical Method for the Analysis of Bridge Trusses.* Greene.

When engineers shall have become convinced of the fact that there is in the continuous girder a theoretical saving of material amounting usually to from 25 to 30 per cent. per truss, and in the extreme case even reaching as high as 50 per cent., as compared with the simple girder; and that the main objection which can be urged—viz., the influence of small variations in level of the supports—has, when properly considered, no force whatever, we shall probably hear less often of designs contemplating many successive and independent spans of considerable length—such as, for instance, for a bridge over the Hudson at Poughkeepsie, consisting of five separate spans of 525 ft. each.

The present work contains the only complete graphical and analytical presentation of this subject in English professional literature, and should it succeed in directing more general attention to what has been done, will not have been in vain. Almost the whole of Chap. XIII. is entirely new and constitutes an important advance in the treatment of the subject. Those acquainted with the old method will, we think, be pleased with the simplicity and comprehensiveness of the new formulæ.

By their aid, indeed, we can solve with ease problems which could hardly be attempted otherwise. In this connection the *list of literature* upon the continuous girder appended to Chap. XIII. may also be of service.

We notice with pleasure in this direction the admirable little treatise of *Clemens Herschel, C.E.*, upon draw spans.* This subject is at least of *admitted* practical value, and we have treated it with a fullness which, in our opinion, leaves little to be desired. We have borrowed from the above work the conception of the "*Tipper*," or draw with secondary span, which is both new and, as it would seem, most adequately represents the true state of the case, and alluded to the idea, also original with Mr. Herschel, of *weighing off* the reactions at the supports of a continuous girder, instead of measuring the differences of level. In this case, as in that of the continuous girder generally, we have clearly brought out the *method of calculation by apex weights*, and here, indeed, lies the whole secret of thorough practical solution. In fact, from this point of view, the complete solution of a continuous girder for any number of spans,

---

* *Continuous, Revolving Drawbridges.* Little, Brown and Company, Boston, 1875.

equal or unequal, offers no more essential difficulty than the calculation of so many separate simple girders. That this is not exaggeration, but accurate statement of fact, a perusal of Chaps. XII. and XIII. will suffice to prove.

We cannot leave this part of the subject without acknowledging our indebtedness to Mansfield Merriman, C.E., Assistant in Engineering in the Sheffield Scientific School of Yale College, for the formulæ of the latter chapter. *Mr. Merriman* has done for the practical solution of the continuous girder what *Weyrauch* has for its theoretical discussion. We refer the student to the Supplement to Chap. XIII. for a specimen of his method of discussion. His formulæ are simple, entirely free, even in general form, from integrals, and are given in just the shape required in practice. This compactness renders it possible for the engineer to enter upon a couple of pages of his note-book *all* the formulæ required for the thorough calculation of a continuous girder of any number of spans, equal or unequal; and this calculation in any particular case proceeds in a manner precisely similar to that of the simple girder, directly and without reference to authorities, tables, points of inflection, elastic line, methods of loading, or any of the " other paraphernalia with which the subject is usually encumbered."

It will be observed that here and throughout we have nowhere left out of sight analytical processes or methods. The reader who considers the present work as an attempt to *supersede*, or even subordinate analytical investigation, misjudges entirely our aim. So far from this, we indulge the hope that its perusal cannot fail to render familiar the use of *both* methods, to bring out their points of difference and relative advantages, to illustrate the one by the other, to enable the reader to check the results of the one by the other, and in any case apply one or both, or a judicious combination of both, as may in such case be most advantageous or desirable. This will be especially noticed in the discussion of the simple and continuous girder and of the braced arch. (Chaps. XII., XIII., XIV. and XVI., and Appendix.)

As to the use of the work, the practical engineer will find in Chap. I., and that portion of the Appendix relating to this chapter, an easy and simple method of solution applicable to any framed structure having simple reactions, and including thus all varieties of bridge and roof trusses of single span. In the Appendix he will find detailed examples calculated to illustrate every practical point of importance, and also a full exposition of *Ritter's* "*method of moments.*" The principles of this chapter alone will enable him to solve readily, both by calculation and diagram, every case usually arising in practice. In problems involving the moment of inertia of areas in the case of the continuous girder, the braced arch and stone arch, as also the suspension system, he will find Chaps. VI., XII., XIII., XIV. and XVI. of value, and in the perusal of any or all of these he will, it is hoped, find no trouble by reason of logical connection with preceding principles. They are in this respect, as far as possible, complete in themselves. We may also call his attention to Chap. XV., upon the stone arch, though it is to be regretted that the practical importance of the subject in the present age of iron renders the ease with which it is graphically treated of less importance than formerly. For his benefit also frequent practical examples are given in detail, so that in all important applications he can easily select a *parallel case*, and follow it out, step by step, in the case in hand, without studying up the whole process of development in order to place himself in a condition to make use of the methods employed. In regard to the subject of the endurance of iron, the student is referred to the author's translation of the work, by Prof. Jacob J. Weyrauch already alluded to, of the Polytechnic School in Stuttgart, issued by the publishers of "Graphical Statics." Weyrauch's book treats of the strength, dimensions, and calculations of iron and steel, and being thoroughly practical, is heartily commended to the student, and the profession of engineers generally.

For the student much of the practical applications may well be at first omitted. Notably Chaps. VII.–XII., inclusive. Chaps. I.–IV. and XIII.–XVI. will put him in complete possession of the method, and, moreover, enable him to solve with ease *any* structure, including the continuous girder, braced arch, suspension system, and stone arch, as well as all the more ordinary forms of bridge and roof trusses, cranes, etc. Indeed, if the first-named structures, which are of comparatively rare occurrence, are at first omitted, Chaps. I.–IV. alone will constitute a complete course upon framed structures so far as usually taught in our schools at the present day. Afterwards, in practice, and in the solution of the particular problems treated of, he will, in common with the practical engineer, find in the other portions of the work and in the Appendix just such assistance as he needs. We would also call the attention of the mathematician more especially to the investigation in Chap. V., Arts. 47–51, of the effects of a given *recurring system of moving loads*, the analytical treatment of which would be almost impracticable by reason of the complexity of the formulæ obtained, and in this respect certainly worthless, even if possible, but the geometrical treatment of which gives rise to some of the most elegant constructions of the graphical statics; also to the Supplements to Chaps. XIII. and XIV., in which the analytical treatment of the continuous girder and braced arch is given.

Finally, if our purpose in writing these pages is accomplished, the principles and methods here set forth will be found easily acquired, accurate in their results, and amply sufficient for the ready determination of the strains in the various pieces of any framed structure which the civil engineer can legitimately be called upon to design.

With this much of introduction and explanation, we present our work to the engineering profession in America and to American technical colleges, in the hope that the spirit which has led to its production, if not the method of its execution, may win for it a favorable reception.

In this spirit and in this hope we may, we trust, be allowed to appropriate the closing lines of *Culmann's* preface—"*Und nun fahre hin—gern hätte ich dich zum Fundament einer auf wissenschaftlicherer Basis gegründeten Ingenieurkunde gemacht, allein kaum darf ich die Hoffnung hegen, so viel Kraft in mir zu finden, um das Ganze dieses umfangreichen Faches umzuarbeiten: dast ist ein Werk, das mir vor Augen schwebt, wie einer jener alten mittelalterlichen Dome sich vor dem Künstler erhob, der ihn entwarf und der der Hoffnung sich nicht hingeben konnte, ihn je in seiner Vollendung zu schauen.*

"*Doch es mögen dich Andere benutzen und weiter bauen,*" und was ich nicht kann, werden meine Nachgänger vollbringen.

NEW HAVEN, *April 17th,* 1875.

PREFACE TO SECOND EDITION.

THE present Edition of this work has undergone careful revision and correction, and no pains have been spared to render it free from typographical errors, and worthy of the very cordial reception it has thus far met with. For this reception the Author would express his gratitude both to the Profession and to the faculties of the several colleges who have already adopted it as a Text Book.

To these and others who may examine it with a view to introduction, a word or two may be allowed us.

The work was not intended to be read through *in sequence* by any class, but must be used with reference to the degree of preparation of the students, to the time at disposal, and to the relative importance and relation of the various subjects treated. The logical order of presentation requires a certain order in the development of the subject as a whole, but it by no means follows that this order should be preserved by the student, or that he should be acquainted with the whole. In fact, some of the subjects treated of are best taken up by the student at a later period, when better prepared for their comprehension; others are best omitted at least in the first reading, and others again may even be omitted entirely as of minor importance, and the student left to pursue them for himself, as taste or the exigencies of practice may demand. In this latter respect, as well as in the completeness with which the several topics are discussed, the work is intended to serve as a book of reference. As a Text Book, it is designed to teach those students possessing the knowledge of mathematics and mechanics usual to the senior classes in our technical schools and colleges, *how to find the conditions of stability in every kind of structure of common occurrence;* and this not alone by graphical construction, but also *by calculation* as well. Structures of less common occurrence may or may not be then taken up, according to the ability of the class and the time at disposal. With the above end in view, the teacher will find it not merely desirable but even essential to depart from the plan of the work as laid down, and to supplement it largely by various examples illustrative of the general principles and bringing out as clearly and repeatedly as may be necessary the various points noticed in the text.

In the work *four* separate and distinct methods of solution are given for such structures as bridge and roof trusses, and the student should become familiar with all. Thus there are two methods by diagram, viz., by resolution of forces (*Maxwell*) and by the equilibrium polygon (*Culmann*); and two corresponding methods by calculation, viz., by composition and resolution of forces, and by moments (*Ritter*). To give these in such manner and with such emphasis that the student shall be conversant with all, and able to use them with discrimination where in any case they best apply, we recommend the following order of perusal:

FIRST. Chap. I., which gives the first method by diagram (*Maxwell's*), and such of the Appendix as relates to this chapter, is read. The class may then go into the drawing-room, and under the supervision of the teacher actually solve a variety of roof trusses from simple to more complex, both for dead load *and wind force* in each case. In each and every case also the results should be checked by calculation by the method of moments (*Ritter's*), at first thoroughly and in detail, and afterwards only a few test pieces to check the accuracy of the diagram. From Roof Trusses we then pass on to Bridges, and here also a series of selected examples of every class used in practice may be solved, and the method of *tabulation of apex weights* referred to in Art. 12 and Appendix to Chap. I., brought out repeatedly until the student has thoroughly mastered it, and appreciates fully the fact that for each form of truss the strains due to only two weights are really necessary to be found, and that the others may then readily be found directly from these. Here also each example should be checked by calculation by the method of moments. In the case of curved flanges the various lever arms may first be measured directly to scale from the frame, and then trigonometrically computed. At this point the student is then already in possession of *two* independent methods of solution for any kind of framed roof or bridge which occurs in general practice.

There are in fact only two framed structures, the continuous girder and the braced arch remaining, which he is not able to solve. Should it be deemed undesirable to consider these, he can at once pass to Chap. II. and then to the stone arch, Chap. XV. If, however, a knowledge of the above is desired, he is now ready to extend his principles and methods to them also. Thus he has already recognized that *provided only all the outer forces are known*, he can both diagram and calculate *any* framed structure. In such structures as he has hitherto had, these outer forces are either given or are easily found. In the cases now considered they are not all given, and must, therefore, first be found. Once known, however, his way is clear. Recognizing clearly, then, what is aimed at, the supplements to Chaps. VII. and XIII. are first taken, and he is now able to find for the continuous girder the outer forces required. Chap. XIII. will then give exercises in finding these forces, and handling the formula he has just deduced. Finally, Chap. XII. resumes again in the light of his present knowledge the same old two methods with which he is already so familiar, of diagram and calculation, and a few examples actually worked out by both methods complete his mastery of the continuous girder and draw span. He can now pass on to the braced arch, and in Chap. XIV. will find all that he needs. Here he must take at first the formulæ and constructions for finding the outer forces on trust. Afterwards, if deemed desirable, he can follow out the development of these formulæ as given in Supplement to Chap. XIV.

In the case of the parabolic arch, at least, the constructions are so simple that it is well adapted to class instruction. The draw span is of such importance as to render some attention to it, at any rate, desirable in any full course. Thus the student is now able to solve *any case whatever* of framed structure in two ways, by diagram and calculation. The same simple principles have been applied throughout, and formulæ have been called in only in a subsidiary way to determine certain forces which are necessary to be first known before these principles can be applied.

Thus Chap. I., Appendix to Chap. I., Supplements to Chaps. VII. and XIII., then Chaps. XIII., XII., with Appendix, and XIV. with Appendix, form by themselves and in this order a complete and systematic course.

If, however, it is deemed undesirable to consider the continuous girder and braced arch, the course indicated in the preceding paragraph may be omitted, and then Chap. I., with Appendix, constitutes a course as thorough as can be desired, the student taking first those examples given in the book, and then such others as the teacher may select, and always solving in two ways, by diagram and calculation. He may then take Chaps. II.-V., and now possesses a *second* method of diagram (Culmann's) by which he may check any or all of the previous cases. In general, a few examples of application to bridge girders, drawing the parabolas for total load (moments) and moving load (shear), will be sufficient. He is now ready to pass on to the Stone Arch, where again suitable problems should be proposed by the teacher and solved under his supervision. The remainder of the work, including moment of inertia and continuous girder treated by the second method of diagram, will in general be found unnecessary and rather advanced, except in a very full course.

The course recommended is then as follows, in the order given, chapters in brackets being omitted or not, at option of teacher:

Chap. I. and Appendix, Supplements to Chaps. VII. (and XIII., Chaps. XIII., XII. and Appendix, XIV. and Appendix), Chaps. II.-V., XV. (XVI., VI., XI.). It will be seen that the order recommended for class instruction is quite different from that of the work itself.

We would ask teachers examining the book with a view to adoption, to look it over *in the order above given.*

With regard to the space occupied in these pages by the continuous girder, and to the opinions held by the writer on that subject, a word of explanation may also not be out of place, in view of the discussion which has arisen since the publication of the first edition of this work.

It is, of course, understood that we claim, in any case, superiority for the continuous girder over the simple, only where circumstances render such a construction advisable, and base that claim upon admitted theoretical results as yet not satisfactorily tested by practice. We do not claim that one should supersede the other. Such a claim represents an extreme view, in our opinion absurd, and which certainly is not advocated in these pages, though there have not been wanting those who would attribute it to us. Circumstances connected with method of erection, etc., may certainly be imagined in which a continuous girder presents advantages not possessed by the simple. It finds, at least, direct application in the *pivot span*. When such circumstances arise, the engineer, we take it, will not find fault with a full and thorough discussion of the subject, such as here given, and he may even find the method of solution which we recommend of service. The continuous girder has then its proper place, if we can only ascertain it, just like the bow-string girder, arch, suspension system, etc., which also may, under proper circumstances, be preferred to the simple girder, without necessarily detracting from the just merits of the latter. The others have been tested by practice, and their relative advantages pretty clearly settled. When the same shall have been done for the continuous girder, the demonstrations and method of solution here given can also be estimated at their true value. Should practice show the system worthless and theory fallacious, we hold ourselves ready at any moment to strike out the portions affected. Meanwhile, the method of solution here given was at the time of publication entirely new; the formulæ were an advance upon any up' to that time presented; the subject, heretofore difficult, was rendered easy of solution, and brought out in a very striking manner the powers of the graphical method, the value of which is of course unaffected by the fate of the theory of continuity. Until then, practice decides very clearly against the system; it has very properly a place in a work professedly treating of framed structures.

As to the merits of the case, the theoretical saving is certainly large enough to justify careful consideration and trial, and our theoretical results have never been called in question. How much, in any case, of this theoretical gain can be realized practically, it remains for intelligent practice to decide—the full solution is beyond the reach of analysis. To assume, as some do without experiment, that the objections to the system and to the theory, however valid in themselves, do actually cover or more than cover the large gain indicated in some cases by theory, is practically to beg the whole question.

To assume, as some have, that because this large theoretical saving is not found in certain special cases of small span, where indeed it was not to be expected, that therefore it is not to be obtained in other cases, is to evade the question. For almost all ordinary spans, apart from the method of erection, the simple girder need fear no rivalry. But for long spans of large number it is not yet clear how the very large theoretical saving is counterbalanced by the objections hitherto urged, such as varying modulus of elasticity, inapplicability of theory to framed structures, effects of temperature, etc.; nor is it quite clear exactly how the *effect* of such objections is estimated. That such objections have weight is not to be denied. That their weight is sufficient to condemn the system is the very point at issue. This point we decline to decide *either way*, without better reasons than have thus far been advanced. Thus, not pretending to decide it ourselves, we cannot but regard those who do, as hasty. The question now is one for the workshop. So far as theory can go we give it here, in a more perfected and practical shape than ever before given. The mathematicians have done their part, and presented their results. It remains now for the practical engineer and constructor to do their part. Only by concert of action can the truth be attained. "If brought to the test of practice, the theory is found at fault and its results delusive—well. If not, well also. In either case engineering art and science are advanced." Until, however, the question is thus settled, we must regard it as an open one, and trust it will "periodically turn up for settlement" until thus finally disposed of. Unless the theory of flexure, so long accepted as practically valuable and received by practical engineers, can be replaced by a better, we must especially deprecate all attempts to arbitrarily settle the question by fruitless discussion. We cannot thus summarily pronounce upon a theoretical question without theory, and upon a practical question without practice; nor is this the spirit in which progress in knowledge or practice can be made.

It may be well to remember, in this connection, that other nations have brought or are bringing the question to a practical test. England, France, Germany, Austria, Prussia, Spain, Italy—have tried or are trying the case, and have not yet decided it. The skill of American engineers, as exemplified in our present practice, needs no encomium. It is world-wide. They might, indeed, be reasonably expected to succeed even where others fail, and if there be any advantage in the system, they, if any, might be expected to find it.

The reasons why they are not included in the above list may well be worth consideration.

Meanwhile, we cannot think that our discussion of the case at this time impairs the value of the work as a whole, or needs any apology. We therefore again present it to the Profession, in substantially its original form, and hope it will be found worthy of a continuance of the favor which has already been accorded to it.

BETHLEHEM, February, 1877.

GENERAL CONTENTS.

	PAGE
INTRODUCTION	xxxi

PART I.—GENERAL PRINCIPLES.

CHAPTER I.
FORCES IN SAME PLANE—COMMON POINT OF APPLICATION. ILLUSTRATED BY PRACTICAL EXAMPLES 1

CHAPTER II.
FORCES IN SAME PLANE—DIFFERENT POINTS OF APPLICATION. PROPERTIES OF FORCE AND EQUILIBRIUM POLYGON 16

CHAPTER III.
CENTRE OF GRAVITY 29

CHAPTER IV.
MOMENT OF ROTATION OF FORCES IN SAME PLANE IN GENERAL 33

CHAPTER V.
MOMENT OF ROTATION OF PARALLEL FORCES—PRACTICAL APPLICATIONS 36

CHAPTER VI.
MOMENT OF INERTIA OF PARALLEL FORCES—PRACTICAL PROBLEMS.. 61

PART II.—APPLICATION TO BRIDGES.

A. THE SIMPLE GIRDER.

CHAPTER VII.
THE SIMPLE GIRDER OR TRUSS SUPPORTED ONLY AT ENDS 84

SUPPLEMENT TO CHAPTER VII.
THE THEORY OF FLEXURE.
CHAP. I. Methods of calculation 98
CHAP. II. Principles of the calculus 102
CHAP. III. Theory of flexure 110

B. THE CONTINUOUS GIRDER OF CONSTANT CROSS-SECTION.

CHAPTER VIII.
GENERAL PRINCIPLES.. 125

CHAPTER IX.
LOADED AND UNLOADED SPANS.. 141

CHAPTER X.
SPECIAL CASES OF LOADING.. 147

CHAPTER XI.
METHODS OF LOADING CAUSING MAXIMUM STRAINS................. 155

CHAPTER XII.
COMBINATION OF GRAPHICAL AND ANALYTICAL METHODS........... 168

CHAPTER XIII.
ANALYTICAL FORMULÆ... 198

SUPPLEMENT TO CHAPTER XIII.
DEMONSTRATION OF ANALYTICAL FORMULÆ........................ 239

PART III.—APPLICATION TO THE ARCH.

CHAPTER XIV.
THE BRACED ARCH. COMBINATION OF GRAPHICAL AND ANALYTICAL METHODS... 251

SUPPLEMENT TO CHAPTER XIV.
DEMONSTRATION OF ANALYTICAL FORMULÆ........................ 271

    CHAP. I. General considerations and formulæ.................. 271
    CHAP. II. Hinged arch in general................................ 277
    CHAP. III. Arch hinged at abutments only....................... 283
    CHAP. IV. Arch fixed at ends—no hinges......................... 287
    CHAP. V. Influence of temperature............................... 299
    CHAP. VI. Partial uniform loading............................... 304

CHAPTER XV.
THE STONE ARCH... 311

CHAPTER XVI.
THE INVERTED ARCH—SUSPENSION SYSTEM......................... 327

APPENDIX.................... 339

TABLE OF CONTENTS.

INTRODUCTION.

ART.		PAGE
I.	Upon mathematical investigations generally.................	xxxii
II.	Analytical and geometrical mechanics.....................	xxxiii
III.	Geometrical statics.......................................	xxxv
IV.	The graphical calculus...................................	xxxvii
V.	Graphical representation.................................	xl
VI.	Graphical statics...	xli
VII.	The methods and limits of the graphical statics...........	xliii
VIII.	The modern geometry......................................	xlv
IX.	The modern geometry in engineering practice..............	xlviii
X.	Practical significance of the graphical statics...........	xlix
XI.	Literature upon the graphical statics....................	li
XII.	Graphical dynamics.......................................	liv

PART I.—GENERAL PRINCIPLES.

CHAPTER I.

FORCES IN SAME PLANE—COMMON POINT OF APPLICATION.

1.	Notation—Representation of forces by lines	1
2.	Resultant of two forces.......................................	2
3.	Resultant of any number of forces.............................	2
4.	Conditions of equilibrium.....................................	3
5.	Properties of *force polygon*.................................	3
6.	Order of forces in force polygon a matter of indifference	4
7.	Forces acting in same straight line...........................	5
8.	PRACTICAL APPLICATIONS..	5
9.	Braced semi-arch..	6
10.	Roof truss...	8
11.	Diagram for wind force.......................................	8
12.	Application of method to bridges.............................	11
13.	Braced arch..	12
14.	Ritter's "Method of Sections"................................	14

CHAPTER II.

FORCES IN SAME PLANE—DIFFERENT POINTS OF APPLICATION.

16.	Resultant of two forces.......................................	16
17.	Case of forces parallel.......................................	18

ART.		PAGE
18.	Perpendiculars let fall upon components from intersection of outer polygon sides, are inversely as the components	18
19.	Equilibrium polygon	19
20.	Case of a couple—Conditions of equilibrium	20
21.	Properties of a couple	21
22.	Force and equilibrium polygons for any number of forces	23
23.	Influence of a couple	24
24.	Order of forces in force polygon, a matter of indifference	25
25.	Pole upon the closing line—Failing case for equilibrium	25
26.	Relation between two equilibrium polygons with different poles	26
27.	Mean polygon of equilibrium	27
28.	Line of pressures in an arch	28

CHAPTER III.

CENTRE OF GRAVITY.

30.	General method for determination of centre of gravity	29
31.	Reduction of areas	30
32.	Reduction of triangle to equivalent rectangle of given base	31
33.	Reduction of trapezoid to equivalent rectangle	31
34.	Reduction of quadrilateral generally	31

CHAPTER IV.

MOMENT OF ROTATION OF FORCES IN THE SAME PLANE IN GENERAL.

35.	Definition of moment of rotation	33
36.	Culmann's principle	33
37.	Application to equilibrium polygon	34

CHAPTER V.

MOMENT OF RUPTURE—PARALLEL FORCES.

38.	Equilibrium polygon—Ordinates to give the moments of rupture	36
39.	Beam with two equal and opposite forces beyond the supports	38
40.	Beam with two equal and opposite forces between the supports	39
41.	Special cases of importance	41
	1st. Beam—load inclined to axis	41
	2d. Force parallel to axis	42
	3d. Forces in different planes	42
	4th. Combined twisting and bending moments	43
	5th. Application to crank and axle	44
42.	Loading continuous—Load area	45
43.	Beam uniformly loaded	46
44.	Moment curve a parabola	46
45.	Beam continuously loaded and also subjected to action of concentrated loads	48
46.	Influence of moving load	50
47.	Load systems	53

ART.		PAGE
48.	Properties of the parabola included by the closing line.............	54
49.	Application of the above principles	54
50.	Most unfavorable position of load system upon a span of given length...	56
51.	Greatest moment at a given cross-section......................	59

CHAPTER VI.
MOMENT OF INERTIA OF PARALLEL FORCES.

52.	Application of moment of inertia in proportioning any cross-section.	61
53.	Graphical determination of moment of inertia	62
54.	Signification of the area of the equilibrium polygon...............	63
55.	Radius of gyration..	63
56.	Inertia curves—Ellipse and hyperbola of inertia.................'...	66
57.	Construction of curve of inertia	69
58.	Example...	69
59.	Central curve—Central ellipse...................................	71
60.	Centre of action of the statical moments considered as forces.......	73
61.	Cases where the direction of the conjugate axes of the inertia curve can be at once determined...................................	75
62.	Practical applications...	76
	1st. The parallelogram.................................	76
	2d. The triangle.......................................	77
	3d. The trapezoid.....................................	78
	4th. The parabolic segment.......'.....................	80
63.	Compound or irregular areas.....................................	81
64.	Cases where there is no axis of symmetry	83

PART II.—APPLICATION TO BRIDGES.

A. THE SIMPLE GIRDER.

CHAPTER VII.
THE SIMPLE GIRDER.

66.	Forces which act upon a bridge	84
67.	Bridge loading..	85
68.	Shearing force—Moment of rupture, etc.........................	85
69.	Concentrated loads—Invariable in position.......................	87
70.	Concentrated loads—Variable in position........................	88
71.	Position of a given system of loads causing maximum shear........	88
72.	Construction of the maximum shear	89
73.	Maximum moments...	90
74.	Construction of maximum moments.............................	91
75.	Absolute maximum of moments	92
76.	Continuous loading..	93
77.	Total uniform load..	93
78.	Method of loading causing maximum shear.......................	94
79.	Live and dead loads..	95

SUPPLEMENT TO CHAPTER VII.

CHAPTER I.
METHODS OF CALCULATION.

ART.		PAGE
2.	Ritter's method	99
3.	Method by resolution of forces	100

CHAPTER II.
PRINCIPLES OF THE CALCULUS.

4.	Differentiation and integration	102
5.	Powers of a single variable	105
6.	Other principles	106
7.	Illustrations	107
8.	First differential coefficient	109

CHAPTER III.
THEORY OF FLEXURE.

9.	Coefficient of elasticity	110
10.	Moment of inertia	110
11.	Change of shape of axis	112
12.	Beam fixed at one end, load at other	113
13.	Beam as above—Uniform load	117
14.	Beam supported at both ends—Concentrated load	118
15.	Beam as above—Uniform load	119
16.	Beam fixed at one end, supported at other—Concentrated load	120
17.	Beam as above—Uniform load	121
18.	Beam fixed at both ends—Concentrated load	122
19.	Inflection points, etc.	124

B. THE CONTINUOUS GIRDER OF CONSTANT CROSS-SECTION.

CHAPTER VIII.
GENERAL PRINCIPLES.

80.	Mohr's principle	125
81.	Determination of tangents to the elastic curve	127
82.	Effect of the moments at the supports	128
83.	Division of the moment area	129
84.	Properties of the equilibrium polygon	130
85.	Polygon for the positive moment areas	131
86.	Construction of the fixed points, and of the equilibrium polygon	132
87.	Construction of the moments at the supports	135
88.	The *second* equilibrium polygon	136
89.	Determination of moments at the supports	138
90.	Comparison with girder fixed horizontally at both ends	139

TABLE OF CONTENTS. XXV

CHAPTER IX.

LOADED AND UNLOADED SPANS.

ART.		PAGE
91.	Unloaded span	141
92.	Two successive unloaded spans	141
93.	The fixed points	142
94.	Shearing force—Reactions at the supports, and moments in the unloaded spans	143
95.	Loaded spans	143
96.	Two successive loaded spans	144
97.	Arbitrary loading	145

CHAPTER X.

SPECIAL CASES OF LOADING.

98.	Total uniform load	147
99.	Practical examples—Girder of four spans	148
100.	Partial uniform load	149
101.	Concentrated load	153

CHAPTER XI.

METHODS OF LOADING CAUSING MAXIMUM STRAINS.

102.	Maximum shearing force	155
103.	Maximum moments	157
104.	Determination of the maximum shearing forces	159
105.	Determination of the maximum moments	161
106.	Practical simplifications of the method	162
107.	Approximate practical constructions	163
	1st. Beam of two spans—Moments	163
	Beam of two spans—Shearing forces	164
	2d. Beam of three or more spans—Moments	164
	Beam of three or more spans—Shearing forces	165
108.	Method by resolution of forces—Draw spans	166

CHAPTER XII.

CONTINUOUS GIRDER CONTINUED—COMBINATION OF GRAPHICAL AND ANALYTICAL METHODS.

111.	Method of finding shearing forces, when inflection points are known	168
	1st. Loaded span	168
	2d. Unloaded span	169
112.	Determination of inflection points—Inflection verticals	170
113.	Beam fixed horizontally at ends	171
114.	Example	172
115.	Counterbracing	174

ART.		PAGE
116.	Beam fixed at one end, supported at the other	175
117.	Practical construction	178
118.	Beam continuous over three level supports	178
119.	Practical construction	179
120.	The pivot draw with secondary central span	181
121.	Supports not on a level—Reactions	182
122.	Beam over four level supports	184
123.	Practical construction	186
124.	Pivot span—Example	187
125.	Method of passing in construction, from one span to another	190
126.	Method of procedure for any number of spans	192
127.	Example—Central span of seven spans	194
128.	Method of calculation by moments	198

CHAPTER XIII.

ANALYTICAL FORMULÆ FOR THE SOLUTION OF CONTINUOUS GIRDERS.

129.	Introduction	202
130.	Notation	204
131.	Theorem of three moments	205
132.	Example—Total uniform load—Moments	206
133.	Triangle of moments	207
134.	Total uniform load—All spans equal—Reactions	209
135.	Triangle for reactions	209
136.	Clapeyron's numbers	210
137.	Uniform live load over any single span—Moments at supports	211
138.	Triangle for moments—Loaded span	212
139.	Unloaded spans—Moments	213
140.	Practical rule and table for finding the above	214
141.	Reactions at supports—Loaded span	216
142.	Triangle for reactions	217
143.	Reactions for unloaded spans—Tables for	218
144.	Concentrated load in any span—Moments	220
145.	Application of above formulæ	220
146.	Triangle and table for moments	222
147.	Reactions at supports—Concentrated load	223
148.	Shear at supports	223
149.	Recapitulation of above formulæ	225
150.	Continuous girder with variable end spans	228
151.	Application of formulæ for	229
152.	Continuous girder with fastened ends	230
153.	Beam of single span—Fastened at both ends—Fastened at one end, etc.—Examples	230
154.	Tables for moments—End spans variable	233
155.	Continuous girder—All spans different—General formulæ—Examples	237
156.	General method of calculation	241

SUPPLEMENT TO CHAPTER XIII.
DEMONSTRATION OF ANALYTICAL FORMULÆ FOR THE CONTINUOUS GIRDER.

ART.   PAGE
1. Conditions of equilibrium.................................. 243
2. Equation of elastic line.................................... 244
3. Theorem of three moments................................. 245
4. Determination of moments—Supports all on level............. 246
5. Uniform load... 248
6. Formulæ for the "Tipper" [Art. 120]....................... 248
7. Example of two span tipper................................. 249
8. LITERATURE UPON THE CONTINUOUS GIRDER................ 251

PART III.—APPLICATION OF THE GRAPHICAL METHOD TO THE ARCH.

CHAPTER XIV.
THE BRACED ARCH.

157. Different kinds of braced arch............................ 255
158. Arch hinged at both crown and abutments.................. 255
159. Hinged at abutments only—Continuous at crown............ 257
    1st. Parabolic arch...................................... 258
    2d. Circular arch, Tables for solution of................. 259
160. Arch fixed at abutments—Continuous at crown.............. 261
    1st. Parabolic arch...................................... 262
    2d. Circular arch... 264
162. General method of solution................................ 266
163. Analytical formulæ for horizontal thrust, and vertical reactions.... 268
164. Arches with solid web..................................... 270
165. Strains due to temperature, Formulæ for................... 271
166. Effects of temperature.................................... 273

SUPPLEMENT TO CHAPTER XIV.
DEMONSTRATION OF ANALYTICAL FORMULÆ FOR THE BRACED ARCH.

CHAPTER I.
GENERAL CONSIDERATIONS AND FORMULÆ.

1. Fundamental equations.................................... 275
2. Displacement of any point................................. 279

CHAPTER II.
HINGED ARCH IN GENERAL.

3. Notation—The outer forces in general...................... 281
4. Intersection line... 282
5. Parabolic arch—Concentrated load......................... 283
6. Circular arch—Concentrated load.......................... 284
7. Integrals used in above discussion......................... 286

CHAPTER III.

ARCH HINGED AT ABUTMENTS ONLY.

A. Parabolic arc.

ART.	PAGE
8. Horizontal thrust	287
9. Intersection curve	287

B. Circular arc.

10. Horizontal thrust	287
11. Intersection curve	289

CHAPTER IV.

ARCH FIXED AT ENDS.

12. Introduction	291
13. Concentrated load—General formulæ	291

A. Parabolic arc.

14. Determination of H, V and M_o	293
15. Intersection curve	295
16. Direction curve	295

B. Circular arc.

17. Fundamental equations	295
18. Determination of H, V and M_o	297
19. Intersection curve	301
20. Direction segments	301
20 (b). Transformation series	301

CHAPTER V.

INFLUENCE OF TEMPERATURE.

21. General considerations	303
22. Influence of temperature on the arch	303
23. Fundamental equations—General	304
24. Arch with three hinges	305
25. Arch hinged at ends	305
26. Arch without hinges	305

CHAPTER VI.

PARTIAL UNIFORM LOADING.

27. Notation	308

A. Arch hinged at crown and ends.

28. Vertical reaction	308
29. Horizontal thrust	309

B. Arch hinged at ends only.

30. Vertical reaction	309
31. Horizontal thrust—Parabolic arch	309
32. Horizontal thrust—Circular arch	310

C. Arch without hinges—continuous at crown.

33. Parabolic arch. Formulæ for V, H and M	311
34. Circular arch. Formulæ for V, H and M	312

CHAPTER XV.

THE STONE ARCH.

ART.		PAGE
167.	Definitions, etc.	315
168.	Pressure line	315
169.	Sliding of the joints	315
170.	Forces acting upon a cross-section—Neutral axis	316
171.	*Kernel* of a cross-section	317
172.	Position of *kernel* for different cross-sections	318
173.	Proper position of resultant pressure	320
174.	Pressure line—True pressure line	321
175.	Support line	322
176.	Deviation of support from pressure line	323
177.	Dimensions of the arch	324
178.	Construction of the pressure line	326
179.	Practical example	328
180.	Proper depth of arch at crown	329
181.	Increase of depth due to change of form	330
182.	Example	330

CHAPTER XVI.

THE INVERTED ARCH—SUSPENSION SYSTEM.

183.	Methods of construction	331
184.	Rear chains, and anchorages	331
185.	Cable with auxiliary stiffening truss	333
186.	Method of loading causing maximum strains	334
187.	Example	336
188.	Analytical investigation of the forces acting upon the stiffening truss	337
189.	Concluding remarks	339

APPENDIX.

NOTE TO CHAP. VIII. OF THE INTRODUCTION—UPON THE MODERN GEOMETRY.

	NOTE TO CHAPTER I.	350
2.	Bent crane	350
3.	Character of strains in the pieces as indicated by the strain diagram	351
4.	Pieces in equilibrium—Points to be avoided in constructing the strain diagram	351
5.	Roof truss—Method of checking the accuracy of the diagram	352
6.	The French roof truss—Solution apparently indeterminate—Method of solution—Method of calculation by moments	353
7.	Application to bridges—Bowstring girder—Method of tabulation illustrated	354
8.	Strains in the flanges—Table	357
9.	Method of calculation by moments	358
10.	Girder with straight flanges—Howe or Murphy Whipple system of bracing	359

ART.	PAGE
11. Lenticular girder, or system of Pauli	358
12. Remarks upon above system—The most economical system for long spans	363

NOTE TO CHAPTER II.

14. Equilibrium polygon considered as a *frame*	366

NOTE TO CHAPTER V., ART. 51.

15. Beam subjected to given force system—Maximum moment at any cross-section	368

NOTE TO CHAPTER XII.

16. Pivot or draw span—Practical example—Solution by diagram—By method of moments—By resolution and composition of forces	369
17. Relative economy of the continuous girder	378
18. Continuous girder—supports out of level	381

NOTE TO CHAPTER XIV. THE BRACED ARCH.

19. Practical example of the braced arch	386
20. Arch hinged at both abutments and crown	387
21. Arch hinged at abutments only	390
23. Temperature strains for above	395
26. Arch continuous at crown—Fixed at ends	398
27. Temperature strains for above	402
28. Advantage of arch with fixed ends for long spans—Comparison with the St. Louis arch	403

INTRODUCTION.

HISTORICAL AND CRITICAL.*

THE subject of Graphical Statics has, since the appearance of Culmann's work (*Die graphische Statik*, Zürich, Meyer and Zeller, 1866), excited considerable attention, but an accurate and just estimate of its methods and practical value is still wanting. Thus there are some who oppose it; others willingly accept it as an efficient and valuable aid in practical investigations of stability; still others even profess to see in it a future rival of Analytical Statics. This last somewhat remarkable claim seems apparently justified by a passage in Culmann's preface, where it is asserted "that the Graphical Statics will and must extend, as graphical methods find ever wider acceptance—but in such case, however, its treatment will soon escape the hands of the practitioner, and it will then be built up by the geometer and mechanic to a symmetrical whole, which shall hold the same relation to the new geometry that analytical mechanics does to the higher analysis." These various and conflicting opinions find their supporters in technical schools and among engineers throughout Germany.

In the consideration of the subject, we shall endeavor especially to give an objective presentation, but shall also feel at liberty to present our own opinions as well, and generally to venture such reflections as seem suited to throw light upon the matter. For both reasons it will sometimes be necessary to make apparent deviations, in order to point out the various fields in which these new investigations take root, to define their limits, and to decide in what directions and to what extent impulse and sustenance for further development may exist. In such a manner only can we satisfactorily ascertain how far the graphical statics may safely count upon more than a passing recognition and brief existence.

We have therefore to ask of the reader who wishes to obtain a just and accurate estimate of this new and, as we venture to think, highly important subject, patience for the following general considerations.

---

* *Ueber die graphische Statik—zur Orientirung.* By J. I. Weyrauch. Leipzig, 1874.

I.

UPON MATHEMATICAL INVESTIGATIONS IN GENERAL.

Mathematical truths may be attained in two essentially different methods —by synthesis or by analysis, by composition or by resolution. In synthesis, we ascend from particular cases to general ones; in analysis, we descend from general cases to particulars. By synthesis we pass from the simplest or admitted truths, by combination and comparison, to more complicated phenomena. Analysis seeks to refer back such phenomena to their fundamental relations, or to deduce special properties from the general conditions.

The analysis of a phenomenon presupposes, then, an accurate comprehension of all its elements. So far as these last stand in relations of cause and effect to the whole and its parts, or so far as such relations exist between the parts themselves, they may be expressed by equations. Thus the operations which are necessary in analysis become independent of concrete phenomena, and are governed only by the laws of abstract quantities as included by *algebra* in the widest sense of the word. Algebra, then, is not analysis itself, but only its instrument, "*instrument précieux et nécessaire sans doute, parce qu'il assure et facilite notre marche, mais qui n'a par lui même aucune vertu propre ; qui ne dirige point l'esprit, mais que l'esprit doit diriger comme tout autre instrument*" (Poinsot, *Théorie nouvelle de la rotation*, près à l' Acad., 1834). Ordinarily the higher branches of algebra, with which numberless really analytical investigations are connected, are designated as analysis. More properly, all investigations which rest upon equations of condition may be termed analytical investigations.

Synthetic investigation rests mainly upon geometrical conceptions, and attains to the knowledge of phenomena through concrete conditions, which latter may be designated as *space* relations and processes. Hence the usual division into analytical and geometrical methods, even in applied mathematics. We have thus with equal appropriateness an analytical geometry as also a geometrical analysis. When pure geometry (in distinction from analytical) makes use of the symbols and operations of algebra, it is only to express with corresponding generality and more concisely than in words truths attained to by abstraction, and independent of the dimensions of the auxiliary figure; or so to formulate such truths that they may be applied in analytical investigation. Accordingly, such use of algebraic formulæ has as little effect upon the synthetic process as from the above it would seem essential to the *analytic* treatment. In either case, algebra is but the instrument, the *method* lies back of and directs it.

If analytical formulæ and operations are entirely excluded from the more complicated geometrical investigations, we are at once restricted to general laws of metrical relation. There remains only the faculty of abstraction and graphical construction. The power of abstraction alone suffices, indeed, to comprehend in full generality metrical relations in elementary geometry and its simplest applications, but fails when the relations sought must be attained step by step by the application of a number of principles, or in the auxiliary figure by a number of constructions. If, indeed, we take the relation sought directly from the auxiliary figure itself, and even if it were possible to take out the required distances with *absolute* accuracy, still this result obtained would stand to the general law desired only in the same relation that the result of a particular numerical computation does to the more general algebraic formula.

Investigations by the aid of graphical figures *can*, however, make known general relations of *form* and *position*, and have in this respect their special advantage. So far also as by them metrical relations are sought, then, by the exclusion of algebraic formulæ, only the *process* of deduction—the *routine* of construction—remains of general significance. Sciences, then, which proceed in this manner, furnish indeed, with respect to metrical relations, no general laws, but for the deduction of these relations *do* give general *methods*. In this category we may place *descriptive geometry* and the more recent *graphical statics*.

II.

ANALYTICAL AND GEOMETRICAL MECHANICS.

It is hardly necessary in these days to call attention to the advantages of a geometrical treatment of mechanical problems. This, however, was not always the case, and the most important developments of geometrical mechanics belong to the present century. It is to *Poinsot, Chasles, Möbius*, etc., that these developments are due.

By the Calculus of Newton and Leibnitz (1646-1714), and its subsequent development, analysis became such a powerful instrument that the activity of mathematicians was for a long time solely directed towards analytical investigations. The power of analysis was in mechanics carried to its highest point by Lagrange (1736-1813), in his *Méchanique analytique*. He undertook the problem of reducing mechanics to a series of analytical operations: "*On ne trouvera point de figures dans cet ouvrage. Les méthodes que j'y expose ne demandent ni constructions ni raisonnement géométrique ou mécanique, mais seulement des opérations algébriques assujéties à une marche régulière et uniforme*" (*Méchanique analytique*. Paris, 1788.) The principle of virtual velocities formed his point of departure. A number of text-books upon theoretical mechanics still follow the method of Lagrange.

The revival of pure geometrical investigations by *Monge* (1746-1818), the creator of descriptive geometry, and his followers, could not well have been without its influence upon mechanics. In the year 1804 appeared the *Eléments de Statique*, by *Poinsot*, in which, in contrast to Lagrange, we find: "*que tous les théorèmes de la Statique rationelle ne sont plus au fond que des théorèmes de Géométrie.*" This work was the beginning of a series of treatises in which the advantages of the synthetic development and geometrical treatment of mechanics were defended and, by most important results, strikingly demonstrated.

At this time the views as to the best method of treating mathematical problems were sharply opposed. *Carnot* (1753-1823), to whom, however, the modern geometry itself owes no slight impulse, gives the preference to analysis. For synthesis "*est restreinte par la nature de ces procédés ; elle ne peut jamais perdre de vue son objet, il faut que cet objet s'offre toujours à l'esprit, réel et net, ainsi que tous les rapprochements et combinaisons qu'on en fait*" (*Géométrie de position.* Paris, 1803.) That which here Carnot considers as a defect in the synthetic and geometrical method, Poinsot claims as its special advantage: "*On peut bien par ces calculs plus ou moins longs et compliqués parvenir à déterminer le lieu ou se trouvera le corps au bout d'un temps donné, mais on le perd entièrement de vue, tandis qu'on voudrait l'observer et le suivre, pour ainsi dire, des yeux dans tout le cours de sa rotation*" (*Théorie nouv. d. l. rot. d. corps*).

The example of Poinsot found numerous followers. In Germany, *Möbius* followed with his "*Lehrbuch der Statik.*" Mechanics as well as geometry thus received enrichment. Möbius gives the preference always to the synthetic method, and also endeavors to interpret geometrically, analytically deduced formulæ—" because in investigations concerning bodies in space the geometrical method is a treatment of the subject itself. and is therefore the most natural, while by the analytical method the subject is concealed and more or less lost sight of under extraneous signs " (*Lehrb. d. Statik.* Leipzig, 1837.)

Even in analytical operations, geometrical considerations came more and more in the foreground. On all sides the development of *Kinematics*, the theory of motion without reference to its cause, was prosecuted. But, neglecting the cause of motion, there remains only its path; that is, geometry proper (*Kinematical geometry, or the geometry of motion*). The investigations of *Chasles, Möbius, Rodrigues, Jouquière,* and others, may yet be still further pursued; and when by the aid of geometry a certain completeness has been given to the theory of the motion of invariable systems, the geometrical theory of regular variable systems (to which the flexible and elastic belong) will be possible. For the discussion of such branches of mathematics, the synthetic geometry is necessary; for their foundation lies in a theory of the relationship of systems.

The advantage of the synthetic method in mechanics is denied by no one. Wherever it is possible, we obtain more comprehensive conclusions as to the nature of the phenomena, while all the properties of the same follow directly from the simple and known truths premised. In analytical investigations it is necessary, even when definite equations are obtained, to deduce the actual laws singly and in a supplementary manner, although they are indeed all contained in the equations themselves.

It is not, however, always possible to preserve the synthetic process throughout. From the first truth the ways diverge in all directions, and a special ingenuity is often needed to reach the goal. Just here analysis comes to our aid with its rich treasures of developed methods, and here it is most certainly not for geometry to " undervalue the advantage afforded by a well-established routine, that in a certain degree may even outrun the thought itself " (*F. Klein: Vergleichende Betrachtungen über neuere geometrische Forschungen.* Erlangen, 1872, p. 41). Algebraic operations are thus, however, not the chief thing, but only the instrument—a most excellent instrument indeed, which can be almost universally applied, and which, by reason of its connection with an extensive and independent mechanism, often needs only to be set in action in order to work of itself.

Geometrical mechanics, moreover, can never entirely free itself from analytical formulæ and operations. For though it may be both interesting and useful to follow, with *Poinsot*, the body during its entire rotation, yet practically this is of minor interest, and the chief problem remains still, "*à déterminer le lieu ou se trouvera le corps au bout d'un temps donné.*"

In the present day all those familiar with both methods of treatment hold fast the good in each; they supplement each other. Often in the course of the same investigation we must interrupt the general analytical process with synthetic deductions, and inversely. Thus we may well close these considerations with the sentence with which *Schell* begins his " *Theorie der Bewegung und der Kräfte* "—both methods, the analytic and the synthetic, can only, when united, give to mechanics that sharpness and clearness which at the present day ought to characterize all the mathematical sciences.

III.

GEOMETRICAL STATICS.

Statics is a special case of dynamics, though earlier treated as independent of the latter. The principle of *d'Alembert* furnishes the means of passing from one to the other. In *technical* mechanics the distinction is still preserved, and indeed, in view of the distinct branches in which the applications on either side are found, not without propriety.

After the mechanics of the ancients, as comprised in the mathematical collections of *Pappus*, the first great step towards our present geometrical statics was made by *Simon Stevinus* (1548-1603), when he represented the intensity and direction of forces by straight lines. Stevinus himself gave a proof of the importance of his method, in the principle deduced from it, that three forces acting upon a point are in equilibrium when they are proportional and parallel to the three sides of a right-angled triangle.

A main discovery was the parallelogram of forces by *Newton* (1642-1727). The composition of two velocities in special cases was long familiar. *Galileo* made use of it for two velocities at right angles, and examples also occur in *Descartes, Roberval, Mersenne,* and *Wallis,* but the fundamental principle was first established when Newton replaced the theories of special by that of universal causation (*Philosophiæ naturalis principia mathematica.* London, 1687).

Varignon in his "*Projet d'une nouvelle mécanique,*" in the same year (1687), and independently of Newton, applied for the first time the general principle of the composition of motions. From this he passes, in the *Nouvelle mécanique ou statique, dont le projet fut donné en* 1687 (published after his death, Paris, 1725), by means of the axiom that "*les effets sont toujours proportionnels à leurs causes ou forces productrices*" to the composition of forces also.

The Statique of *Varignon* is purely geometrical. He postulates nothing beyond books 1–6 and 11 of Euclid, and even explains the significance of + and − signs. In this work, the first founded upon the parallelogram of motion and of forces, we find also the *force* and *equilibrium* polygons (*Funiculaire*, Section II.), to the application and development of which almost the whole of Graphical Statics is to be attributed. *Varignon* recognized the value of the equilibrium polygon, and gave it as the seventh of the simple machines.

After the great Interim of Geometry, *Monge* wrote a *Traité élémentaire de Statique* (Paris, 1786). The work claims to contain for the first time everything in statics which can be synthetically deduced. In a later edition we learn that synthetical statics must be taken up as preliminary to analytical, just as elementary geometry before analytical geometry. Thus the work of Monge contains the necessary preparation for *Poisson's "Traité de mécanique"* (Paris, 1811).

The greatest influence upon the development of geometrical statics was exercised by *Poinsot*. By the introduction of *force pairs*, he solved in the most elegant manner the fundamental problem of any number of forces acting upon a body (*Eléments de Statique*, Paris, 1804, and *Mémoire sur la composition des moments et des aires dans la mécanique*).

Chasles completed the solution by the proof that the contents of the tetrahedron, which is determined by the resultant forces, is constant, however the forces may be composed.

In the hands of *Möbius*, geometry and geometrical statics were most completely developed.

Of the greatest importance, for later applications, was the introduction of the rule of signs.

The germ of this had existed already in the preceding century.* *Möbius* recognized its significance, extended it to the expression of the contents of triangles, polygons, and three-sided pyramids, and applied it systematically (*Barycentrischer Calcul.* Leipzig, 1827).

A new impulse, extended field of action, and numerous additions were given to geometrical statics by the *Graphical Statics* of *Culmann*.

---

* Möbius alludes to this, and we find, for example, in Kästner (*Geometrische Abhandlungen*, I. Saml., 1790, p. 464), the equation $\overline{AB} + \overline{BA} = o$.

IV.

THE GRAPHICAL CALCULUS.

The most extended applications of statics are in the field of engineering. Here, not only general properties of form and position are required, but in a large number of cases numerical relations are also necessary. General results of the latter character can, as we have seen, only be embraced by algebraic formulæ (I.). The pure graphical theory of construction is therefore in this respect lacking in completeness, as it is unable to furnish *general* metrical relations.

The practical engineer has almost always, however, to do with *special* problems; dimensions and acting forces are numerically given. Geometry in such cases could give no general relations, because the results desired are the consequences of the special proportions of the figure. In any determinate case, however, we may obtain a result *holding good for that case*, and it only remains to show how generally to obtain such a result. The graphical calculus treats of such methods, and so, although not exclusively, does graphical statics. As soon now as practical use is made of the actual proportions of the figure, everything depends upon the exactness of the drawing. One condition for the application of the graphical method is, therefore, skill in geometrical drawing—a requisition, indeed, which the practical engineer can most readily meet.

The idea at bottom of the graphical calculus is simple. The modifications of numbers in numerical calculations correspond always to similar modifications of the quantities represented by these numbers. The measure of a quantity can be as well given by a line as a number, by putting in place of the numerical the linear unit. In order for a graphical calculus, then, the modifications of lines answering to corresponding numerical operations are necessary, and these are furnished by geometry. They consist of graphical constructions, and rest upon the known properties of geometrical figures. The scale furnishes the means of converting directly any numerical quantity into its corresponding linear representation, and inversely any graphically obtained result can be at once transformed into numbers.

The graphical determination of desired or computable numbers is naturally nothing new. From the "*Traité de Gnomonique*" of *de la Hire* (1682) to the "*Géométrie descriptive*" of *Monge* (1788), many examples are to be found. The graphical calculus, however, goes further than this. It aims to found a method, a *routine*, which shall not only apply to bodies in space, but which shall also, like the arithmetical or algebraic calculus, be independent of concrete relations and of general application. It seeks further to obtain its results (products and powers) in the shape of lines convertible by scale into numbers. (Hence the important part which area transformation plays in the graphical calculus.) Such was the problem which *Cousinery* proposed, and whose solution he attempted in his "*Calcul par le trait*" (*Ses Éléments et ses applications*. Paris, 1839).

Cousinery applied the graphical calculus to powers, roots, proportion and progression; to the measure of lines, surfaces, cubes, graphic interpolation, and the stability of retaining walls. The presentation is naturally by no means complete, and labors also under a prolixity and minuteness of detail to which the results obtained are by no means commensurate. It sounds somewhat comic when Cousinery, in his "*Calcul par le trait*," claims the then already-existing graphical solutions of *Poncelet* ("*Mémoire sur la stabilité des revêtements, in Mémorial de l'off du génie*") as an elegant example of the application of his graphical calculus.

While Cousinery thus sought to apply geometry in a direction where until then analysis had held sway, he acted in entire accordance with the spirit of his age, though without making use of those means for aid which lay at his disposal. "Without effect upon him," says Culmann, "were the researches of Steiner, already published in 1832, as well as those of his predecessor; and instead of simply premising the elementary principles of the modern geometry, he laboriously sought to deduce them independently by the aid of perspective." The works, at least, of the French predecessors of Steiner were, at any rate, well known to Cousinery. In his preface we read: "Peut-être même nos efforts eussent-ils été complètement infructueux, sans les ressources que nous ont procurées et les annales de M. Gergonne et les travaux de M. Brianchon, et ceux plus récents de M. Poncelet. Nous avons envers M. Chasles une obligation encore plus droite, car outre les précieux documents que renferme son '*Histoire des méthodes en géométrie*,' nous avons à lui faire agréer un témoignage particulier de reconnaissance pour la manière dont il a bien voulu mentionner nos premiers essais sur le système de projection polaire."

Why Cousinery made use of perspective and not of the modern geometry, is easily understood. The development of geometry at that time, as today, proceeded in various almost independent directions, and Cousinery himself had the pleasure of seeing his "*Géométrie perspective*" (Paris, 1828) designated by the reporters for the Academy, *Fresnel* and *Matthieu*, as new and ingenious, as well as favorably noticed by *Chasles*.[*] He sought, therefore, naturally to develop and render fruitful his own method, so much the more as the true significance and value of the various growing branches of geometry could not then, as now, be correctly estimated. Accordingly, the Ingénieur-en-chef, *B. E. Cousinery*, wrote avowedly for his colleagues, and did not feel justified in directly premising a knowledge of the newest investigations, more especially of his own.

We have noticed the above somewhat in detail, because it bears directly

---

[*] Its newness, at least, is not without doubt. According to *Fiedler*, the principles are completely given in *Lambert's* celebrated work, "*Die freie Perspective*" (Zürich, 1759). Poncelet also takes issue with the estimation of the "*Géométrie perspective*" by *Chasles* ("*Traité des propr. proj.*," II., éd. 1865, p. 412).

upon a point of our discussion; for the introduction of the modern geometry in the graphical method by *Culmann*, is still, thirty years after Cousinery, a chief hindrance to its rapid spread.*

After Cousinery, no one occupied himself with the graphical calculus till Culmann gave it a place in his *Graphische Statik*. The presentation is here far better, and especially shorter. The rule of signs, which was unknown to Cousinery, is at once brought out. Instead of such long and tedious applications as the graphical interpolation, a few examples from engineering practice are given, among which we may especially notice earth-work calculations. In the extensive earth works of roads, canals, and railways, the method shows not only most plainly the extent and best arrangement of transport, but also allows, with the aid of the planimetre, the cost of transport to be determined.

As to the rest, it would appear as if the graphical calculus should play an important part in engineering practice. This circumstance, as well as the interesting problems which present themselves in connection, has gained for the *Arithmography* many friends. Several publications have since sought to win for it a wider recognition without furnishing anything essentially new. [*II. Eggers:* "Grundzüge einer graphischen Arithmetic," Schaffhausen, 1865. *J. Schlesinger:* "Ueber Potenzcurven," Zeitschr. d. österr. Arch. u. Ing. Ver., 1866. *E. Jäger:* "Das graphischen Rechnen," Speier, 1867. *K. von Ott:* "Grundzüge des graphischen Rechnens und der graphischen Statik," Prag, 1871.]

Recently the method of the graphical calculus has been applied to Differentiation and Integration. A treatise by *Solin* shows the first exact, so far as possible in a construction, the last approximate only (" Ueber graph. Integr. ein Beitrag z. Arithmographie, Abhand. d. königl. böhm. Gesellsch. d. Wissenbach." VI. Folge, 5 Bd. Separate reprint by Rivnác, Prag, 1871). It is to be remarked, also that examples of double integration and differentiation were given by *Mohr* in 1868. The graphical construction of the elastic line, and the determination of the moments at the supports of a continuous girder, are essentially examples in point (*Mohr:* "Beitrag zur Theorie der Holz und Eisenconstructionen," Zeitschr. d. Hannöv. Ing. und Arch. Ver., 1869; or *W. Ritter:* "Die elastische Linie," Zürich, 1871.)

As to the importance of the graphical calculus as an independent study or discipline, it is, as we believe, often exaggerated. The theoretical value is but little, for graphical constructions, as given by the graphical calculus, offer in no respect anything new. That which pertains to practical applications may be easily based directly upon geometry, and is nowhere found as a consequence of the method itself. If it is considered advisable to call special attention to a few general points before making such applications, all that can be desired can be easily presented in ten or a dozen pages octavo.

---

* See Preface; also Chaps. VII. and VIII. of this Introduction.

V.

GRAPHICAL REPRESENTATION.

Graphical representation, in the widest sense of the word, includes every visible result of writing or drawing. The written sentence is the graphical representation of a thought—the drawn line the graphical indication of an idea. In such generality we naturally do not here regard graphical representation. In a narrower sense we understand the graphical representation of the diversity or dependence of numerical quantities. In this sense we cannot speak of the graphical representation of pure geometry. This last was introduced into analysis by *Vieta* (1540-1603). Here the figure merely aids the conception, while the equation embraces the characteristics of the phenomena (I.), and ensures the independent character of the drawn lines. Thus the clearness of geometry is combined with the fruitfulness of analysis.

If the graphical representation is constructed from a number of suitably chosen and calculated values, the intermediate values can be directly measured and, by means of the scale, reconverted into numbers. The graphical representation, then, replaces numerical tables. Illustrative examples often occur in practice. We instance, for example, the graphical representation of maximum moments and shearing forces in the continuous girder. If the several values are *calculated* from a formula, their graphical union gives a simultaneous view—a *picture*—of the law which the formula represents. If these values are merely known—*observed*, for example their graphical combination may enable us to deduce the law which connects them. Thus the graphical representation is of assistance in the deduction of empirical formulæ, and indirectly in the discovery of exact relations. Illustrations of such application occur frequently in applied mathematics, especially in astronomy and meteorology.

In this connection we may also remark that graphical representation plays also an important part in statistics. By its aid a comprehensive view is obtained of a series of separate results. Or it may be applied to still higher problems—for example, from comparison of simultaneous but different series of observations to determine an inner connection.

In engineering practice, graphical representations have in recent times notably multiplied. All graphical constructions, so far as they do not depend upon analytical formulæ, and therefore are not directly given by geometrical laws, are nothing more than consequences of graphical representation.

VI.

GRAPHICAL STATICS.

The few text-books upon graphical statics and the more numerous works upon its applications, afford us no definition, and can afford none, because neither the method nor scope of this new study are anywhere sufficiently indicated.

If, following *Culmann*, we speak of it in contradistinction to the applications of a *pure* graphical statics, we may define it somewhat as follows: *Graphical Statics comprises the theory of those geometrical constructions which occur in the graphical solution of statical engineering problems;* it treats further of the general relations deducible from such constructions. This limitation, so far as it does not follow from the preceding, we shall seek in the course of these remarks still further to establish.

Graphical representations of analytically obtained results have, as has been already noticed, long been used in engineering practice. They served also the purposes noticed in the preceding chapter. Often also certain values, whose analytical determination is somewhat complicated, have been sought by graphical constructions. Examples of this may be found in many text-books upon the theory of structures, and we notice only, as one of the most notable of recent date, the construction of lever arms and limits of loading in *A. Ritter's* "Theorie und Berechnung eiserner Dach und Brückenconstructionen" (Hannover, 1862). *Poncelet* applied analysis in general to practical investigations, but sought in several complicated cases to elucidate the deductions of formulæ by geometrical constructions, and to deduce graphical solutions from analytical relations. This procedure found considerable acceptance, and the investigations of *Poncelet* were afterwards resumed upon more general assumptions by *Saint Guilhem* (*Mémoire sur la poussée des terres avec ou sans surcharge, Ann. des ponts et chauss.*, 1858, sem. 1, p. 319).

The first, however, to give pure geometrical determinations of stability in structures was *Cousinery*. He gave a number of examples as applications of his graphical calculus, but his ideas appear to have found in France little acceptance. On the other hand, the graphical construction of the curve of pressure in the arch by *Mery* (*Mémoire sur l'équilibre des voûtes en berceau—ann. d. ponts et chauss.*, 1840, sem. 1, p. 50) was extensively used, and has since been extended by *Durand-Claye* to iron arches also (*Ann. d. ponts et chauss.*, 1867, sem. 1, p. 63, and 1868, sem. 1, p. 109). Special prominence was given to graphical investigations of stability by Culmann's "Graphische Statik" (first part, Zürich, 1864, entire work, 1866; second edition, 1st part, 1875.)

This work of Culmann must be considered as original in all those parts relating to structures. Poncelet and Cousinery, beyond the general idea, furnished only unessential contributions. Culmann recognized the fruitfulness of the relations between the force and equilibrium polygon, upon which most of the practical solutions depend. He developed these relations, applied them in the theory of moments by the introduction of the closing line (*Schluss Linie*), and, accepting the rule of signs, obtained general points of view for the discussion of the most diverse figures which could arise in the same problem. In this and in many other respects even geometrical statics can profit from Culmann's work, as, for instance, in the investigation of the projective relations between the force and equilibrium polygon.

The fundamental importance of the force and equilibrium polygon was also recognized by those who, after Culmann, occupied themselves with the graphical method. Here we may notice two works of special influence upon the development of the graphical statics—those of *Mohr* and *Cremona*. The idea of *Mohr*, that the elastic line is an equilibrium polygon or curve (" Beitrag zur Theorie der Holz und Eisenconstructionen." Zeitschr. d. Hannov. Ing. und Arch. Ver., 1868) is of special significance for graphical statics.

That from it *Mohr* obtained the graphical determination of the moments at the supports of a continuous girder, is an example both useful as well as interesting. Already it has been endeavored to utilize the same idea in other cases (*Fränkel:* " zur Theorie der Elastischen Bogenträger," Zeitschr. d. Hannov. Ing. u. Arch. Ver., 1869, p. 115), and by it an impulse has been given to similar investigations.

Cremona has kept more especially in view the geometrical side of graphical statics. Starting from the theory of reciprocal polyhedrons, he gave the reciprocal relations between the force and equilibrium polygon with a generality and elegance to be expected from this distinguished Italian mathematician (*Le figure reciproche nelle statica grafica.* Milan, Länger, 1872). By this investigation the theoretical development of the graphical statics is essentially anticipated.

It was under the most unfavorable circumstances that Culmann introduced his graphical statics in the engineering department of the Zürich Polytechnic in the year 1860. It was finally, indeed, admitted as a regular study, but not the geometry of position which he premised. It was not till 1864 that this last was given in a series of lectures by Reye, and then the time at disposition for both courses was insufficient. Meanwhile the method spread, crept into the construction department of the engineering school, and wherever it came, even in the other departments of the Polytechnic, gained friends. Finally, at the present time, it is to be found, together with the modern geometry of position, upon which it was based, in every Polytechnic throughout Germany.

According to the above given definition of graphical statics, the methods of the graphical calculus, as far as applied in statical investigations, may also be regarded as belonging to graphical statics, and justly so; for these methods follow directly from geometrical principles, and can be applied by any one acquainted with geometry, without being collected under the special name of the "graphical calculus." Thus, for instance, *Bauschinger*, in his "Elemente der graphischen Statik" (München. 1871), disregards entirely the graphical calculus, and also cuts loose from the modern geometry; he develops the elementary principles of the subject in a logical and easily comprehended, if not purely geometrical manner, and thus brings the subject within the reach of those persons for whom it seems so especially designed. The work is remarkable for clear presentation, but expressly avoids all special investigations and practical applications, for which it is merely intended to prepare the way. In the present work, also, a similar plan is pursued, but all such applications as are of most value to the engineer or mechanic find likewise a place. Thus, combining the method of presentation of *Bauschinger* and the practical applications of *Culmann*, it has been endeavored to make it a practical manual, as well as a text-book of elementary principles—to serve the wants of the practical engineer, and also meet the requirements of the engineering student. How far this twofold design has been realized, the judgment of the reader must decide.

VII.

THE METHODS AND LIMITS OF THE GRAPHICAL STATICS.

The most perfect method of the graphical statics is the synthetic or geometric, since in geometrical statics the solution must always, when possible, rest upon pure mechanical or geometrical reasoning. Culmann presents his graphical statics to practitioners " as an attempt to solve by the aid of the modern geometry such problems pertaining to engineering practice as are susceptible of geometrical treatment."

The graphical statics, however, is not in and of itself the product of endeavors to make the modern geometry of service in applied mechanics; *graphical solutions* merely were required. How to obtain these, was another question. Thus it is that Poncelet's solutions consist almost entirely of graphical representations of analytical relations; that Cousinery avoided all use of formulæ; that Culmann made use of the new geometry wherever it was possible; that Bauschinger and others make use only of the ancient geometry; and that the latest graphical solutions—in a certain degree, those of Mohr also—entirely in the spirit of Poncelet's, rest again upon analysis. The pure geometric solution is, indeed, desirable, but is not always attainable.

If now we review all the cases in which direct and exclusively geometrical solutions are not possible, we see at once that this occurs when it is required to make use of the physical properties of bodies, as elasticity, cohesion, etc. Why? The actual condition of a body after equilibrium is attained, is a consequence of the motion of a variable system of points. The theory of the motion of variable systems has, however, by no means, as yet, been brought to practical efficiency (II.). We are therefore obliged to start from an hypothetical condition or state of the body (in the theory of flexure, for instance, we rest upon the assumption that all plane cross-sections made before the action of the outer forces remain plane after their action). To deduce now from this general condition the special relations necessary for solution, demands an essentially analytical process (I) Hence the dependence of the graphical solutions in such cases upon analytical relations—relations which, when the body is assumed to be rigid, as in the arch, in frame work, or the simple girder, no longer exist.

The sphere of action of an independent graphical statics is, then, confined to those problems which, under the assumption of inflexibility, are determined by a sufficient number of conditions. Beyond this point we have chiefly graphical *interpretations* only.

It has been already noticed that graphical statics, without the application of algebraic operations, can furnish no general laws (IV.). From relatively simple figures, indeed, here and there, general formulæ of metrical relations have been derived, as is, in fact, not theoretically impossible (I.), but such formulæ were always previously known. Such a result holds, in general, immediately good only for that form of figure which has been discussed, or, according to the terminology of *Carnot*, only for the existing "primitive figure," and must be proved or transformed for all "correlative figures" which can occur in accordance with the conditions of the problem. When the graphical investigation is guided by analytical operations, it is these last which render possible the deduction of general metrical relations.

Thus, in the theory of structures, there remains subject to pure graphical treatment only the general relations of form and position. Here we have the elegant deductions upon unfavorable loading, and here the graphical method often attains its end in a more elegant manner than the analytical. A complete exploration and development of such form and place relations, without a *geometry of position*, would evidently be impossible (IX.). The scientific future of the graphical statics, therefore, rests essentially upon the influence of the modern geometry. To endeavor to separate the higher geometry from the graphical method would be as unwise and fruitless as the attempt to exclude the higher analysis from analytical investigations. As, however, for certain purposes an elementary presentation of analytical theories relating to engineering practice will ever be acceptable, so also an elementary development of graphical methods is not without justification, the more so as long as the modern geometry itself is not sufficiently well known.

Culmann says of the graphical statics: "It includes, thus far, only the general part which we need in the investigation of problems in construction, but it must and will extend, as graphical methods find ever wider acceptance. Then, however, it will escape the hands of the practitioner, and must be built up by the geometer and mechanic to a symmetrical whole, which shall bear the same relation to the new geometry that analytical mechanics does to the higher analysis." Such an estimation does not appear to be entirely correct. It is *geometrical* statics (or mechanics) for which the above relation may subsist, and to this, indeed, Culmann's valuable work has itself greatly contributed. It was, moreover, developed quite independently of and much earlier than *graphical* statics (III.). In this respect, therefore, the spread of graphical methods is of less importance than that of geometrical views and knowledge; for when practical calculations are disregarded, and the deduction of *general* truths alone occupies us, then, first of all, we must exclude from the drawn figure all *special* relations—that is, strike out of graphical statics the essentially graphical part. A truth comprehended only in the abstract holds good for *all* figures which can be drawn in accordance with the given conditions.

We place, then, in one line *geometry* and *geometrical statics* (mechanics). From geometry we obtain a method of construction, or *descriptive* geometry, which finds its practical applications in architecture and machine drawing. From geometrical statics we obtain also a construction method or routine—viz., *graphical* statics—which finds its practical applications in the graphical calculation of structures and machines. Both descriptive geometry and graphical statics have still, with reference to these practical ends, to develop and make use of the general relations which subsist between the geometrical constructions to which they give rise, and thus each, according to its means, contribute to the discovery and spread of geometrical and mechanical truths.

From this co-ordination of descriptive geometry and graphical statics we must not, however, infer an equal importance; for, while in geometrical drawing we have always to represent an ideal image, and the graphical method is therefore directly suggested, we have for statical calculations the analytical process also at our disposal, and everything depends then upon the relative advantages and disadvantages of the graphical and analytical methods. We have thus noticed all the most important points which occur in a theoretical consideration, and there only remains to make a comparison from a practical standpoint (X.).

VIII.

THE MODERN GEOMETRY.

Geometry treats of figures or constructions in space. These figures and their properties are not always regarded and treated in equal extent and generality.

Geometrical knowledge found its origin in practical needs, and the ancients confined themselves almost exclusively to special investigations of individual figures and bodies of definite form, such as presented themselves to the eye. In the *phorisms* of Euclid (-285), according to Pappus (end of the fourth century), the mutual relations of the circle and straight lines were, indeed, given with a certain degree of completeness, but these have not come down to us.

Properties thus determined had naturally only a limited significance, and could neither count upon permanence nor give satisfactory conclusions. Investigators sought, therefore, assistance where it was best afforded, in analysis. This was, in the sixteenth century, by the algebra of *Vieta* (1540–1603), notably enriched.

From this period geometry, for a long time, served merely as an aid to analysis, interpreting graphically its results (V.). From this union the greatest advantages were derived, as analysis led to the infinitesimal calculus of Newton and Leibnitz, and geometry to the analytical geometry of Descartes (1596-1650).

But the extension and generality which geometrical truths received by this great creation of Descartes was essentially due to analysis. *Desargues* (1593-1662) and *Pascal* (1623-1662) extended pure geometrical considerations, and made the first step towards the modern geometry when they regarded the conic sections as projections of the circle, and deduced the properties of the first from those of the last. Then *De la Hire* (1640-1718), *Le Poivre* (1704) and *Huygens* (1629-1695) occupied themselves with geometrical investigations. While the two first developed the methods of Desargues and Pascal, Huygens and, later, Newton (1642-1727) applied pure geometry in optics and mechanics. Soon, however, the Calculus of Newton and Leibnitz (1684 and 1687) showed itself so wonderfully fertile in analytical geometry, that geometry proper was put in the background. Only a few, as *Lambert* (1728-1777), still regarded it with favor.

Then appeared *Monge* (1728-1777), and gave the impulse to a complete revolution in geometrical views, and to the reconstruction of the science upon a new basis. In his *Leçons de Géométrie descriptive* (Paris, 1788), all those problems previously treated in a special and uncertain manner in stereotomy, perspective, gnomonics, etc., were referred back to a few general principles, and, without the aid of analysis, the most important properties of lines and surfaces were deduced. While descriptive geometry taught the relations between bodies in space and drawn figures, it strengthened the power of abstraction; introducing into geometry the transformation of figures, it gave to its deductions an advantage till then possessed only by analysis; and while, finally, it owed its comprehensive results to the application of projections, it pointed the way for the further development of geometry itself.

Meanwhile, in the field of analytical geometry, the conclusion had been reached that the desired truths admitted of a still more general comprehension. All properties had been obtained only with respect to and by means of a determinate co-ordinate system. But already *Godin* (1704-1760) had announced "*que l'art de découvrir les propriétés des courbes est à proprement parler, l'art de changer le système de co-ordonnées*" (*Traité des propriétés communes à toutes les courbes*). This idea *Carnot* seized upon (1753-1823), and in the sixth chapter of his *Géométrie de position* (Paris, 1803) he sought to obtain a more general comprehension of figures by analysis, and to avoid the indeterminancy of this last by the introduction of the idea of position, and by many solutions after the method already pointed out by Liebnitz and d'Alembert.

Now began a veritable race in the condensation and promulgation of geometrical truths, in which the pure geometrical method obtained the palm. The scholars of Monge—*Brianchon, Servois, Chasles, Poncelet, Gergonne*—working with him and in his spirit, filled the *Annales des mathématiques* and the *Correspondance sur l'école polytechnique* with new results—the two last named discovering the general law of *reciprocity* or *duality*.

The foundation proper of the modern geometry was laid by Poncelet in his *Traité des propriétés projectives des figures* (Paris, 1828): *"Aggrandir les resources de la simple Géométrie, en généraliser les conceptions et le langage ordinairement assez restreints, les rapprocher de ceux de la Géométrie analytique, et surtout offrir des moyens généraux, propres à démontrer et à faire découvrir, d'une manière facile, cette classe de propriétés dont jouissent les figures quand on les considère d'une manière purement abstraite et indépendamment d'aucune grandeur absolue et déterminée, tel est l'objet qu'on s'est spécialement proposé dans cet ouvrage."*

The new ideas found in Germany especially fruitful soil. *Möbius, Plücker, Steiner, Grassman*, and many others, proceeding in part from entirely different points of view, opened out an abundance of new directions which have not yet been thoroughly explored, and which, in union with other investigations, have caused a thorough change in our conceptions of space relations, whose latest phases are indicated by the names of *Riemann, Helmholtz* and *Lie-klein*.

In this development period, also, still existed the two parties in analytical and synthetic, or pure geometry. *Plücker* held the analytical relations as the most general, and which were with advantage to be illustrated and interpreted geometrically; while *Steiner* recognized in the space figure itself the true object and most efficient aid of investigation. Both directions—the modern analytic and synthetic—lead naturally to the same results. With reference to the methods, however, they diverge the nearer the ideas and transformations of geometry approach the generality and ease of the algebraic method, thus rendering possible an abandonment of this last. Thus, while analytical geometry, through the theory of determinants of Hesse, came into ever closer connection with analysis—a direction in which English and Italian investigators—as *Salmon, Cayley, Cremona*—brilliantly assisted, the Erlangen Professor *von Staudt* cut loose from algebraic formulæ and metrical relations, and gave us the geometry of position (*Nürnberg*, 1847, Beitr. z. Geom. d. Lage).

After von Staudt, the strict geometry of position remained a long time disregarded, while the synthetic geometry of Steiner has enjoyed, without intermission till the present day, a special preference on the part of mathematicians. One reason may indeed be that mathematicians take little interest in an independence of geometry to which analysis can lay no claim; but another, still more potent, is the extremely condensed, almost schematic presentation of von Staudt, which has not exactly an encouraging effect upon every one.

Culmann gave the impulse to a change in this respect. In his graphical statics he rests directly upon the work of von Staudt, and, with something more than boldness, assumes a knowledge of the geometry of position among all practical men. Such a course was not indispensable for the foundation of his method, and impeded the spread of the graphical statics; but by it the geometry of position gained. This last had next, of necessity, to be introduced into the Zürich Polytechnic, and thus arose the first, until now, only complete text-book upon the subject, the "Geometrie der Lage," by Reye (Hannover, 1868), as the direct result of the graphical statics of Culmann.

Since then, the modern geometry has been introduced into all technical institutions throughout Germany, and thus placed at the disposal of the arts and sciences.

As, according to its founder, Poncelet, it reaches the highest range of speculation, so also in the most practical relations it acts to simplify and condense: "*Peu à peu les connaissances algébriques déviendront moins indispensables, et la science, reduite à ce qu'elle doit être, à ce qu'elle devrait être déjà, sera ainsi mise à la portée de cette classe d'hommes, qui n'a que des moments fort rares à y consacrer.*"

[For illustrations of the method of the modern geometry, the reader may consult the Appendix to this chapter.]

IX.

THE MODERN GEOMETRY IN ENGINEERING PRACTICE.

One who should infer that a science created thus from its very inception with reference to the needs of practice* must have found access, above all, in technical circles, would be much mistaken. As Culmann sent out his graphical statics, deep silence prevailed, and if the modern geometry appeared here and there in the lecture plan of one and another polytechnic, it was, without doubt, due to the zeal of some enthusiastic *privat docent* who had undertaken the thankless task of holding forth to empty benches.

Whence came this indifference to a discipline proceeding from the *Ecole polytechnique?* It is hard, indeed, to find a sufficient reason. We often hear it said that by reason of the colossal extension which engineering sciences have experienced, students are already overburdened. Most true! and it is just here that the modern geometry comes to our assistance. It is precisely to this that the learned critic of Monge, *Dupin*, alludes : "*Il semble que dans l'état actuel des sciences mathématiques le seul moyen d'empécher que leur domaine ne devienne trop vaste pour notre intelligence, c'est de généraliser de plus en plus les théories que ces sciences embrassent, afin qu'un petit nombre des verités générales et fécondes soit dans la tête des hommes l'expression abrégée de la plus grande variété des faits particuliers.*"

The modern geometry in its present form starts with a small number of elementary constructions whose properties are first set forth, and then, proceeding from these by combination and comparison, it covers the entire department of space. The engineer, during and after his preparation, has to do with space problems, with geometrical principles and constructions;

---

* Poncelet himself set upon the title-page of his work: "*Ouvrage utile à ceux qui s'occupent des applications de la Géométrie descriptive et d'opérations géométriques sur le terrain.*"

"how many superfluous definitions and demonstrations could not be spared, if they were already completely comprehended and recognized by the scholar as parts of a higher whole" (*Culmann*—"Die Graphische Statik"). At no very distant day it will no longer be possible to read a scientific work upon applied mathematics without familiarity with the principles of the modern geometry.* Permitting pure graphical applications, without the aid of analytic symbols, it forms the common point of view for descriptive geometry, practical geometry, and graphical statics.

Descriptive geometry existed before the modern, and this last has sprung from it. Now, reversely, the geometry of position comes to the aid of descriptive geometry, and offers in return its most fruitful principles and efficient aid. Thus in *descriptive geometry* we may refer to the works of *Pohlke*, *Schlesinger*, and *Fiedler*. The effect of the geometry of position in this direction to simplify and condense may be seen from the work of *Staudigl* ("Ueber die Identität von Constructionen in perspective, schiefer und orthogonaler Projection"), where it is proved that "all problems of the descriptive geometry, in which neither linear nor angular measure are considered—therefore all problems which belong to the geometry of position—can in similar manner and by precisely similar constructions be solved as well in perspective as in oblique and orthagonal projection." In shades and shadows and in geometrical drawing, *Burmeister* and *Paulus* owe to the modern geometry the simplicity of their constructions.

In the department of *practical* geometry also, in geodesy, perspective, surveying, we mark the influence of the modern geometry in the works of *Müller* and *Spangenberg*, of *Franke* and *Baur*.

In mechanics and physics, we see it again in the works of *Lindemann*, *Burmeister* and *Zech*.

X.

PRACTICAL SIGNIFICANCE OF THE GRAPHICAL STATICS.

We have already remarked (VII.) that the importance of graphical statics is in great part dependent upon its advantages as compared with the analytical method, and have reserved for this place a comparison from a practical point of view.

Here, first of all, we have to notice the independence of the graphical construction of the regularity or irregularity of the given relations. Whether the forces are equal or not, whether they act at equal or varying distances, even their relative position, are matters of indifference. Centre of gravity, central ellipse, kernel—for all, even the most irregular figures, are found in similar manner, with equal ease, even when exact analytical solutions are hardly conceivable. Thus a process, a routine almost mechanical is rendered possible in many investigations of stability, without losing sight of interior relations; for in the repeated and independent compositions of the forces we always perceive the origin, connection and

---

* Well illustrated in Gillespie's *Land Surveying*. New York, 1870.

reason of the result obtained, which, in the substitution of numbers in a formula, is not always the case.

With this advantage goes hand in hand a disadvantage. This very regularity of the process is a consequence of its special, we might almost say numerical, character (I.). In a numerical analytical example greater or less regularity has also but little effect. This numerical character has also for consequence that we can never attain to general laws and relations (IV., VII.).

The practical engineer becomes with time ever more familiar with the dividers and rule, while facility in analytical operations gradually disappears. A graphical construction once completed is not easily forgotten, or a single glance at a similar figure suffices to recall the whole process. It is indeed easy in clearly given formulæ to substitute special numerical values; but formulæ unfortunately are not always clearly given, in some cases *cannot* be so given, without presuming upon the thorough familiarity of the reader with the processes involved; these and the very many and various systems of notation in use leave to the constant, easily acquired and remembered graphical solution many advantages.

But here we may remark that graphical solutions can only be easily acquired, retained or quickly recovered when the constructions are based upon methods *purely geometric*, and not when they are simply the interpretation of previously obtained analytical results. In the latter case we must recall the process of development of the formula as well as the graphical construction, and the method is thus too often confusing instead of simple.

Often it is desired to make visible the results of an investigation, as in the case of the arch, where the graphical method is especially advantageous, and has in France been long used (VII.).

Errors relating to the mutual relation of strains are more easily discovered in graphical solutions than in analytical, as a certain law of regularity is always visible, which breaks abruptly for an error in construction. By calculation, on the other hand, we can more easily select any one place in the structure, and determine the strain there independently of the others.

As to which of the two methods demands the least time is a matter of minor importance. In a construction costing from thousands to millions, it matters little whether the calculations require one or several days, more or less, if only the results are clear and correct. It is a question also which can hardly be decided in favor of one or the other, dependent as it is upon elements other than those pertaining to the methods themselves— such as varying individual skill and capacity in either direction. The declaration which is already sometimes encountered, that the numerical calculation of a continuous girder requires about three times as much time as the graphical solution, sounds questionable. Why not at once furnish the statement with *decimal places?* In general, for ordinary cases, the analytical solution requires less time; for irregular and more complicated cases, the graphical.

The exactness of the graphical solution is sufficient, but it, too, depends upon the care and skill of the draughtsman. The greater the forces and dimensions with which one works, the better the results obtained. The scales should not, then, be taken too small.

It is hoped that these considerations, now drawing to a close, will suffice to give the reader clear ideas upon the nature and origin, advantages and disadvantages, of the graphical statics. The determination whether he will enter more fully into the subject—it may be, even take part in its development (there is abundance of room for workers), and in this case the choice of direction may thus be facilitated.

The graphical statics is certainly suited, especially in extended applications of the geometry of position, to furnish many new points of view, and in a practical respect it can often greatly simplify. Whoever has really studied the new methods must admit this.

On the other hand, the importance of the graphical statics is sometimes exaggerated. It appears out of place when in works designed for practice graphical solutions are given of problems which any reasoning being can almost solve in his head.

Such solutions may find a place in special text-books upon the subject, where they may, indeed, be desirable for completeness.

If it is desired to make two independent investigations of stability, as for large and important constructions is always desirable, it will be found of advantage, if a suitable graphical solution exists, to make the first determination graphically. Nothing more ensures a conviction of the correctness of an investigation than a correspondence of the graphical and calculated results.

XI.

LITERATURE UPON GRAPHICAL STATICS.

We have already referred in VI. to the most important contributions in the branch of graphical statics, and now annex a list of the literature upon the subject so far as known to us.

Where several works treat of the same subject, we have allowed ourselves a brief critical notice. Opportunity is thus given to those who would take part in the development of graphical statics to make themselves acquainted with all existing works, and at the same time the practical man is enabled in any case that may come up to inform himself as to where assistance may best be sought. A short remark to specify the contents may in this respect often help in the right direction. The succession is in each division chronologically arranged

Although the literature of the subject would seem from the following tolerably extensive, still the number of pure geometrical solutions in which no analytical formulæ appear is much less. Publications upon the subject would, moreover, beyond doubt, be still more numerous were it not for the difficulty and cost of production of lithograph plates.

I. TEXT-BOOKS UPON GRAPHICAL STATICS.

Culmann, K.—"Die graphische Statik." With Atlas of 36 Plates. Zürich, Meyer and Zeller, 1866. [I. Part, 1864: Elements and Graphical Investigations of Structures. Also, second edition, first volume, 1875, with 17 Plates. General Principles, second volume, to follow shortly.]

Bauschinger—"Elemente der graphischen Statik." With Atlas of 20 Plates. München, 1871. [Without the aid of modern geometry, and without practical applications. Admirable exposition of the Principles.]

Reuleaux.—An outline of the graphical statics is to be found in "Der Constructeur," by Reuleaux, third ed. Braunschweig, 1872.

Levy—"La Statique Graphique et ses Applications." Paris, 1874. With Atlas of 24 Plates. [Principles and several applications; clear and elegant exposition of the subject.]

Otto, K. von—"Die Grundzüge des graphischen Rechnens und der graphischen Statik." Prag, 1872, pp. 107. [English translation, by G. S. Clarke. Small rudimentary treatise.]

Favaro, Antonio—"Lezioni di Statica Grafica." Padua, 1877, pp. 650. [Containing introduction to Geometry of Position.]

II. PAPERS UPON THE GRAPHICAL STATICS.

Most—"Ueber eine allgemeine Methode, geometrisch den Schwerpunkt beliebiger Polygone und Polyeder zu bestimmen." Archiv d. Math. und Phys., IL. (1869), p. 355. [Also applicable to curve areas, without equilibrium polygon.]

Culmann, K.—"Ueber das Parallelogram und über die Zusammensetzung der Kräfte." Vierteljahrsschr. d. Naturforsch. Ges. zu Zürich, 1870. [Correspondence of the graphical statics with the Statics of Plücker.]

Mohr—"Beitrag zur Theorie der Holz- und Eisenconstructionen." Zeitschr. d. Hannöv. Arch. u. Ing. Ver., 1870, p. 41. [Relation between the neutral axis and centre of strains.]

Grunert, J. A.—"Ueber eine Graphische Methode zur Bestimmung des Schwerpunktes eines beliebigen Vierecks." Arch. d. Math. u. Phys., LII. (1871), p. 494. [Simple and brief. Compare also L., p. 212.]

Cremona, B.—"Le figure reciproche nelle statica grafica." With 5 Plates. Milan, 1872. German translation in Zeitschr. d. Ost. Arch. u. Ing. Ver., 1873, p. 230. [Force and equilibrium polygon as reciprocal figures.]

Du Bois, A. J.—"The New Method of Graphical Statics." Van Nostrand's "Engineering Magazine," Vol. XII., Nos. 74, 75, 76, 77, 78. [General properties of force and equilibrium polygons, with practical applications to bending moments, and several important mechanical problems. Also, Maxwell's Method applied to bridges, roof trusses, etc.] Separate reprint, 1875. Van Nostrand, New York.

Weyrauch, J. J.—"Ueber die graphische Statik—zur Orientirung." Leipzig, 1874. [Historical and critical.]

Favaro, Antonio—"La Statica Grafica." Venice, 1873. [Historical and critical.]

III. APPLICATION TO THE SIMPLE GIRDER.

Culmann, K.—"Der Balken." Third chap. of d. graph. Statik, 1866. [Contains also the construction of the inner forces.]
Vojácek—"Graphische Bestimmung der Biegungsmomente an kurzen Trägern." Zeitschr. d. Vereins Deutsch. Ing., 1868, p. 503. [Graphical interpretation of analytical relations.]
Cotterill, J. H.—"On the Graphic Construction of Bending Moments." "Engineering," 1869 (VII.), p. 32. [Equilibrium polygon for the simple truss, with references to Reuleaux and Culmann.]
Winkler, E.—"Einfache Träger," "Theorie der Brücken," "Aeussere Kräfte gerade Träger." Wien, 1872. [Simultaneous presentation of analytical and graphical methods.]
Ott, K. von—"Wirkung paralleler Kräfte auf einfache Träger mit Gerade Längenachse." In die Grundzüge d. graph. Rechnens u. d. graph. Statik. Prag, 1872, p. 28. [The most elementary principles pertaining to composition of forces in a plane are prefaced.]

IV. APPLICATION TO THE CONTINUOUS GIRDER.

Culmann, K.—"Der continuirliche Balken." Fourth chap. of the Graph. Statik, 1866. [With examples—the moments at the supports are analytically determined.]
Mohr—"Beitrag zur Theorie der Holz- und Eisenconstructionen." Zeitschr. d. hannöv. Arch. u. Ing. Ver., 1868, p. 19. [Completion of Culmann's method—the moments at the supports are graphically determined.]
Lippich—"Theorie des continuirlichen Trägers Constanten Querschnitts." Wien, 1871. Separate reprint from Förster's Bauzeit., 1871, p. 103. [Graphical method, together with elementary analytical.]
Ritter, W.—"Die elastische Linie und ihre Anwendung auf den continuirlichen Balken." Zürich, 1871. [Mohr's method—given as a supplement to the Graph. Statik of Culmann.]
Winkler, E.—"Continuirliche Träger. Theorie der Brücken—aeussere Kräfte gerade Träger." Wien, 1872. [The Mohr-Culmann method, together with analytical.]
Solin, J.—"Geometrische Theorie der continuirlichen Träger." Mitth. d. Arch. u. Ing. Ver. in Böhmen, 1873.
Greene, Chas. E.—"Graphical method for the analysis of Bridge Trusses; extended to Continuous Girders and Draw Spans." New York, 1875. [Moments at supports found by successive approximation, or balancing of moment areas.]

V. APPLICATION TO FRAME WORK.

Culmann, K.—"Das Fachwerk." Fifth chap. Graph. Statik, 1866. [Most general form of parallel truss, suspension truss, Pauli's truss, roof trusses.]

Keck, W.—"Ueber die Ermittelung der Spannungen in Fachwerk trägern mit Hülfe der graphischen Statik." Zeitschr. d. hannöv. Arch. u. Ing. Ver., 1870, p. 153. Separate reprint, Hannover, 1872. [Presentation of the method with reference to practice.]

Jenkin—"On the Practical Application of Reciprocal Figures to the Calculation of Strains in Frame-work. Transact. of the R. Soc. of Edinburgh, 1870, (XXV.) p. 441.

Maxwell, Prof. Clerk—"Reciprocal Figures, Frames, and Diagrams of Forces." Trans. of R. Soc. of Edinburgh, 1869-70.

Unwin—"Iron Bridges and Roofs." London, 1869. [Application to roof trusses, wind force, etc.]

Ranken, F. A.—"The Strains in Trusses." New York, Appleton, 1872. [Examples of simple trusses drawn to scale.]

Bow, Robert H.—"Economics of Construction in Relation to Framed Structures." London, 1873. [Application of Maxwell's Method only to roof trusses, etc.]

Ott, K. von—"Das Fachwerk." In Grundzüge d. graph. Rechnens u. d. graph. Statik. Prag, 1872. [Roof trusses, truss fixed at one end and free at the other, bridge trusses.]

Reuleaux—"Hilfslehren aus der Grapho statik." Second chap. of the Constructeur, third ed., 1872. [Compound truss, roof trusses, etc.]

Schäffer—"Graphische Ermittelung der Ordinaten des Schwedler'schen Trägers." Zeitschr. für Bauwesen, 1873, p. 237. [Proceeding from the equation for the same.]

Heuser—"Graphische Ermittelung der Ordinaten des Schwedler'schen Trägers." Zeitschr. f. Bauwesen, 1873, p. 523. [Preceding method simplified—another by means of equilibrium polygon.]

VI. APPLICATION TO THE IRON ARCH.

Culmann, K.—"Der Bogen." Sixth chap. der graph. Statik, 1866. [Contains also the inverted or suspended arch. The arch as a rigid body.]

Durand-Claye, A.—"Sur la vérification de la stabilité des arcs métalliques et sur l'emploi des courbes de pression." Ann. d. ponts et chauss., 1868, sem. 1, p. 109. [Mery-Durand pressure curves, but with reference to the absolute resistance of the material.]

Fränkel, W.—"Zur Theorie der elastischen Bogenträger." Zeitschr. d. hannöv. Arch. u. Ing. Ver., 1869, p. 115. [Following out Mohr's idea of the equilibrium polygon as elastic line.]

Mohr—"Beitrag zur Theorie der elastischen Bogenträger." Zeitschr. d. hannöv. Arch. u. Ing. Ver., 1870, p. 389. [Criticism of the preceding method, and giving another.]

Vála—"Beiträge zur graphischen Berechnung elastischer Bogenträger mit Kämpfergelenken." Mitth. d. Arch. u. Ing. Ver., in Böhmen, 1873.

Greene, Chas. E.—"Graphical Analysis of Roof. Trusses." Chicago, 1877.

VII. APPLICATION TO THE ARCH.

Cousinery, E. B.—"Application des procédés du calcul graphique à divers problèmes de stabilité." Fourth chap. of Calcul par le Trait. Paris, 1839. [With especial reference to the strength of abutments—pure graphical treatment.]

Mery—"Mémoire sur l'équilibre des voutes en berceau." Ann. d. ponts et chauss., 1840, sem. 1, p. 50. [Geometrical determination of every possible pressure curve.]

Culmann, K.—"Der Bogen." Sixth chap. of Graph. Statik, 1866. [Containing also arch centerings; exact discrimination of support and pressure line.]

Durand-Claye, A.—"Sur la vérification de la stabilité des voûtes en maçonnerie et sur l'emploi des courbes de pression." Ann. d. ponts et chauss., 1867, sem. 1, p. 63. [Reference to relative resistance of material.]

Harlacher, A. R.—"Die Stützlinie im Gewölbe." Tech. Blätter, 1870, p 49. [Practical method by inscription of support line, according to Culmann.]

Heuser—"Zur Stabilitätsuntersuchung der Gewölbe." Deutsche Bauzeit. 1872, p. 365. [Also methods for unsymmetrical form and load.]

VIII. APPLICATION TO RETAINING WALLS.

Poncelet, J. V.—"Mémoire sur la stabilité des revêtements et leur fondation." Mem. de l'off. du Génie, 1838 (XIII.); separate reprint, Paris, 1840. [First analytical graphical theorie.]

Cousinery, E. B.—"Application des procédés du calcul graphique à divers problèmes de stabilité." Fourth chap. of "Calcul par le Trait," 1839. [Pure graphical, without formulæ.]

Saint Guilhem—"Mémoire sur la poussée des terres avec ou sans surcharge." Ann. d. ponts et chauss., 1858, sem. 1, p. 319. [Further development and generalization of Poncelet's Theory.]

Rankine—"Manual of Civil Engineering." London, fourth ed., 1865. [Containing graphical construction of pressure parallel to earth surface upon vertical wall.]

Culmann, K.—"Theorie der Stütz- und Futter-Mauern." Eighth chap. of Graph. Statik, 1866. [With use of equilibrium polygon, pressure upon tunnel arches.]

Holzhey, E.—"Beiträge zur Theorie des Erddrucks und graphische Bestimmung der Stärke von Futter-Mauern." Mitth. über Gegenst. d. Artill. und Geniewesens; separate reprint, with two plates, Wien, 1871. [Point of application of earth pressure for complicated contour.]

Mohr—"Beiträge zur Theorie des Erddrucks." Zeitschr. des hannöv. Arch. u. Ing. Ver., 1871, p. 344. [Point of application of earth pressure and new analytical theory.]

Winkler, E.—" Neue Theorie des Erddrucks." Wien, 1872. [Containing graphical methods according to the old theory.]
Häseler, C.—" Beiträge zur Theorie der Futter- und Stütz-Mauern." Zeitschr. d. hannöv. Arch. u. Ing. Ver., 1873, p. 36. [Graphical determination of earth pressure according to Culmann.]

MISCELLANEOUS APPLICATIONS.

Reuleaux—" Die graphische Statik der Axen und Wellen." Published by polytech. Ver. in Zürich, 1863. [Autograph copy of lectures.]
Culmann, K.—" Der Werth der Constructionen." Seventh chap. of Graph. Statik, 1866. [Best and cheapest systems under given conditions, especially for bridges.]
Reuleaux—" Graphostatische Berechnung verschiedener Axen, Kranpfosten, Kurbeln," in the Constructeur, third ed., 1872.

Scattering graphostatical constructions are to be met with in many textbooks upon construction, especially since the appearance of Culmann's work, a second edition of which is in course of preparation, and expected soon to appear.

XII.

GRAPHICAL DYNAMICS.

The scientific or practical value of graphical solutions once recognized, there remains no reason for limiting them to statical problems only, and endeavors in the above direction are already forthcoming. We limit ourselves to a passing notice.

First, we have an attempt to employ graphical constructions in the theory of the overshot and breast-wheel (*Seeberger*, " Arbeitung der Theorie der oberschlächtigen Wasserräder auf graphischen Wege." Civil Ing. 1869, p. 398, and 1870, p. 339). We cannot here notice the value of the solutions given, but the very sparing applications of geometry hardly justify the title of the work.

A short article, which gives the graphical determination of the force at every position of a moving point, may also be noticed. (*Rapp*, " Zur graphischen Phoronomie," in Zeitsch. f. Math. u. Phys., 1872, p. 19.)

The genuine foundation of a graphical dynamics has been first attempted by Pröll ("Begründung graphischer Methoden zur Lösung dynamische Probleme," in Civil Ingenieur, 1873). From the fact that the effects of forces in dynamics are measured by the changes of velocity of any point or points of a machine system, Pröll concluded that it must be possible to represent these force effects by geometrical relations, such as kinematic geometry teaches.

His investigations, since published in independent form ("Versuch einer graphischen Dynamic," with 10 plates, 1874), fall into three parts. The first part treats of the action of the " outer forces " in machines whose motion is in a plane, the outer forces being also in this plane. In the second part he subjects to graphical treatment the action of outer forces upon a free movable material point. The third part, finally, considers the motion of rigid invariable systems acted upon by given forces.

In the course of the development extended use is made of analytical formulæ. The work is but the beginning of the future structure, but this beginning will be thankfully received by all those with whom graphical methods have found acceptance.

PART I.
GENERAL PRINCIPLES.

CHAPTER I.
FORCES IN THE SAME PLANE—COMMON POINT OF APPLICATION.

1. Notation, etc.—In order that a force may be " given " or completely determined in its relations to other forces, we must know not only its *intensity*, but also its *direction*, and the position of its *point of application*. These three being known, the geometrical expression of our knowledge is very simple. We have only to assume a certain length as the unit of force, and then any force is at once given by the *length, direction, and position of a straight line*. This method of force representation is so obvious, that it is in fact used in mechanics, even where the treatment itself is essentially analytical.

Unless expressly stated, all the forces with which we have to do, will be considered as lying and acting in the *same plane*. Graphically then, any force is completely determined by a straight line, the beginning of which represents the point of application, and the length and direction of which give the intensity and direction of the force.

We shall indicate a force in general by the letter **P**, its point of application by **A**. When we have several forces we represent the points of application by A_1, A_2, A_3, etc., and the *ends* of the corresponding lines by P_1, P_2, P_3, etc. The direction in which a force is supposed to act is thus unmistakably indicated.

When, however, lines representing several forces are laid off one after another, the beginning of each at the end of the preceding, it will be sufficient to put 0 at the beginning of the first, and 1, 2, 3, etc., at the end of each. No confusion can arise, as each force acts and reaches from the point indicated by the figure which is one *less* than its index, to the point indicated by that index.

When, finally, we designate a force by the two letters or figures which stand at the beginning and end, we shall always indicate by the *order* in which the letters or figures are written, the direction of action of the force, first naming the point of application, and then the end.

A force due to the composition of several forces, as P_1, P_2, P_3, we denote by $P_{1\cdot3}$ or $R_{1\cdot3}$. Thus $R_{1\cdot3}$ denotes the *resultant* of the forces $P_1, P_2,$ and P_3.

2. Parallelogram of Forces.—If two forces, P_1 and P_2, given in direction and intensity by the lines OP_1 OP_2 [Fig. 1, Pl. 1], have a common point of application O, the resultant $R_{1\cdot2}$ is found by the well known principle of the "parallelogram of forces," by completing the parallelogram as indicated by the dotted lines, and drawing the diagonal. OR then gives the resultant of the forces P_1 and P_2. If this resultant acts in the direction from O to R, as indicated by the arrow, it *replaces* P_1 and P_2; that is, it produces the same effect as both forces acting together. If it were taken as acting in the opposite direction—*i.e.*, from O outwards, *away* from R—it would hold the forces P_1 and P_2 in *equilibrium*.

Now, we see at once that it is *unnecessary to complete the parallelogram*. It is sufficient to draw from the end of the force P_2 the line $P_2 R$ *in the same direction that* P_1 *acts in*, and make it equal and parallel to P_1. The point R thus found is the end of the resultant R, or is a point upon the direction of the resultant prolonged through O.

As to the *direction* of action of the resultant—if we follow round the triangle from O to P_2 and from P_2 to R and R to O —*i.e.*, if we follow round *in the direction of the forces*—the direction for the resultant from R to O thus obtained is, as we have already seen, the direction necessary for *equilibrium*.

3. If, instead of two forces, we have three or more, as P_1, P_2, P_3, P_4 [Fig. 2] we still have the same construction. Thus completing the parallelogram for P_1 and P_2 we find $R_{1\cdot2}$. Completing the parallelogram for $R_{1\cdot2}$ and P_3, we find $R_{1\cdot3}$, and again, with this and P_4 we obtain $R_{1\cdot4}$. Again, we see it is unnecessary to complete all the parallelograms. We have only to draw lines $P_1 R_{1\cdot2}, R_{1\cdot2} R_{1\cdot3}, R_{1\cdot3} R_{1\cdot4}$, parallel to the forces $P_2 P_3$ and P_4 respectively, and equal in length to the intensities of these forces, and then, no matter what may be the number of forces, the *line drawn from the point of beginning to the end of the last line laid off* will give the intensity and position of the resultant. As to direction, the same holds good as before.

If the end of the last line laid off as above, should coincide with the point of beginning, there is, of course, *no* resultant, and the forces themselves are in equilibrium.

4. The polygon formed by the successive laying off of the lines parallel and equal to the forces, we call the "*force polygon.*" Hence we have the following principles established :

If any number of forces having a common point of application and lying in the same plane, are in equilibrium, the "force polygon" is closed.

If the "force polygon" is not closed, the forces themselves are not in equilibrium, and the line necessary to close it gives the resultant in intensity and direction.

This resultant, if considered as acting in the direction obtained by following round the "force polygon" with the forces, will produce equilibrium—acting in the opposite direction, it replaces the forces.

The resultant thus found in intensity and direction can be inserted in the *force diagram* at the common point of application.

5. Thus, required the position, intensity, and direction of the resultant of the forces P_1, P_2, P_3, P_4, P_5.

These forces are given in position, direction, and intensity by the *force diagram*, Fig. 3 (*a*). The resultant of all these forces must have of course the same point of application A as the forces themselves—it remains to find then its relative position and the direction of its action, so that we may properly insert it in the *force diagram*.

We have simply to draw the *force polygon*, Fig. 3, (*b*) by laying off successively $O\ P_1$, $P_1\ P_2$, etc., equal, parallel, and in the same direction as the forces P_1, P_2, etc., as given by Fig. 3 (*a*). Then the line $P_5\ O$ necessary to close the force polygon gives the intensity of the resultant, and in order to replace P_{1-5} it must act in the *direction from* O *to* P_5; *i.e.*, contrary to the order of the forces. If then in Fig. 3 (*a*) we draw $A\ R_{1-5}$ equal and parallel to $O\ P_5$, we have the resultant applied at the common point of application A, and given in position, intensity and direction.

Moreover, it is evident that *any diagonal* of the force polygon as R_{3-4} [Fig. 3 (*b*)] is the resultant of P_{3-4}, and acting in the direction from P_4 to P_2, it holds P_{3-4} in *equilibrium*. But it is also the resultant of P_1, P_2, P_5, and R_{1-5}, and acting in the same direction as before, it *replaces* these forces. The force polygon thus shows that the force which *replaces* P_1, P_2, P_5, and $R_{1.5}$, at the same time holds P_3 and P_4 in *equilibrium*, just as it should do.

If, on the other hand, we had originally only P_1, P_2, $R_{3.4}$, P_5, and $R_{1.5}$ forming a system of forces in equilibrium, we could *decompose* $R_{3.4}$ into two components by simply assuming any point as P_3 [Fig. 3 (*b*)] and drawing $P_3 P_4$, $P_3 P_2$. Then following round this new polygon in the direction of the forces, or, what amounts to the same thing, taking the direction of the components $P_3 P_4$, *opposed* to the direction of $R_{3.4}$ for equilibrium, we obtain the direction of action of P_3 and P_4 as shown by the arrows in Fig. 3 (*b*). These forces inserted in Fig. 3 (*a*), *in the place* of $R_{3.4}$ and in these directions, will not disturb the equilibrium.

Hence, *any diagonal in the force polygon, is the resultant of the forces on either side, holding in equilibrium those on one side and replacing those on the other, according to the direction in which it is conceived to act.*

Also, *any force or number of forces may be decomposed into two others in any desired direction, by choosing a suitable point in the plane of the force polygon and drawing lines from this point to the beginning and end of the force or force polygon.*

6. It matters not in what Order we lay off the Forces in the Construction of the force Polygon.—Thus, in Fig. 1, whether we draw from the end of P_2 the line $P_2 R_{1.2}$ equal and parallel to P_1 or from the end of P_1 the line $P_1 R_{1.2}$ equal and parallel to P_2, in either case we obtain the same resultant and the same direction for the resultant. But by a similar change of two and two, we can obtain any order we please. For example, we lay off in Fig. 3 (*c*) the same forces in the order $P_3 P_2 P_1$, $P_5 P_4$, and obtain precisely the same resultant, in the same direction as before. For, the resultant of P_3 and P_2 must be the same as that of P_2 and P_3 in the first case. The resultant of $R_{3.2}$ and P_1 must then be the same in both polygons, and so on.

Generally, then, no matter what the order in which the forces are laid off, the line necessary to close the force polygon is the resultant of the forces, and the diagonals of the force polygon give us the resultants of the forces on either side.

By assuming a point at pleasure, and drawing lines from this point to the beginning and end of any side of the force polygon, and taking the direction of these lines *opposed* to the direction of that side, we can *decompose* any force in the force polygon into its components. Thus the force polygon gives us complete information as to the action of the forces.

7. If the Forces act in the same straight Line, the force polygon of course becomes a straight line also, and the resultant is the sum or difference (algebraic sum) of the forces.

Thus, if we have P_1, P_2, P_3, all acting at the point A, as shown by the force diagram Fig. 4 (*a*), we form the force polygon by laying off from 0, Fig. 4 (*b*), the intensity of P_1, from the end of this line $P_1 P_2$ equal to $A P_2$ and from P_2, $P_2 P_3$ equal to $A P_3$. Then the line necessary to close the polygon is evidently $0 P_3 = P_1 + P_2 - P_3$. A single force acting then at A in the direction of and having the intensity represented by the line $0 P_3$ would replace P_1, P_2, and P_3. If acting from P_3 to 0, it will produce equilibrium.

If we again choose an arbitrary point as C [*we shall hereafter call this point the "pole" of the force polygon*], and draw lines $S_0 S_3$ from this *pole* to the beginning and end of the force polygon, we can decompose the resultant into two forces in any required direction. If the resultant is supposed to act down, then the arrows show the direction in which these components must act in order to *replace* the resultant. If then at A we draw lines parallel and equal, we have these components in position, direction, and applied at the common point of application.

8. Practical Applications.—Simple and even self-evident as all the preceding may seem, we have already acquired all the principles requisite for a rapid, accurate, and very elegant method of finding by diagram the strains in the various members of all kinds of framed structures, such as roof trusses, bridge girders, cranes, etc., no matter how complicated the structure, or what special assumptions are made as to the loading, provided only, that all the exterior forces are known. A complicated or unsymmetrical arrangement of parts increases greatly the labor of calculation, but has no effect upon the ease or accuracy of the graphical method. The method moreover checks its own accuracy, does not accumulate errors, and shows in one view the *relation* of the strains to each other, and the variations which would be caused by a change in the manner of load distribution, or in the form of construction.

As this method is not as well known as it deserves to be, it will perhaps be of advantage to pause for a moment in the development of our subject, and make this direct application of the principles already established.

BRACED SEMI-ARCH.

9. Stoney, in his "Theory of Strains," Vol. I., page 123, gives the following example of a "braced semi-arch," represented by Fig. 5, Pl. 1. The dimensions are as follows: projecting portion, 40 ft. long, 10 ft. deep at wall. Lower flange, circular, with a horizontal tangent 2 ft. below the extremity of girder. Radius of lower flange, 104 ft. Load uniform and equal to one ton per running foot supposed to be collected into weights of 10 tons at each upper apex, except the end one, which has only 5 tons.

Fig. 5 shows the arch drawn to a scale of 10 ft. to an inch.

This scale is too small in this case to ensure good results; in general the larger the scale to which the *frame* can be drawn, the better; but for the purpose of illustration it will answer well enough. With a large scale for the *frame diagram*, a scale of 10 tons to an inch will in general be found to answer well. Fig. 5 (*a*) gives the strains in the various members to a scale of 10 tons to an inch and Fig. 5 (*b*) 20 tons to an inch; the first for the load at the extremity alone, the second for a uniform load.

Fig. 5 (*a*) is thus obtained. We first lay off the weight, 5 tons, to scale, in the direction in which it acts; *i.e.*, *downwards*. Now this weight and the strains in diagonal 1, and flange **A**, are in *equilibrium;* therefore by article (4) the force polygon must close. Drawing lines therefore from the ends of the line representing the weight of 5 tons, parallel to these pieces and prolonging them to their intersection, we obtain the strains in **A** and 1. Commencing with the beginning of the weight line and following *down* around the triangle thus formed, we find that **A** acts from right to left, as shown by the arrow. **A** acts then *away from the apex;* it is therefore in *tension.* Diagonal 1 acts *towards* the apex and is hence *compressed.*

We pass now to apex *a*, of the frame. Here we have the strains in **E** and diagonals 1 and 2, and these three strains hold each other in equilibrium. The strain in 1 we have already, and know it to be *compressive*. We have then simply to draw lines from 0 and b parallel to **E** and 2, and follow round the triangle, to obtain the intensity and quality of the strains in **E** and 2. We must remember that as 1 is in compression, and we are *now* considering apex a, we must follow round from o to b in Fig. 5 (a), and so round. We thus find 2 acting *away* from apex a and therefore in *tension*, and **E** acting towards this apex, and hence *compressed*.

Pass now to apex c. We have the strains in **A** and 2 in equilibrium with **B** and 3. [No weights are supposed to act except the one at the end.] But **A** and 2 we already have. We draw 3 and **B**. Diagonal 2 has been found to be in tension. With reference to apex c it must therefore act *away* from c; *i.e.*, from d to b in the force polygon. This is sufficient to give us the hint how to follow round. We pass from d to b for 2, from b to e for **A**, then from e to **B** and from **B** to d for **B** and 3. **B** is therefore *tension* and 3 *compression*. And so we proceed. For the next apex g, we have **E** and 3 in equilibrium with **F** and 4. We draw parallels to **F** and 4 so as to close the polygon of which we have already two sides, **E** and 3, given, and remembering that as 3 is in compression, it must therefore act *towards* g, we follow round the completed polygon with this to guide us, and find 4 *tension* and **F** *compression*. Thus we go through the figure, and when all is ready we can scale off the strains. The strains in the lower flanges it will be observed all radiate from o. The upper flanges are all measured off on the horizontal e **C**, and the diagonals are the traverses between. We see at once that however irregular the structure, we can always easily and readily determine the strains at any apex, *provided no more than two unknown strains are to be found*. If more than two pieces, the strains in which are unknown, meet at an apex, we can evidently form an indefinite number of closed polygons. The problem is indeterminate, and the structure has unnecessary or superfluous pieces.

Fig. 5 (b) gives the strains for a uniform load, taken, for convenience of size, to a scale of 20 tons to an inch. Here until we arrived at apex c of the frame the strains are evidently the same as before. Observe the influence of the weight at c. Here we have the strains in **A** and 2 given in the diagram, in equilibrium with **B**, 3 *and the known weight* acting at c; viz.,

10 tons. We lay off therefore 10 tons *downward* from *e*, Fig. 5 (*b*), and follow *down* from *e* around the polygon. We thus find B tension and 3 compression. Then 4 and F are found as before for apex *g*, 4 tension and F compression; and then we come to the next apex and the *next weight*. This is laid off downwards from the end of the preceding, and then we follow round, finding C tension and 5 compression; and so on.

10. As another example, let us take the

ROOF TRUSS,

given in Fig. 6, Pl. 2. This truss is given by Stoney, Vol. I., page 128. Dimensions: span, 80 ft.: rise of top and bottom flanges, 16 and 10 ft. respectively. Radii, 58 and 85 ft. The figure shows two different kinds of bracing. In the left-hand part the extreme bay of the lower flange is half as long again as the others. The upper flange is divided into 4 equal bays. In the right-hand section, both flanges are divided into 4 equal bays, and every alternate brace is therefore nearly radial. Each upper apex in both cases is supposed to sustain a weight of one ton.

The strains in the various pieces are given in Fig. 6 (*a*).

We form the force polygon by laying off the weights from 0 to 7 and then laying off the reactions 3.5 apiece, upwards, we come back to 0, and the force polygon is closed as it should be, since the sum of the reactions must be equal and opposite to the sum of the weights. Starting then with the reaction at the left support A, we go through from apex to apex in a manner precisely similar to the previous case. The operation is so simple that it is hardly necessary to detail it again, but we recommend the reader to go over it with the aid of Fig. 6 (*a*), lettering the figure as he proceeds. The dotted part gives the strains for the right-hand half.

DIAGRAM FOR WIND FORCE.

11. It is of considerable importance to investigate the influence of a partial load, such as that caused by the wind blowing on one side of the roof, and this by the aid of our method we can easily do.

From the experimental formulæ of Hutton,[*]

---

[*] *Iron Bridges and Roofs.* Unwin. p. 120.

$$P_n = P \sin i^{\,1.84\cos i - 1}$$
$$P_h = P \sin i^{\,1.84\cos i}$$
$$P_v = P \cot i \sin i^{\,1.84\cos i}$$

where **P** is the intensity of the wind pressure in lbs. per sq. ft. upon a surface perpendicular to its direction, i is the inclination of any plane surface to this direction; P_n is the normal pressure, P_h the horizontal component of this normal pressure, and P_v its vertical component.

That is, if the wind blows horizontally, P_h is the horizontal and P_v the vertical component of the pressure on the roof. If we take **P**=40 lbs., which probably allows sufficient margin for the heaviest gales, we have the following values of the normal pressure and its components for various inclinations of roof surface:

Angle of Roof	Lbs. per square foot of surface.		
	P_n	P_v	P_h
5°	5.0	4.9	0.4
10°	9.7	9.6	1.7
20°	18.1	17.0	6.2
30°	26.4	22.8	13.2
40°	33.3	25.5	21.4
50°	38.1	24.5	29.2
60°	40.0	20.0	34.0
70°	41.0	14.0	38.5
80°	40.4	7.0	39.8
90°	40.0	0.0	40.0

The load at each joint may be taken as equal to the pressure of the wind striking a surface whose area is equal to that portion of the roof supported by one bay of the rafter, and inclined at the same angle as the tangent to the rib at the joint. Thus we can calculate P_1, P_2, P_3, P_4, (Fig. 6), resolve these forces into their horizontal and vertical components, and find the reactions at the supports as well as the horizontal force at the left abutment, which in our construction is supposed to be fixed. Should the wind be supposed to blow from the right side, the strains would be entirely different, and it would be necessary to form a second diagram. Each piece must be proportioned to resist the strains arising in either case. The forces $P_{1\text{-}4}$ and their horizontal and vertical components, as also the reactions, being known, we can now form the force polygon.

Thus in Fig. 6 (*b*), we lay off the forces $P_{4\text{-}1}$, make *a c* equal to the vertical reaction at **A**, $a\,b =$ the sum of the horizontal components, or the horizontal force at **A**, and $o\,b$ the vertical reaction at the right support. This last line should close the force polygon and bring us back to o.

Now starting at the left support, we have the vertical reaction $a\,c$, the horizontal force $a\,b$, and the wind force $\mathbf{P}_1$, in equilibrium with **A** and **E**. Closing the polygon by lines parallel to **A** and **E**, we obtain the strains in these pieces, **E** tension and **A** compression. At the next apex we have **A** and $\mathbf{P}_2$ in equilibrium with 1 and **B**. Completing the parallelogram, we find 1 compression and **B** compression. At the next apex 1 and **E** are in equilibrium with 2 and **F**, and we find **F** and 2 tension and so on. The upper flanges are in compression and start from the ends of the forces $\mathbf{P}_1$, $\mathbf{P}_2$, etc. The lower flanges radiate from b. If we were to carry out the construction for the rest of the frame, the upper flanges after **D** would radiate from o.

A comparison of Fig. 6 (*a*) and (*b*) shows that whereas under uniform load the strain in 1 is *tension*, for wind force the same brace is in *compression*. In fact in the first case *all* the braces are in tension, while in the second 1, 3, and 5 are compressed, and 3 and 5 quite severely. The strains in the bracing generally are much greater in the second case.

Were we to consider the wind as blowing from the other side, or what is the same thing, suppose the right end fixed and the left supported on rollers, then the horizontal reaction $a\,b$ will be applied at the right abutment. In this case the lower flanges will radiate from a instead of b, and the first upper flange will start from o. Supposing the first two lines of this new diagram drawn, as indicated by the dotted lines, and following round from b to o, and so round to a and back to b, it may easily happen that the last upper flange is in *tension* and the last lower flange in *compression*; that is, a *complete reversal of the ordinary condition of strain*.

For an excellent presentation of the above method, we refer the reader to *Iron Bridges and Roofs*, by W. C. Unwin, pp. 128–140. The above method is there referred to as "*Prof. Clerk Maxwell's Method*," and as such is known and used in England.*

---

* Phil. Mag., April, 1864, and a Paper read before the British Association for the Advancement of Science, by Prof. Maxwell, in 1874.

BRIDGES.

12. For *bridges* the strains due to a uniform load are of course easily found. In most cases a *rolling load* can be managed also, *without* making a separate diagram for each position of the load. Thus, if we diagram the strains for the load at the first and last apex, the strains due to intermediate loads will be multiples or submultiples of these, provided all the bays are equal. A calculation for a simple Warren girder of small span, and a consideration of the reaction for each position of the load, will at once illustrate what is meant. [Compare Stoney, *Theory of Strains.* Pp. 99–111, Vol. I.]

Thus Stoney, in his *Theory of Strains*, Vol. I., p. 99, gives the girder represented in Fig. 7, Pl. 2, span 80 ft., depth of truss, 5 ft., 8 equal panels in upper flange, 7 in lower.

For the first weight of 10 tons, P_1, the strains are given by Fig. 7 (*a*) to a scale of 10 tons to an inch. We form first the *force polygon* by laying off from *o*, 10 tons, to P_1. From the end of this line we lay off upwards the reaction at right abutment $= \frac{1}{8}$ of 10 tons, or 1.25 tons; and then the reaction at the left abutment $= \frac{7}{8}$ of 10 tons, back to *o*, thus closing the force polygon. [Note.—In *any* structure which holds in equilibrium outer forces, the force polygon must close. If it does not, there is no equilibrium, and *motion* ensues (see Art. 20).] Commence now with the reaction at *a* in the frame diagram, Fig. 7, because here we have a known reaction, *a o* (force polygon), and only *two* unknown strains to be determined. Drawing lines parallel to **A** and 1, we obtain the strains in **A** and 1. Then pass on to apex *b*. With the now known strain in 1, we can determine 2 and **E**.

Passing now to the next apex, we have **A** and 2 known, and *also the weight* P_1. Join therefore P_1 and **E** [Fig. 7 (*a*)] by lines parallel to **B** and 3. **B** and 3 are both in compression. We find diagonal 2 also in compression, and 1 in tension. That is, *both the diagonals under the weight are compressed*, as evidently should be the case. From 4 on we have tension and compression alternately.

Fig. 7 (*b*) gives the strains due to the *last* position of the load P_7. The strains in the diagonals are evidently all equal, and alternately tension and compression.

Now it is not necessary to construct more than these two diagrams. From these two alone we can determine the strains for any intermediate weight. Thus scaling off the strains in Fig. 7 (*a*) and (*b*), we can tabulate them under P_1 and P_7, as shown by the table.

DIAGONALS.	P_1	P_2	P_3	P_4	P_5	P_6	P_7	C +	T −
1............	−12.4	−10.6	− 8.9	− 7.1	− 5.3	− 3.5	− 1.8	......	−40.6
2............	+12.4	+10.6	+ 8.9	+ 7.1	+ 5.3	+ 3.5	+ 1.8	+40.6	......
3............	+ 1.8	−10.6	− 8.9	− 7.1	− 5.3	− 3.5	− 1.8	+ 1.8	−37.2
4............	− 1.8	+10.6	+ 8.9	+ 7.1	+ 5.3	+ 3.5	+ 1.8	+37.2	− 1.8
5............	+ 1.8	+ 3.5	− 8.9	− 7.1	− 5.3	− 3.5	− 1.8	+ 5.3	−26.6
6............	− 1.8	− 3.5	+ 8.9	+ 7.1	+ 5.3	+ 3.5	+ 1.8	+26.6	− 5.3
7............	+ 1.8	+ 3.5	+ 5.3	− 7.1	− 5.3	− 3.5	− 1.8	+10.6	−17.7
8............	− 1.8	− 3.5	− 5.3	+ 7.1	+ 5.3	+ 3.5	+ 1.8	+17.7	−10.6

Now the reaction at the left abutment due to P_6 is *twice* as great as that due to P_7. Hence the values in the column for P_6 will be twice as great; in the column for P_5 three times as great, and so on. For similar reasons the strain in 5 for P_2 will be twice that for P_1. In column P_2, then, from 5 down we multiply the strains in P_1 by 2. In P_3 from 7 down by 3. Thus we fill out the table of strains completely, and find the maximum tension and compression. A similar procedure will give the flanges.*

APPLICATION TO AN ARCH.

13. For a "braced arch" (Stoney, p. 136) as represented in Fig. 5 (*c*) Pl. 2, the strains in every piece due to any load are in similar manner easily found by first finding the components of the load acting at the abutments, and then proceeding as above. Thus for a load P_2, the left half of the arch is in equilibrium with the forces acting upon it; viz., a horizontal and a downward force at *a*, and a horizontal and an upward force at **A**. The resultant of the forces at *a* must then pass through

---

* The reader not familiar with the above method of tabulation will find it further illustrated in Art. 7 of the Appendix. He cannot do better than to refer to it here and now.

a and A, and be equal and opposite to the resultant at A. The resultant at the right abutment must pass through that abutment, and also through the intersection of P_2 with $A\,a$. So for any other force, as P_6, we have simply to draw $B\,a$ to intersection with P_6, and then $P_6\,A$. We can now decompose P_6 or P_2 along the resultants through the abutments thus found. Thus resolving P_2 along $A\,a$ and $P_2\,B$, Fig. 5 (*e*), we find the force acting at apex a. This force resolved into A and 1 gives the strains on these pieces both compressive. Passing then to the next apex, we obtain the strains in 2 and E. Then to the next, and we get 3 and B, compression and *tension* respectively, and so on, as shown by diagram, Fig. 5 (*e*), which, it will be seen at once, is similar to Fig. 5 (*a*), already obtained for the "semi-arch," except that the strain in A is less than for the semi-arch and compressive, while $B\,C$ and D are in tension. The reason is obvious. At a [Fig. 5 (*c*)] the resultant lies *between* A and 1, and therefore causes compression in both, while it passes outside of the arch entirely, to the right of the apex for diagonals 3 and 4, and hence causes tension in $B\,C$ and D. Fig. 5 (*d*) gives the strains due to P_6. Here the resultant or reaction at A is first found and resolved into 9 and H, and then we go through the frame as before. We see that 4 and 5 under the load are both compressed, that E and F are in tension and G and H, as also the entire upper chord, in compression. The work checks from the fact that the line closing the polygon formed by E and 2 should be exactly parallel to and give the strain in diagonal 1, or A and 1 should be in equilibrium with the resultant through a [see Fig. 5 (*d*)].

In every case of the kind we first, then, have to draw the *frame diagram*. Then lay off the *force polygon* which should close. Finally we construct the *strain diagram*. The frame diagram should be taken to as large a scale as possible consistent with reasonable size, and the scale for the force and strain diagrams as *small* as possible, consistent with scaling off the strains to the requisite degree of accuracy. A small frame diagram does not give with the proper accuracy the relative positions and inclinations of the various pieces, so as to ensure the proper direction for the lines of the strain diagram. A slight deviation from parallelism causes sometimes considerable variation. Nevertheless with practice, care, and proper instruments the accuracy of the method is surprising ; even in complicated structures, the variation resulting from performing the operation *twice* being inappreciable. Every symmetrical frame gives also a symmetrical strain diagram, and the accuracy of the work is tested at every point by this double symmetry, and finally by the end or last point of the second half, exactly coinciding with the last point of the first half. Thus in Fig. 6 (*a*), if we had but one system of triangulation carried through the frame, the strain diagram for the right half would be precisely similar and symmetrical to that already found for the first, and the end of the last line would fall, or should fall, *precisely* upon the point *b* of the first. If it does not, and the error is too great to be disregarded, then by checking corresponding points in each half, we can find the point where the error was committed. In any case *errors do not accumulate*. Thus, armed with straight edge, scale, triangle, and dividers, we can attack and solve the most intricate problems, without calculation or tables, with ease, accuracy, and great saving of time.

METHOD OF SECTIONS.

14. The results obtained by the above method are best checked in general by Ritter's "method of sections," or the use of moments.* This consists in supposing the structure divided by a section cutting only *three* pieces. We can then take the intersection of *two* of these pieces as a centre of moments, and the sum (algebraic) of the moments of all the exterior forces, such as reaction, loads, etc., upon one of the portions into which the structure is divided by the section, with reference to this centre of moments, must be balanced by the moment of the strain in the third piece, with reference to this same point. Thus in Fig. 6, Pl. 2, required the strain in **D**. Take a section through **D**, 7 and **H** (right half of Fig.), and let *a* be the centre of moments. The moments of the strains in 7 and **H** are then, of course, zero, since these pieces pass through *a*. The moment of the strain in **D** with reference to *a* must then be balanced by the sum of the moments of all the outer forces acting upon the portion to the left (or right) of the section.

Thus, strain in **D** multiplied by its lever arm with respect to *a*, is equal to moment of reaction at **A**, minus sum of the moments of loads between **A** and *a*, all with reference to *a*. If

---

* *Dach- und Brücken-Constructionen.* Ritter. Hannover, 1873.

we take the direction of rotation of the forces on the left of the section when in the direction of the hands of a watch as *positive*, and find the moment of strain in **D** *negative*, it shows negative rotation about a, and the strain in **D** to resist this rotation must act *away* from b, or be tensile. If the resultant rotation of the outer forces is on the other hand positive, the strain in **D** must act *toward* b, and **D** is therefore compressed.

This method of calculation, it will be observed, is both simple and general. It can be applied to any structure, when the outer forces are completely known, and only three pieces are cut by the ideal section.

15. It is unnecessary to give here further applications of our graphical method. The reader can easily apply it for himself to the "bowstring girder," bent crane, etc., and satisfy himself as to its accuracy, and the ease with which the desired results are obtained.

Enough has been said to indicate the many important applications which even at the very commencement of our development of the graphical method we are enabled to make, and here we shall close our discussion of forces lying in the same plane and having a common point of application. As we pass on to forces having *different* points of application, we shall have occasion to develop new principles and relations not less fruitful and useful in their practical results.*

---

* We refer the reader here to the *Appendix* to this chapter for further illustrations of the application of the above principles, as well as for information upon several points of considerable practical importance. We would also remind him here once for all, that the *Appendix* to this work was NOT intended to be disregarded, but has been thought desirable in order to avoid encumbering the general principles with too much of detail in the text. We earnestly request him to neglect no reference to it which may be made in the text.

He will do well in the present case, after first making himself familiar with the above points, to solve for himself with scale and dividers a number of similar problems, checking his results always by the method of moments. He will thus in a very short time master the method, and be able to solve readily and accurately every problem of usual occurrence in practice. Though the method is very simple, *actual practice with the drawing board is here indispensable.*

CHAPTER II.

FORCES IN THE SAME PLANE—DIFFERENT POINTS OF APPLICATION.

16. Resultant of Two Forces in a Plane—Different Points of Application.—Heretofore we have considered forces having a common point of application, and have seen that in any case the direction and intensity of the resultant is easily found by *closing the force polygon.*

But suppose we have two forces P_1 P_2 having *different* points of application A_1 A_2; required the position and direction of the resultant [Pl. 3, Fig. 8].

Any force acting in a plane may be considered as acting at any point in its line of direction.

P_1 and P_2 may then be supposed to act at their common point of intersection a, and through *this* point the resultant should pass. The case reduces therefore to a common point of application. The resultant is given in intensity and direction as before by the force polygon (*b*), and its position is determined by the point of intersection a. At this point, or at *any point* in the line through a, parallel to 0 2, the resultant may be supposed to act.

But the direction of the forces may not intersect within reasonable limits, or the forces may be supposed parallel to each other, so that they *may not intersect at all*. In any case the *force polygon* will still give the intensity and direction of action of the resultant, but its *position* in the plane of the forces remains yet to be determined. Now we have seen [Art. 5] that we can decompose a force into two components in any desired directions, by choosing a "*pole*" and drawing lines to the beginning and end of the force in the force polygon. Let us choose then a *pole* C [Fig. 8 (*b*)] and decompose the resultant thus into two forces given in intensity by the lines 0 C and 2 C. The forces P_1 P_2 being supposed to act at the points A_1 A_2 in the common plane, at what point in the plane and in what direction must the resultant 0 2 be applied to keep this plane and hold the forces *in equilibrium?* The direction of action of the resultant is given at once from the force polygon [Art. 5 (*b*)]. It must act in a direction from 2 to 0, and must be equal to 2 0, taken to the scale of force. Now at *any* point in the line of direction of P_1, as for instance 1, let us suppose the component given by C 0 to act. What is then the resultant of P_1 and C 0 ? A glance at the force polygon gives us 1 C, because this line closes the polygon made by C 0, 0 1 and 1 C. At 1 then, the three forces S_0 (parallel and equal to C 0) S_1 (parallel and equal to 1 C) and P_1 are in equilibrium, and there is no tendency of the point 1 to move. But 1 C or S_1 may be considered as acting in the plane at *any* point in its line of direction; therefore at 2 its intersection with P_2 prolonged. Suppose at 2, S_2 or 2 C to act. We see at once from the force polygon that 2 C, C 1 and P_2 are in equilibrium. There is therefore no tendency of the point 2 to move, and the two forces P_1 P_2 are then in equilibrium with C 0, 1 C, C 1 and 2 C. But since the resultant of C 0 and 2 C or of S_0 and S_2 is also the resultant of the forces, and since it must therefore act through the point of intersection of S_0 and S_2; we have only to *prolong these lines to intersection b.* Through this point the resultants $R_{1.2}$ must pass and acting *downwards* (from 0 to 2) as indicated in the Fig., it *replaces* P_1 P_2. Acting upwards it would hold them in equilibrium. We thus easily find the point 2 in the plane at which 2 C or S_2 must be applied, *when* C 0 or S_0 *acts at* 1, and S_0 S_2 are thus found in proper relative position. The position, intensity, and direction of the resultant are thus completely determined.

Had we taken any *other* point than 1, as the point of application of C 0, we should have found a different corresponding point for application of 2 C, but in any case the prolongations of 2 C and C 0 would intersect upon the line *a b*, prolonged if necessary. The same holds true for any position of the *"pole"* C. This construction is evidently general whatever the position or whatever the number of the forces. We may thus obtain any number of points along the line *a b*; that is, the resultant also, may act at *any* point in its line of direction.

[NOTE.—*That b is a point in the resultant of* P_1 *and* P_2 *can be proved in a method purely geometrical. In the two "complete quadrilaterals"* 0 1 2 C *and* 1 *b* 2 *a, the five pairs of corresponding sides* 0 1 *and a* 1, 1 2 *and a* 2, 2 C *and b* 2, C 0 and $b\,1$, $C\,1$ and $1\,2$, *are parallel each to each, therefore the sixth pair $0\,2$ and $a\,b$ must also be parallel; b is therefore a point of the resultant passing through a, parallel to $0\,2$.*]

17. The above Construction holds good equally well for Parallel Forces.—By means of it we find in Pl. 3, Fig. 9 (*a*) and (*b*) and Fig. 10 (*a*) and (*b*), the resultant of a pair of parallel forces, in the first case, both acting in the same direction; in the second, in opposite directions.

In both cases we have simply to choose a pole **C**, and draw S_0 S_1 and S_2. Then taking any point c in the line of direction of P_1, as a point of application for S_0, draw through this point S_1, thus finding d, the point of application for S_2. S_0 and S_2 prolonged, intersect upon the resultant, whose intensity, direction, and position thus become fully known.

18. Property of the Point b.—It is plain that thus a point of intersection b, through which the resultant must pass, can always be found, provided S_0 and S_2 do not fall together in the force polygon, or intersect without the limits of the drawing. By properly choosing the position of the pole **C**, this can always be avoided if the points 2 and 0 in the *force polygon do not themselves coincide*, i.e., if the force polygon does not close.

The point b, Figs. 8, 9, and 10, which by reason of the arbitrary position of the *pole* may lie anywhere upon the resultant, has a remarkable property. If we draw a line $m\,n$ through this point parallel to S_1, and let fall from it perpendiculars p_1 and p_2 upon P_1 and P_2, then in all three cases, and therefore generally, the triangle $c\,m\,b$ is similar to $0\,C\,1$, and $d\,b\,n$ is similar to $1\,C\,2$. Hence we have the proportions—

$0\,1 : 1\,C :: c\,m : m\,b$, and
$1\,C : 1\,2 :: n\,b : n\,d$.

From these proportions we find

$0\,1 : 1\,2 :: c\,m \times n\,b : m\,b \times n\,d$.

Now the triangles $c\,m\,b$ and $d\,n\,b$ have the same height above the base $m\,n$; the bases $m\,b$ and $b\,n$ are therefore proportional to their areas. But their areas are equal to half their sides $c\,m$ and $n\,d$ multiplied by p_1 and p_2 respectively. Hence we have from the above proportion, since $c\,m = n\,d$,

$0\,1 : 1\,2 :: n\,d \times p_2 : n\,d \times p_1$ or
$0\,1 : 1\,2 :: p_2 : p_1$.
or $P_1 : P_2 :: p_2 : p_1$.

That is, *the perpendiculars let fall from any point of the*

CHAP. II.] DIFFERENT POINTS OF APPLICATION. 19

resultant upon the components, are to each other inversely as the components. Regarding any point of the resultant as a centre of moments, the moments of the forces then are equal, and of course the forces themselves are inversely as their lever arms.

19. Equilibrium Polygon.—If we consider the forces P_1 P_2, Figs. 8, 9, and 10, held in equilibrium by their components C 0, 1 C, and 2 C, C 1, which act parallel to the lines S_0 S_1 and S_2; then regarding the line S_1 or $c\,d$ as part of the material plane in which the forces act, C 1 and 1 C balance one another, and cause either tension or compression in $c\,d$. Suppose the resultant R is to act so as to cause equilibrium, or prevent the motion of the plane due to P_1 and P_2. Then R must act upwards in Figs. 8 and 9, and downwards from 2 to 0 in Fig. 10. In Figs. 8 and 9 then, S_0 and S_2 act *away* from c and d (Art. 4), and in Fig. 10 *towards* c and d. Following round the force polygon, we find in the first two cases $c\,d$ in *tension*, in the last $c\,d$ in *compression*.

In the first two cases, the points of application c and d of S_0 P_1 and S_2 P_2 if connected by a *string* stretched between c and d will be perfectly fixed and motionless; while in the latter case, the string must be replaced by a *strut*. In case of three or more forces the polygon or broken line which we thus obtain, by choosing a pole, drawing lines to the beginning and end of the forces in the force polygon, and then parallels to these lines intersecting the lines of direction of the forces in the *force diagram*, we call the "*string*" or "*funicular polygon*," or the "*strut polygon*," according as the forces act to cause tension or compression along these lines. We can apply to both cases the general designation of polygon of equilibrium or "*equilibrium polygon.*"* The perpendicular let fall from the pole C *upon the direction of the resultant* in the force polygon, we call the "*pole distance*" and shall always designate it by H. The straight line joining the points c and d, or the beginning and end of the equilibrium polygon, we call the "*strut*" or "*tie line*" or generally the "*closing line*" and designate it by L. The convenience and application of these terms and conceptions will soon appear. In the present case of only *two* forces, the equilibrium polygon becomes a straight line and coincides with L, or $c\,d$.

[NOTE.—We repeat that in order to determine the *quality* of

---

* The term "equilibrium polygon" is preferred to "funicular," as it expresses the idea generally, without implying either tension or compression

the strain in cd, we have only to follow round the force polygon in *the direction of the forces*, and then refer to the force diagram. Thus Fig. 9, at c, P_1 S_0 and S_1 act, and are in equilibrium. The corresponding closed figure is given in the force polygon (*a*). S_0 acts away from c, P_1 acts downwards from 0 1. Continuing this direction we find S_1 acting from 1 towards C. *Reversing* this direction (Art. 4), we find that the resultant which *replaces* S_0 and P_1 acts from C to 1. Referring now to the force diagram (*b*), and transferring this direction to the point c, we find this resultant acts to pull c *away* from d or *contrary* to the direction of the force 1 C which replaces S_2 and P_2. The strain in cd is therefore tension.

A much better way of arriving at the same result is to consider the triangle cbd as a jointed *frame* which holds in equilibrium the forces P_1 P_2 and $R_{1.2}$. Then the strains in any two pieces cd, cb, meeting at a point, *are in equilibrium with the force or forces acting at that point*.

We have then the force P_1 acting at apex c, decomposed into strains along cb and cd (Art. 5) represented by C 0 and 1 C in the force polygon. All three are in *equilibrium*. P_1 acts *down*. Follow *down* then from 0 to 1 from 1 to C and C to 0. Refer back now to apex c of the frame and transfer these directions. The strain in cd acts *away* from the apex c and is therefore in tension, while the piece cb would be in compression, since the direction of C 0 is *towards apex c*.

See also "practical applications" of the preceding chapter for illustrations of this. In the same way follow round 0 1 C Fig. 10 (*a*) and refer to (*b*) and S_0 is in *compression*.]

20. Case of a Couple.—In Article 18 we remarked that the pole can always be chosen in such a position as to give S_0 and S_2 intersecting within desired limits, *provided* that S_0 and S_2 or the point 0 and 2 do not *coincide*. This case however actually happens, with a pair of equal and opposite forces—that is, with a *couple*.

Thus in Fig. 11, Pl. 3, we have two equal and opposite forces P_1, P_2.

The force polygon closes: *therefore the resultant is zero*. S_0 and S_2 are parallel, hence their point of intersection in the equilibrium polygon is *infinitely distant*. By changing the position of the pole, we see that S_0 and S_2 may take *any* positions in the plane.

Two forces therefore which form a couple *cannot be replaced by a single force*. Their resultant is an indefinitely small force situated in any position in the plane of the forces, at an infinite distance.

21. Conditions of Equilibrium.—If then, similarly to Art. 4, any number of forces lying in the same plane and having different points of application, are in *equilibrium*, the force polygon always *closes*.

For this reason, as already repeatedly seen in the practical applications of our last chapter, the force polygon formed by the exterior forces must always close.

But inversely, if the force polygon closes, *it does not follow that the forces are in equilibrium*—a couple *may* result.

To determine whether this is the case inspect also the "equilibrium polygon." If this *also closes* [*i.e.*, if S_0 and S_n intersect] the forces are in equilibrium. If this does *not* close [*i.e.*, if S_0 and S_n are *parallel*] there is no single resultant, but the forces can be replaced by a *couple*, and this couple, as we have seen, may have *any* position in the plane.

Thus if we suppose in Fig. 11, Pl. 3, P_1 and P_2 decomposed into their components S_0, S_1, and S_1, S_2, the compressive strains in S_1 at c and d are equal and opposite [see (*a*)]. We have then S_0 and S_2 remaining, which again form a couple which must have the same action as the first.

Hence we see that *one couple can be replaced by another without changing the action of the forces.*

It is easy to determine a simple relation between any two couples.

If from c we lay off $c\,a$ equal to $o\,1$, and $c\,o$ equal to Co, we have $o\,a$ parallel to $C\,1$ or S_1, and therefore to $c\,d$. Join $a\,d$ and $o\,d$. The triangles $c\,d\,a$ and $c\,d\,o$ having a common base $c\,d$ and their vertices o and a in a line parallel to $c\,d$, are equal in area. The side $c\,a$ of one is known, and the opposite apex lies in the line of the force P_2. Its area is then $c\,a = P_1$ multiplied by half of the perpendicular distance of P_1 from P_2, and is therefore completely determined. So also for the other triangle, one side of which $o\,c$ is one force of the new couple, and the opposite apex of which lies in the other force S_2.

Hence—*a couple can be turned at will in its plane of action, and the intensity and direction of its forces can be changed at will if the area of the triangle the base of which is one of the* new forces, and whose opposite apex lies in the other force, is constant; or when the product of the intensity of the forces into their perpendicular distance remains the same. The direction of rotation, of course, must also remain the same.

We shall see further on the significance of this area, or of this product—so much is clear, that a couple (or infinitely small, infinitely distant force) is completely determined in its plane when the direction of rotation is given, and the area of the triangle or value of the product to which it is proportional, is known. The couple itself can be replaced by any two parallel equal and opposite forces whatever, if only the triangle having one force as base, and the opposite apex in the other, has a given constant area.*

22. Force and Equilibrium Polygons for any Number of Forces in a Plane.

In Pl. 3, Fig. 12 (b) we have the forces P_{1-5} acting in various directions and at different points of application. P_2 and P_3 form a *couple;* that is, are equal, parallel, and opposite in direction. Required the position, intensity and direction of action of the resultant.

First, form the *force polygon*, Fig. 12 (a), by laying off the forces to scale one after the other in proper direction. Thus we have 0 1, 1 2, 2 3, 3 4, 4 5 in Fig. 12 (a) parallel respectively to P_1 P_2 P_3, etc., in Fig. 12 (b). The line necessary to close the polygon, 0 5, is the resultant in intensity and direction. In intensity because the length of 0 5 taken to the scale of force, gives the intensity of the resultant; in direction because acting from 5 to 0 it produces equilibrium, while acting in the opposite direction, from 0 to 5, it replaces the forces.

We have, therefore, only to find the *position* of the resultant in the plane of the given forces in Fig. 12 (b). Hence:

Second, choose anywhere a *"pole"* as C, and draw the lines or rays, or *"strings"* S_0 S_1 S_2 S_3 S_4, etc. S_0 and S_5 are evidently components of the resultant, since they form with it a closed figure in the force polygon.

Third, form the *equilibrium polygon a b c d e o'*, Fig. 12 (b), as follows:

Draw a line parallel to S_0 intersecting P_1 (produced if necessary) at any point as *a*. From this point draw a line parallel

* *Elemente der Graphischen Statik.* Bauschinger. München. 1871. Pp. 11, 12.

to S_1 to intersection with P_2 (also produced if necessary) at b. From b parallel to S_2 to c, then parallel to S_3 to d, and finally parallel to S_4, to intersection e with P_5. Through this last point draw a line parallel to the last ray S_5. Now S_0 and S_5 are components of the resultant 0 5 [Fig. 12 (*a*)] and are found in proper relative position. Produce them, therefore, to intersection o'. Through this point the resultant must pass. Drawing then through o', a line parallel to 0 5, we have the resultant in proper position, and acting in the direction indicated in the figure, it produces *equilibrium*.

Any other point than a, upon the direction of P_1, assumed as a starting point, would have given a different point o'; so also for any other assumed position of the *pole* C. But in every case we shall obtain a point upon the line of direction of $R_{1.5}$ already found. The reader may easily convince himself of this by making the construction for different poles, and points of beginning.

Now the polygon or broken line, $a\ b\ c\ d\ e$, we call the *equilibrium polygon*—that is, *it is the position which a system of strings or struts, $S_0\ S_1\ S_2$, etc., would assume under the action of the given forces at the assumed points of application.*

Thus P_1 acting at a, is held in equilibrium by the forces along S_0 and S_1, P_2 acting at b, by S_1 and S_2 and so on. If we join any two points in the line of direction of S_0, and S_5, as $m\ n$ by a line, we have then a *jointed frame*, which acted upon at the apices $a\ldots e$ by the forces $P_1\ldots P_5$, and at m and n by S_0 and S_5 is in *equilibrium*.

For S_0 acting at m, we see from the force polygon may be *replaced* by a force a 0 parallel and opposed to the resultant R and a force C a acting along the line L. In like manner S_5 may be *replaced* by a C and 5 a parallel and opposed to the resultant. The two forces a C and C a being equal and opposed balance each other through $m\ n$, while the sum of 0 a and 5 a is equal and opposed to the resultant 0 5. There is, therefore, equilibrium, and m and n may be considered as the *points of support* of the frame acted upon by the forces $P_1\ldots P_5$ at the apices $a\ldots e$, a 0 and 5 a being the upward *reactions* at the points of support.

As to the *quality* of the strains in the different pieces; as before the reaction at m, viz., a 0, is in equilibrium with the strain in $m\ n$ and $m\ a$. Following round, then, in the force polygon from a to 0, 0 to C and C to a, and referring back to the *frame*, we find strain in $m\,n$ acting *towards* apex m, therefore *compressive;* strain in $m\,a$ acting *away* from m, therefore *tensile*. In like manner $S_1\,S_2\,S_3$ are in tension, while S_4 or $d\,e$ and S_5 or $e\,n$ are compressed.

Hence we may *fix* any two points of the equilibrium polygon by joining them by a line. The forces acting at these points are at once found by drawing from C in the *force polygon* a parallel to this line to intersection with resultant. Thus a C (since we have taken $m\,n$ parallel to S_1) is the force in $m\,n$ and $a\,0$, $5\,a$, are the forces opposed to the resultant at m and n.

23. Influence of a Couple.—Among the forces in Fig. 12 there are two, P_2 and P_3 which are equal, parallel and opposite, the direction of rotation being as indicated by the arrow. Examining the equilibrium polygon, we see that the influence of the couple is to *shift S_1 through a certain distance parallel to itself,* to S_3. Now suppose the forces composing the couple *were not given,* but the *value* of the couple known, from the direction of rotation and the area of the triangle $A_2\,P_2\,P_3$, which has its base equal to one of the forces and a height equal to their perpendicular distance. In this case the lines 1 2, and S_2 in the *force polygon,* would disappear, but we can none the less find the point d, and from this point continue the polygon by drawing S_4 and S_5, and thus find the same points e and o' as before. To do this we have simply to apply the principle deduced in Art. 21, that one couple *can be replaced by another provided the area of the triangle is constant.*

In the present case we must replace the given couple by another whose forces are S_1 and S_3, having the same direction of rotation.

Lay off then from a, $a\,i$ equal by scale to S_1 as given in the *force polygon.* Describe upon S_1 the triangle $a\,g\,h$ equal to the given area $A_2\,P_2\,P_3$. Draw $g\,i$, and then through h, $h\,k$ parallel to $g\,i$. The point k is upon the line of direction of S_3, or in other words the area of the triangle $i\,k\,a$ is equal to $a\,g\,h$. The proof is easy. The two triangles $i\,g\,h$ and $i\,g\,k$ are equal, since they have the same base $i\,g$, and height. But if from the triangle $a\,i\,g$ we subtract $i\,g\,h$, we obtain $a\,g\,h$. If from the *same* triangle $a\,i\,g$ we subtract $i\,g\,k$, which is *equal* to $i\,g\,h$, we obtain $i\,k\,a$. Equals subtracted from equals leave equals. Hence $i\,k\,a$ is equal to $a\,g\,h$.

If then through k we draw a line parallel to S_3 and produce it to d, we have the same point as before, and thus from d, can continue the polygon.

[*Note* that the direction of rotation *shows the side of* S_1 *upon which the point k must fall*. S_1 acts away from a [from 1 to C in (a)] hence for rotation as shown by the arrow, g must fall above S_1, and S_1 is shifted *upwards*.

24. Order of Forces Immaterial.—As in the case of a common point of application, so also here, the *order in which the forces are laid off is immaterial*. To prove this for two forces is sufficient, as by continued interchange of two and two, we can obtain any desired order.

Let the two forces be P_4 and P_5 (Fig. 13, Pl. 3) existing either alone, or in combination with others preceding and following.

Taking the forces first in the order P_4 P_5, we have the *equilibrium polygon* S_3 S_4 S_5, (b) giving the point a in the resultant. Taking them now in reverse order, P_5 P_4, we have the polygon S_3 S'_5 S'_4 giving the same point a in the resultant. The resultant in the force polygon (a), viz., 0 5, is of course unchanged in intensity and direction in either case. It is required to prove that in the second case the last string S'_4 is not only parallel to S_5 in the first, but *coincides with it*.

This is easy. The resultant of P_4 P_5 goes through a, the intersection of S_3 and S_5. The same resultant in the second case must also pass through the intersection of S_3 and S'_4. But S_3 is the same in position and direction in both cases. If the second point of intersection does not coincide with a, still it must lie somewhere upon S_3. Hence as the resultant must pass through both points, it must *coincide* with this last line; viz., S_3. But this is not possible, as the resultant must also pass through d, the point of intersection of the forces, or when these do not intersect must be parallel to them. As therefore S'_4 must be parallel to S_5 (shown by the force polygon), the intersections in each case must coincide, as also the lines S'_4, S_5 themselves, and the polygon from e on has the same course in either case.

25. Pole taken upon closing line.—We have seen (Art. 20) that when any number of forces are in equilibrium *both* the force and equilibrium polygon must close. There is one exception to this statement. Since the pole may be taken anywhere, suppose it taken somewhere *upon the line closing the force polygon*. This line, as we know, is the resultant, and holds the other forces in equilibrium. But now the equilibrium polygon evidently *will not close*. On the contrary, the first and last strings will be parallel. This position of the pole should then in general be avoided. For any other position of the pole our rule holds good; viz.,

If the force polygon closes as also the equilibrium polygon, the forces are in equilibrium. If the equilibrium polygon however does not close, the forces cannot be replaced by a single force but only by a couple. The forces of this couple act in the parallel end lines of the equilibrium polygon, and are given in intensity and direction of action by the line from the pole to the beginning of the force polygon [beginning and end coinciding].

26. Relation between two equilibrium polygons with different poles.—We may deduce an interesting relation between the two equilibrium polygons formed by choosing different poles, with the same forces and force polygon.

Thus with the forces P_1 P_2 P_3 P_4, we construct the force polygon Fig. 14 (*a*), Pl. 4. Then choose a pole C and draw S_{0-4}, and thus obtain the corresponding equilibrium polygon S_0 *a b c d* S_4 Fig. 14 (*b*). Choose now a second pole C'. Draw S'_{0-4} and construct the corresponding polygon S'_0 *a' b' c' d'* S'_4. [In our figure *c* and *c'* fall accidentally nearly together.]

Join the two poles by a line C C'. Then—*any two corresponding strings of these two polygons intersect upon the same straight line* M N *parallel to* C C'. Thus S_0 and S'_0 intersect at *g*, S'_1 and S_1 at *k*, S'_2 and S_2 at *l*, S'_3 and S_3 at *n*, S'_4 and S_4 at *m*—and *all* these points *g, k, l, n* and *m*, lie in the same straight line M N parallel to the line C C' connecting the poles.

The proof is as follows.* If we decompose P_1 into the components S_0 S_1 and S'_0 S'_1, these components are given in intensity and direction by the corresponding lines in the force polygon. If we take the two first as acting in opposite directions from the two last, they hold these last in equilibrium. The resultant therefore of any two as S_0 and S'_0 must be equal and opposed to that of the remaining two, S_1 and S'_1, and both resultants must lie in the same straight line. This straight line must evidently be the line *g k* joining the intersections of S_0 S'_0

---

* *Elemente der Graphischen Statik.* Bauschinger. München, 1871. Pp. 18–19.

and $S_1 S'_1$. But from the force polygon we see at once that the resultant of S_0 and S'_0 is given in direction and intensity by C C', and this is also the resultant of S_1 and S'_1. The line joining g and k must therefore be parallel to C C'. For the second force P_2 we can show similarly that the line joining k and l is parallel to C C'. But k is a common point of *both* lines—hence $g\ k$ and l lie in the same straight line parallel to C C'.

[NOTE.—*The pure geometric proof is as follows: The two complete quadrilaterals* 0 1 C' C *and* $g\ k\ a'\ a$ *have five pairs of corresponding sides parallel, viz.,* 0 1 *and* $a\ a'$, $a\ a'$ 1 C' *and* $a'\ k$, C 0 *and* $a\ g$, o C' *and* $a'\ g$, 1 C' *and* $a\ k$; *hence the sixth pair are also parallel, viz.,* C C' *and* $g\ k$. *In like manner for* 1 2 C C' *and* $l\ k\ b'\ b$ *and so on.*]

We can make use of this principle in order from one given equilibrium polygon $S_0\ a\ b\ c\ d\ S_4$ and pole, to construct another, the direction of C C' being known. For this purpose, having assumed the position of the first string S'_0 we draw through its intersection g with S_0 a line **M N** parallel to C C'. The next string must therefore pass through the intersection a' of S'_0 and P_1 and through the point k, of intersection of the second string of the first polygon and the line **M N**. It is therefore determined. The next side must pass through b' and l, and so on.

[*Note.* Observe that the intersections r and r' of the first and last lines of both polygons must lie in a straight line parallel to 0 4, the direction of the resultant.]

27. Mean polygon of equilibrium.—Since the pole may have any position, let us suppose it situated in *one of the angles* of the force polygon. It is evident that the first line of the corresponding equilibrium polygon, then *coincides with the first force*. If now the pole be taken at the *beginning of the first force* in the force polygon, then the first side of the corresponding equilibrium polygon will coincide with the first force, and the last line *will be the resultant itself in proper position*.

Take for instance, the pole at o in the force polygon, Fig. 15 (*a*), Pl. 4. The first side S_0 reduces to zero. The next S_1 coincides with 0 1. In (*b*) therefore P_1 is the first side of the equilibrium polygon. The next side S_2 corresponds with S_2 in (*a*). Thus we obtain the polygon $a\ b\ c\ d\ e$, the last side of which S_7, *is the resultant itself.* That is, S_2 is the resultant of P_1 and P_2, S_3 of $P_{1\text{-}3}$, S_4 of $P_{1\text{-}4}$ and so on. Every line in the polygon then is the resultant of the forces preceding, and we call such a polygon the *mean polygon of equilibrium*.

If we wish to find the mean polygon for P_{3-7} we have only to take the new pole C' at 2 in the force polygon (a). According to the preceding Art., each side of the new polygon must pass through the intersection of the corresponding side of the first with the line S_2 which passes through a and is parallel to $C C'$. Thus S'_4 must pass through b' and o. S'_5 through c' and n, and so on. S'_7 is the resultant of P_{3-7}, and since S_2 is the resultant of P_{1-2}; S_7, the resultant of P_{1-7}, must pass through the intersection m of S'_7 and S_2.

We observe here again the influence of the *couple* P_5 and P_6. S_4 and S'_4 are simply *shifted* through certain distances, without change of direction, to S_6 and S'_6; and as we have seen above, knowing the direction of rotation, and the moment of the couple, we might have omitted it in the force polygon and still obtained S_7 and S'_7 as before.

28. Line of pressures in an arch.—The practical application of the above will be at once seen in the consideration of an arch. Thus with the given horizontal thrust applied at a given point of the arch, and the forces P_{1-5}, we construct the force polygon $C \, o \, 5$, and then the line of pressures $a \, b \, c \, d$. [Fig. 16, Pl. 4.]

Required with another thrust $H' = o \, C'$ acting at another point, and the same forces P_{1-5}, to construct the corresponding line of pressures. To do this we have only to lay off $o \, C'$ equal to the new horizontal thrust, then choose a point of the force line, as 3, as a *pole* and draw the corresponding polygon, $k \, o \, p \, k$; the point of intersection, k, is a point upon the line $m \, n$ parallel to $o \, C$, and upon this line will be found the intersection of corresponding sides of the two polygons. Thus from the intersection of the side $a \, p$ of the first polygon with $m \, n$, draw a line to o and we have a'. From the intersection b of the second line of the first polygon draw a line to a', and we have $b' \, a'$, and so on.

29.—The preceding articles comprise all the most important principles of the Graphical Method which can be deduced independently of its practical applications. Future principles will be best demonstrated, and at the same time illustrated, by considering the various special applications of the method, and to these applications we shall therefore now proceed.

CHAPTER III.

CENTRE OF GRAVITY.

30. General Method.—One of the most obvious applications of our method, as thus far developed, is to the determination of the *centre of gravity* of areas and solids. We shall confine ourselves to areas only, merely observing that all the principles hitherto developed apply equally well to forces in space. The forces being given by their orthographic projections upon two planes after the manner of descriptive geometry, the projections upon each plane may be dealt with as forces lying in that plane, and thus the projections of the force and equilibrium polygons, the resultant, etc., determined.

A body under the action of gravity may be considered as a body acted upon by parallel forces. The resultant of these forces being found for one position of the body [or the body being considered as fixed, for one common direction of the forces] may have its point of application anywhere in its line of direction.

For a new position of the body [or another direction of the forces] there is another position for the resultant. Among all the points which may be considered as points of application of these two resultants there is *one* which remains unchanged in position, whatever the change in direction of the parallel forces. This point must evidently lie upon *all* the resultants, and is therefore given by the intersection of any two.

It is hardly necessary to give illustrations of the method of procedure.

Generally, we divide up the given area into triangles, trapezoids, rectangles, etc., and reduce the area of each of these figures to a rectangle of assumed base. The heights of these reduced rectangles will then be proportional to the areas, and hence to the force of gravity acting upon them; *i.e.*, to their *weights*. Consider then these heights as forces acting at the centres of gravity of the partial areas. Construct the *force polygon* by laying them off one after the other. Choose a pole and draw lines from it to the beginning and end of each force. These lines will give the sides of the *funicular* or *equilibrium polygon*. Anywhere in the plane of the figure, draw a line parallel to the first of these pole lines (S_0). Produce it to intersection with the first force (P_1), prolonged if necessary. From this intersection draw a parallel to the second pole line (S_1), and produce to intersection with second force (P_2). So on to last pole line, which produce to intersection with first pole line. Through this point the resultant must pass, and of course it must be parallel to the forces.

Now suppose the parallel forces all revolved say 90°, the points of application remaining the same. Evidently the new force polygon will be at right angles to the first, as also the new pole lines, each to each. It is unnecessary then to form the new force polygon. The directions of the new pole lines are given by the old, and this is all that is needed.

Anywhere then in the plane of the figure, draw a line (S'_0) perpendicular to the first pole line (S_0) previously drawn, and prolong to intersection with new direction of first force (P_1'). Through this point draw a perpendicular (S_1') to second pole line, to intersection with new direction of second force (P_2') and so on. We thus find a point for new resultant, parallel to new force direction. Prolong this resultant to intersection with first and the centre of gravity is determined.

[NOTE.—If the area given has an *axis of symmetry*, that can of course be taken as one resultant, and it is then only necessary to make one construction in order to find the other.]

The given area of irregular outline must, as remarked above, be divided by parallel sections into areas so small that the outlines of these areas may be considered as practically straight lines. The forces are then taken as acting at the centres of gravity of these areas. This division will give us generally a number of triangles and trapezoids.

It is therefore desirable to reduce graphically to a common base the area of these triangles and trapezoids, and for this purpose the following principles will prove of service:

32. Reduction of Triangle to equivalent Rectangle of given Base.—Let b be the base and h the height. Then area $= \dfrac{b\,h}{2}$. Take a as the given reduction base, and let x represent the height of the equivalent rectangle. Then

$$ax = \frac{bh}{2}, \text{ or } \frac{h}{a} = \frac{x}{\frac{1}{2}b}.$$

Now a, b, and h being given, it is required to find x graphically.

Let **A B C** be the triangle, and **D** the middle of the base. [Fig. 17, Pl. 5.] Lay off **A E** $= h$ and **A F** $= a$. Draw **F D**, and parallel to **F D** draw **E** x. Then **A** x is the required height.

$$\text{For}: \frac{\mathbf{A}\ x}{\mathbf{A D}} = \frac{\mathbf{A E}}{\mathbf{A F}} \text{ or } \frac{x}{\frac{1}{2}b} = \frac{h}{a}.$$

As to the *centre of gravity* of the triangle, it is at the intersection of the lines from each apex to the centre of the opposite side; since these are *medial* lines

33. Reduction of Trapezoid to equivalent Rectangle.—In the trapezoid **A B C D**, Fig. 18, Pl. 5, draw through the middle points of **A D** and **B C** perpendiculars to **D C**, and produce to intersections **E** and **F** with **A B** produced.

Then lay off **F** $g = a =$ the given reduction base, and draw g **E** intersecting **D C** in x. Then **H** x is the required height.

$$\text{For } \frac{\mathbf{E F}}{\mathbf{F}\ g} = \frac{\mathbf{H}\ x}{\mathbf{H E}} \text{ or } \frac{\mathbf{E F}}{a} = \frac{x}{\mathbf{H E}};$$

hence $a\ x =$ **E F** $\times$ **H E** $=$ area.

To find the *centre of gravity*, draw a line through the middle points of the parallel sides **A B** and **D C**. Prolong **A B** and **C D** and make **C** $a =$ **A B** and **A** $b =$ **C D** and join a and b. Then the intersection of $a\,b$ with the axis of symmetry gives the centre of gravity.

The construction for the reduction of a *parallelogram* is precisely similar. [Fig. 18 (*b*).]

The points **F** and **E** here coincide with **A** and **B**, and we have

$$\frac{\mathbf{A}\ x}{h} = \frac{\mathbf{A B}}{\mathbf{B}\ g}, \text{ or } a\ x = h \times \mathbf{A B} = \text{area}.$$

The same construction also holds good, of course, for a *rectangle* or square. The centre of gravity in each case is at the intersection of two diameters, since these are axes of symmetry.

34. Reduction of Quadrilaterals Generally.—In general

any quadrilateral may be divided into two triangles which may be reduced separately, or into a triangle and trapezoid.

It is also easy to reduce any quadrilateral to an *equivalent triangle*, which may then be reduced by Art. 32 to an equivalent rectangle of given base.

Thus we reduce the quadrilateral **A B C D** [Fig. 18 (c)] to an equivalent triangle by drawing **C C$_1$** parallel to **D B** to intersection **C$_1$** with **A B**, and joining **C$_1$** and **D**. The triangle **D B C$_1$** is then equal to **D B C**, and hence the area **A D C$_1$** is equal to **A B C D**. The triangle **A D C$_1$** can now be reduced to an equivalent rectangle of given base by Art. 32.

The *centre of gravity* of the quadrilateral may be found as follows:

Draw the diagonals **A C** and **B D** and mark the intersection **E**. Make **A E$_1$** = **C E** and **B E$_2$** = **D E**, also find the centres **O$_1$** and **O$_2$** of the diagonals **A C** and **B D**. Join **O$_2$ E$_1$** and **O$_1$ E$_2$**; the intersection **S** of these two lines is the centre of gravity required.

The above is sufficient to enable us to find the centre of gravity of any given area of regular or irregular outline. The method may be applied to finding the centre of gravity of a loaded water-wheel (as given in *Der Constructeur*, Reuleaux, Art. 47), and many similar problems. The reader will have no difficulty, following the general method indicated in Art. 30, in making such applications for himself. The method itself is so simple that it is unnecessary to give here any practical examples in illustration. We shall, moreover, have occasion to return to the subject in the consideration of *moment of inertia of areas*.

We pass on therefore to the *moment of rotation of forces in a plane*.

CHAPTER IV.

MOMENT OF ROTATION OF FORCES IN THE SAME PLANE.

35. The "Moment" of a Force about any Point is the product of the force into the perpendicular distance from that point to the line of direction of the force. The importance and application of the "*moment*" in the determination of the strains in the various pieces of any structure will be evident by referring to Art. 14, where Ritter's "method of sections" is alluded to. In general, when the moments of all the exterior forces acting upon a framed structure are known, the interior forces, or the strains in the various pieces, can be easily ascertained.

As we shall immediately see, these moments are given directly in any case by the "*equilibrium polygon.*"

36. Culmann's Principle.—If a force **P** be resolved into two components in any directions as b **C**, b **C**$_1$ (Fig. 19, Pl. 5), and these components be prolonged, it is evident that the *moment* of **P** with reference to any point as a situated anywhere in the line $c\,d$ parallel to **P**, is **P** × $b\,a$. But if from C we draw the perpendicular **H** to **P**, then by similar triangles,

$$\mathbf{P} : \mathbf{H} :: c\,d : b\,a\,;$$

$$\mathbf{P} \times b\,a = \mathbf{H} \times c\,d.$$

That is, *the moment of* **P** *with respect to any point a is equal to a certain constant* **H** *multiplied by the ordinate c d, parallel to* **P** *and limited by the components prolonged.* The constant **H** we call the "*pole distance.*"

This holds good for any point whatever, and we have only to remember that if we assume the ordinates to the right of **P** as positive, those to the left are negative.

We can choose the *pole* **C** where we please, and thus obtain various values for **H**, but *for any one value the corresponding ordinates are proportional to the moments.*

The above principle is due to *Culmann*, and will be referred to hereafter as *Culmann's principle*.

37. Application of the above to Equilibrium Polygon. —Let P_{1-4} be a number of forces given in position as represented in Fig. 19 (*a*) Pl. 5. By forming the *force* polygon Fig. 19 (*b*), choosing a pole C, and drawing S_0 S_1, S_2, etc., we form the *equilibrium polygon a b c d e f*, Fig. 19 (*a*).

The resultant of the forces P_{1-4} acts in the position and direction given in the Fig. Now, as we have seen in Art. 22, regarding the broken line *a b c d e* as a system of *strings*, we may produce equilibrium by joining any two points as *a* and *f* by a *line*, and applying at *a* and *f* the forces S_0 and S_4. Let us suppose this line *a f* perpendicular to the direction of the resultant. Since we can suppose the broken line or polygon fastened at any two points we please, this is allowable, and does not affect the generality of our conclusion.

Then the compression in the line *a f* is given by H, the "*pole distance*," or the distance of the pole C from the resultant in the force polygon. We have therefore at *a* the force H and $V_1 = H\ 0$ acting as indicated by the arrows. At *a* then V_1 acting up, H and S_0 acting away from *a*, are in equilibrium, or V_1 is decomposed into H and S_0, as shown by the *force polygon*.

According to *Culmann's principle* then, the moment of V_1 with reference to any point, as *m* or *o*, is equal to $H \times o\ m$. Therefore H being known, the ordinates between *a f* and S_0 are proportional to the moment of V_1 at any point. V_1 acting upwards gives positive rotation (left to right) with respect to *m*.

At the point *b*, P_1 may be replaced by a force 0 K parallel to R and a force K 1 along S_1 [see force polygon]. This we see at once from the force polygon where 0 K and K 1 make a closed polygon with P_1, and taken as acting from 0 to K and K to 1, *replace* P_1. But these two forces are in equilibrium with S_1 and S_0, or 1 C and C 0 [see force polygon], and since K 1 and 1 K balance each other, all the forces acting at *b* may be replaced by S_0, 0 K and K C. We have then at *b* the force 0 K resolved into components in the directions S_0 and S_1.

By *Culmann's principle*, therefore, the moment of 0 K about any point as *m*, is proportional to the ordinate *n m*, and since 0 K acts downward this moment is negative. Hence the *resultant* moment at *m* or *o* of the components at *a* and *b* parallel to **R**, is proportional to the ordinate *o n*.

So for any *point, the ordinate included by the polygon a b c d e f, and the closing line a f, to the scale of length multiplied by the "pole distance"* **H** *to the scale of force, gives the moment at that point of the components parallel to the resultant.*

The practical importance and application of this principle will appear more clearly in the consideration of parallel forces in the next Chapter.

CHAPTER V.

MOMENT OF ROTATION.—PARALLEL FORCES.

38. Equilibrium Polygon.—Since the forces acting upon structures are generally due to the action of gravity, these forces may be considered as parallel and vertical, and in all practical cases therefore, we have to do with a system of parallel forces.

Given any number of parallel forces P_{1-5}, Pl. 6, Fig. 20; required to find the direction, intensity and position of the resultant, and the moment of rotation at any point.

1st. Draw the *force polygon*. In this case it is, of course, a straight line.

2d. Choose a *pole* O, and draw the lines S_0, S_1, S_2, etc.

3d. Draw the string or equilibrium polygon *a b c d e f*. Considering this polygon as a system of *strings*, the forces will be held in equilibrium if we join any two points, as *a* and *g*, by a strut or compression piece, and apply at *a* and *g* the upward forces V_1 and V_2.

4th. Prolong *a b* and *f g* to their intersection *o*. Through this point the resultant must pass. It is of course parallel and equal to the sum of the forces.

Now, if *a g* is assumed horizontal, the perpendicular H to the force line, or the "*pole distance*," divides the resultant 0 5 into the two reactions V_1 and V_2 (Art. 22).

All the forces in the equilibrium polygon have the same horizontal projection H, in the force polygon.

Let *a g* represent a *beam* resting upon supports at *a* and *g*. We have then at once the vertical reactions V_1 and V_2 or *k* 0 and 5 *k*, which, in order to cause equilibrium, must act *upwards*.

For the moment at any point, as *o*, due to V_1, we have, by Culmann's principle, *m o* multiplied by H. The triangle formed by *a b*, *a g*, and P_1, gives then the moment of rupture at any point of the beam as far as P_1. For a point *o*, beyond P_1, the moment due to V_1, must be *diminished* by that due to P_1, since these forces act in opposite directions, and rotation from left to right upon the *left* of any point is considered positive. We see at once from the force polygon that P_1 is resolved into S_0 and S_1 or into $a\,b$ and $b\,c$. Hence the moment at o due to P_1 is $m\,n$ multiplied by H. The total moment at o is then $m\,o - m\,n = n\,o$, multiplied by H.

Hence we see that *the ordinates to the equilibrium polygon from the closing line $a\,g$, are proportional to the total moments; while the ordinate at any point between any two adjacent sides of this polygon, prolonged, represents the moment at that point of a force acting in the vertical through the intersection of these two sides.*

[The reader should make the construction, changing the *order* in which the weights are taken, and thus satisfy himself that the order is a matter of indifference. As to the *direction* of the reactions V_1, V_2, it must be remembered that $a\,b$ is to be *replaced* by V_1 and H, hence V_1 must be opposed to $C\,0$, the direction obtained by following round in the force polygon the triangle $0\,1\,O$. Force and distance *scales* should also be assumed. Thus the ordinates to the equilibrium polygon scaled off say in inches, and multiplied by the number of tons to one inch, and then by the "pole distance" taken to the assumed scale of distance, will give the moments of any point.]

The resultant of *any* two or more forces must pass through the intersection of the outer sides of the equilibrium polygon for those forces (Art. 16). Thus, the resultant of P_1 and P_2 must pass through the intersection of $a\,b$ and $c\,d$. Of V_1 and P_1, through the intersection of $a\,g$ and $b\,c$; of $P_1\,P_2$ and P_3, through intersection of $a\,b$ and $d\,e$, and so on. In every case the intensity and direction of action of the resultant is given directly by simple inspection of the force polygon.

Thus from the force polygon we see that the resultants $k\,2$ and $k\,3$ of $V_1\,P_1\,P_2$ and $V_1\,P_1\,P_2\,P_3$, act in *different* directions. Their points of application are at the intersection of $c\,d$ and $d\,e$ respectively with $a\,g$, or upon either side of d in the equilibrium polygon. At d the ordinate and hence the moment is greatest, and at this point the tangent to the polygon is parallel to $a\,g$. If we had a continuous succession of forces; if $a\,g$, for instance, were continuously or uniformly loaded; the equilibrium polygon would become a *curve*, and the tangent at d would then coincide with the very short polygon side at that point.

The points of application of the resultants of all the forces right and left of d are then at the intersection of this tangent with $a\,g$, or *at an infinite distance.*

At d then we have a *couple*, the resultant of which is as we have seen (Art. 20), an indefinitely small force acting at an indefinitely great distance. That is, with reference to d, the forces acting right and left cannot be replaced by a single force.

Hence generally: at the point of maximum moment ("*cross section of rupture*"), the resultant of the outer forces on either side reduces to an indefinitely small and distant force, the direction of which is reversed at this point, and the point of application of which changes from one side to the other of the equilibrium polygon.*

The "cross section of rupture" then, is that point where the weight of that portion of the girder between it and the end is equal to the reaction at that end, or where the resultant changes sign.

The *value of the moment* at this point, is therefore *equal to the product of the reaction at one end into its distance from the point of application of the equal resultant of all the loads between that end and the point.*

Thus for a beam uniformly loaded with w per unit of length, the reaction at each end is $\dfrac{w\,l}{2}$. From the above, the cross section of rupture is then at the middle. The point of application of the resultant of the forces acting between one end and the middle is at $\dfrac{l}{4}$, hence the maximum moment is $\dfrac{w\,l}{2} \times \dfrac{l}{4} = \dfrac{w\,l^2}{8}$.

39. Beam with Two Equal and Opposite Forces beyond the Supports.—The ordinates to the equilibrium polygon thus give, as it were, a *picture* or simultaneous view of the change and relative amount of the moments at any point. The point where the moment is greatest, *i.e.*, where the beam is most strained, is at once determined by simple inspection.

Let us take as an example a beam with two equal and opposite forces *beyond* the supports. Thus, Fig. 21, Pl. 6, suppose the beam has supports at **A** and **B**, the forces being taken in the order as represented by $P_1\ P_2$. We first construct the force

---

* *Die Graphische Statik.*—Culmann, p. 127.

polygon from 0 to 1, and 1 to 2 or 0. Next choose a pole **C**, and draw S_0 S_1 and S_2. Draw then a parallel to S_0 till intersection with *first force*, P_1, then parallel with S_1 to *second force*, P_2, then parallel to S_2 or S_0 to intersection with vertical through support **B**, and finally draw the *closing line* **L**. A line through **C**, parallel to **L**, gives as before the vertical reactions. Following round the force polygon, we find at **A** the reaction *downwards*, since S_0 acts from **C** to 0 and is to be *replaced* (Art. 4) by **L** and V_1; at **B** reaction *upwards*, since P_2 acts up, and following round, S_2 acts from 0 to **C**. *Both* reactions are equal to *a* 0. At **A** then the support must be above, and at **B** below the beam. The shaded area gives the moments to pole distance **H**. Had we taken the pole in the perpendicular through *o*, S_0 would have been parallel with the beam itself. This is, however, a matter of indifference. The moment area may lie at *any* inclination to the beam. We also see here again the effect of a *couple* (Art. 23). S_0 is simply *shifted* through a certain distance to S_2, parallel to S_0, and therefore the moment at any point between P_2 and **B** is *constant*. This is generally true of any couple, as we have already seen, Article 21, and may be proved analytically as follows:

Let the distance between the forces be $a =$ **A B**, Fig. 22. Then for any point *o*, we have $P \times (a + B\,o) - P \times B\,o = P\,[a + B\,o - B\,o] = P\,a$. For *o'* between **A** and **B**, $P \times A\,o' + P \times o'\,B = P\,[A\,o' + o'\,B] = P\,a$.

So also for any point to the left, the same holds true.

Graphically the proof is as follows:

Decompose both forces into parallel components, Fig. 23. Then for any point, as *o*, we have the moment $M = H \times m\,n - H \times m\,p$ or $M = -H \times n\,p$. But $n\,p$ is the constant ordinate between the parallel components **A** n and **A** p.

We see, therefore, by simple inspection, that the distance of P_1 and P_2 from the support **B**, Fig. 21, has no influence whatever upon the moment or strain in **A B**, provided the distance between the points of application remains the same, and that the moment at all points between P_2 and the support **B** is constant and a maximum. From **B** and P_2 the moments decrease left and right, and become zero at **A** and P_1.

40. Beam with Two Equal and Opposite Forces between the Two Supports.—Let the beam **A B**, Fig. 24, Pl. 6, be acted upon by the two equal and opposite forces P_1 P_2.

Construct the force polygon 0 1 2. Choose a pole C and draw C 0, C 1, C 2. Parallel to C 0, draw the first side of the equilibrium polygon to intersection with *first force* P_1; then parallel to C 1 to *second force* P_2; then parallel to C 2 to d. Join d and 0. Parallel to this draw C a in force polygon. Then 0 a is the vertical reaction at A, which acts upwards, since it must with C a *replace* C 0; and C 0, when we follow round from o to 1 and 1 to C, acts from C to 0.

We have the same vertical reaction at B, but here, since we must follow from 1 to 2 and 2 to C, C 2 acts from 2 to C, hence following round, the reaction at B is downward. The shaded area gives the moments to pole distance H, as before.

We see at once that at a certain point e the moment is zero. Left and right of this point the moment is positive and negative. At the point itself we have a *point of inflection*, and here, since the moment is zero, there is *no longitudinal strain*. At b and c the moments are greatest; here the beam is most strained, and at these points, therefore, are the "cross sections of rupture." Here again, if we had taken the pole C in the perpendicular through a, the closing line of the polygon o d would have been horizontal. It is, however, indifferent at what inclination a d may lie, but we may if we wish make it horizontal *now*, and then lay off from its new intersections with P_1 and P_2 along the directions of these forces, the ordinates already found at b and c, and join the points thus obtained with the ends of o d (*i.e.*, with its intersections with the verticals through the supports). The ordinates of the new polygon thus found will be for any point the same as before, and will also be perpendicular to the beam.

[NOTE.—Had we taken the forces precisely as above but in *reverse order*, the force line would be reversed, and we should have 0 and 2 in place of 1, and 1 in place of 0 and 2; that is, in place of C 1 we should have C 0 and C 2. Constructing then the equilibrium polygon by drawing a line parallel to new C 0 to intersection with new P_1, then parallel to new C 1 to intersection with new P_2, then parallel with new C 2 to intersection with vertical through B, and finally joining this last point with intersection of the first line drawn (C 0) with vertical through A, we have at first sight a very different equilibrium polygon. This new polygon will consist of *two* parts. If the ordinates in one of these parts are considered positive, those in the other must be negative. The *difference* of the ordinates in these two portions for any point, will give the same result as above. This, by making the above construction, the reader can easily prove.]

41. Many other problems will readily occur, which may in a similar manner be solved. The weights may have any position, number and intensities desired; in any and every case we have only to construct with assumed pole distance the corresponding equilibrium polygon, and we obtain at once the moments at every point. By the use of convenient scales, numerical results may be obtained which may be checked by calculation, and the practical value and accuracy of the method thus demonstrated.

The above principles will be sufficient for the solution of any such problem which may arise, and we shall therefore content ourselves with the above general indication of the method of procedure, and pass on to the consideration of a few cases where the above needs slight modification, and which, from their practical importance, and the ease with which they may be treated graphically, seem worthy of special notice.

1ST. BEAM OR AXLE—LOAD INCLINED TO AXIS.* [Fig. 25, Pl. 6.]

We have here simply to draw the "closing line" **A C** parallel to the beam or axle. From d draw d **B** parallel to the force **P**, then draw **A B** in *any* direction at pleasure, and join **B C**. We have thus the equilibrium polygon **A B C**, the ordinates to which, as d **B**, *parallel to the force* **P**, will give the moments, *provided we know the corresponding pole distance*.

But this can easily be found. As we have already seen, the force polygon being given, the equilibrium polygon may be easily constructed. Inversely, the equilibrium polygon being given, the force polygon may be constructed. Thus from **A** draw **A** c equal and parallel to **P**, and then draw c **C**$_1$ parallel to **B C**. **A** a and b c are the vertical reactions **P**$_1$ and **P**$_2$; a b is the horizontal component of the force which must be resisted at one or both of the ends; and the moments at any point are given by the ordinates parallel to **P** multiplied by the perpendicular distance from **C**$_1$ to **A** c. If we suppose the force **P**, as in the Fig., as causing two opposite vertical forces, instead of acting directly upon the axis, we have only to prolong **A B** to **B**$_1$ and join **B**$_1$ **B**$_2$, and then the ordinates of **A B**$_1$ **B**$_2$ **C** parallel to **P** or **A** c, multiplied by **H** (perpendicular distance from **C**$_1$ to **A** c) will give the moments.

---

* See *Der Constructeur*, Reuleaux.

42   MOMENT OF ROTATION—PARALLEL FORCES. [CHAP. V.

2D. FORCE PARALLEL TO AXIS. [Fig. 26, Pl. 6.]

We have an example of this case in the "bayonet slide" of the locomotive engine.

We have here two pairs of forces, the reactions V_1 and V_2 and the forces over B_1 and B_2. The points of application of these last change of course periodically, but for any assumed position the moments are easily found. Thus draw $A\ B_1$ at pleasure, and $C\ B_2$ parallel to it, and join $B_1\ B_2$ and $A\ C$, and we have at once the equilibrium polygon. To find the corresponding *force polygon*, suppose P_1 applied at b, and join b with the other support. Make $b\ c$ equal to P then $c\ d = V_2$. Lay off then $A\ a = c\ d = V_2$ and draw $a\ C_1$, which is the *pole distance*. Draw $C_1\ e$ parallel to $B_1\ B_2$. Then $A\ e$ and $e\ A$ are the forces acting over B_2 and B_1, and $A\ a$ is the reaction V_1. The case is, indeed, precisely similar to that in Art. 40.

[NOTE.—The moment area should properly be turned over upon $A\ C$ as an axis, so that $A\ a$ should be laid off and e fall *below* A. This can, however, cause no confusion.]

The application of the method to car axles,* crane standards, and a large number of similar practical cases in Mechanics is obvious. The formulæ for many of these cases are too complex for practical use; in some, no attempt at investigation of strain is ever made, the proportions being regulated simply by "Engineering precedent" or rules of thumb. Those familiar with the analytical discussion of such cases will readily recognize the great practical advantages of the Graphical Method.

3D. BEAM OR AXLE ACTED UPON BY FORCES LYING IN DIFFERENT PLANES.

The analytical calculation in such a case for instance is of considerable intricacy, but by the graphical method, on the contrary, the difficulty of investigation is scarcely greater than before.

Thus, let Fig. 27, Pl. 7, represent a beam acted upon by two forces P_1 and P_2 *not in the same plane*.

First, we draw the force polygons $A\ O_1\ M$ and $D\ O_2\ 2$ for the forces P_1 and P_2, having both the *same* pole distance $G\ O_1 = O_2\ H$, the pole O_2 being so taken that the closing lines of the

---

* *Der Constructeur*, Reuleaux, pp. 215–222.

corresponding polygons **A** b' **D** and **A** c'' **D** coincide. This is easily done, as if the closing line of the second polygon for any assumed position of O_2 (O_2 H being equal to G O_1) does not coincide with **A D**, the ordinate at c'' can be laid off from **C** and **A** c'' **D** thus found in proper position, and then the pole O_2 can be located. It will evidently be at the intersection of the vertical O_2 O'_2 with c'' **D**.

The two force polygons being thus formed, we construct the polygon **A C″ D** by drawing lines **B B″**, **E E″**, **C C″**, etc., so that their angles with the vertical shall be equal to the angle between the planes of the forces, and making them equal to the ordinates **B** b'', **E** e'', **C** c'', etc., respectively. Join b' **B″**, e' **E″**, f' **F″**, c' **C″**, etc., and lay off the ordinates **B** b, **E** e, **F** f, **C** c, etc., respectively equal. The ordinates to the polygon thus obtained, viz.: **A** $b\ e\ f\ c$ **D** multiplied by the *pole distance* O_1 **G** or O_2 **H**, give the moments at any point. **A** b and c **D** are straight lines, $b\ e\ f\ c$ is a curve (hyperbola). If we drop verticals through O_1 and O_2, and draw the perpendiculars O_1' **M**, O'_2**K**; **A M** is the reaction R_1, and **D K** the reaction R_2, both measured to the scale of the force polygon. Their directions are found by the composition of **A G** and **H 2** and **D H** and **G M** respectively, under the angle of the forces.

4TH. COMBINED TWISTING AND BENDING MOMENTS.

In many constructions pieces occur which are subjected at the same time to both bending and *twisting* moments. Both can be represented and given by moment areas. Thus, Fig. 28, Pl. 7, represents an axle turning upon supports at **A** and **B** and having at **C** a wheel upon which the force **P** acts tangentially. We have then a moment of torsion $M_t = P\ R$ and reactions $P_1 = P\ \dfrac{s}{a+s}$ and $P_2 = P\ \dfrac{a}{a+s}$; s being the distance of **P** from **B**, and a of **P** from **A**.

Let the bending moments be represented by the ordinates to the polygon a **C** b; then laying off $a\ o$ equal to **P** and drawing o **O** parallel to $b\ c$, we find the *corresponding pole distance* **O** k, and the reactions P_1 and P_2 equal to $k\ a$ and $o\ k$ respectively.

Now, in the force polygon **O** $a\ o$ thus found, at a distance from **O** equal to **R**, draw a line $m\ n$ parallel to **P**. This line $m\ n$ evidently gives *for the same pole distance* the moment of torsion $P \times R$. Laying off $C'C_1 = b \; b', = m \; n$, we have the torsion rectangle $C_1 \; b' \; b \; C'$.

Now the combined moment of torsion M_t and bending M_b is $\frac{5}{8} M_b + \frac{3}{8} \sqrt{M_b^2 + M_t^2}$.* We make then $C' \; C_0$ equal to $\frac{3}{8} \; C' \; C_1 = \frac{3}{8} \; m \; n$ and $C \; C_2$ equal to $\frac{5}{8} \; C \; C' = M_b$, and draw $C_2 \; b$. Then any segment of any ordinate, as $f f_2$ is $\frac{5}{8}$ of $f f'$. Revolve now $C' \; C_0$ with C' as a centre, round to C'_0 and join $C'_0 \; C_2$. Then $C_2 \; C'_0$ is equal to $\frac{3}{8} \sqrt{M_b^2 + M_t^2}$, and therefore with C_2 as centre revolving $C_2 \; C'_0$ to C_3, we find the point C_3, $C \; C_3$ being equal to $\frac{5}{8} M_b + \frac{3}{8} \sqrt{M_b^2 + M_t^2}$. In the same way we find any other point as f_3, by laying off $f' f'_0$ equal to $f' f_0$, joining f_2 and f'_0 and making $f_2 \; f_3$ equal to $f_2 \; f_0$. The line $C_3 \; f_3 \; b_0$ thus found is a hyperbola, and the ordinates between it and $b \; C$ give the combined moments [for pole distance $O \; k$] at any point.

[NOTE.—We suppose the axle to turn freely at **A**, and the working point or resistance beyond **B**; hence the moments left of the wheel are given by the ordinates to a **C**.]

5TH. APPLICATION TO CRANK AND AXLE.

The above finds special and important application in the case of the crank and axle.

Thus in Pl. 8, Fig. 29, let **E D C B** be the centre line of crank and shaft. Lay off a **P** equal to the force **P** acting at **A**, choose a pole o and draw $o \; a$ o **P** and the parallels $o \; a$ and a **E**. Join **E** and d and draw $o P_1$ parallel to **E** d. Then $P P_1$ is the downward force at **E** and $P_1 \; a$ the upward reaction at **D**. The ordinates to **E** $d \; a$ to pole distance o **P**, give the bending moments for the shaft. Make a **F** equal to the lever arm **R**, then **F G** is the moment **P R**, and we unite this as above with the bending moments and thus find the curves $c' \; d' \; e'$ the ordinates to which give the combined moments at every point of the shaft [see 4th].

For the arm **B C**, make the angle a_0 **B C** equal to **D** $a \; d$, and then the horizontal ordinates to a_0 **B** give the bending moments for the arm. Make $C \; c_0$ equal to $C \; c$ and we have the *torsion rectangle* $C \; c_0 \; b_0$ **B**, and as in the previous case we unite the two and thus find the curve $b_0 \; h$ **F**, the horizontal ordinates to which from **B C** give the required combined moments, to

---

* *Der Constructeur*, Reuleaux, p. 52, Art. 18.

pole distance o P. Thus $h'h_0 = \frac{1}{8} H h_0$, $H i = \frac{1}{8} B b_0$, and $H h = h_0 h' + h' i = \frac{1}{8} M_b + \frac{1}{4} \sqrt{M_b^2 + M_t^2}$.

The application of the method when the crank is *not* at right angles to the shaft, as also when the crank is double, and generally in the most complicated cases, is equally simple and satisfactory. Our space forbids any more extended notice of these applications, and we must refer the reader to *Der Constructeur*, by *F. Reuleaux, Braunschweig*, 1872, for further illustrations and applications of the method to the solution of various practical mechanical problems.

42. Continuous Loading—Load Area.—Thus far we have considered only concentrated loads. But whatever may be the law of load distribution, if this law is known, we can represent it graphically by laying off ordinates at every point, equal by scale to the load at that point. We thus obtain an *area* bounded by a broken line, or for continuous loading, by a curve, the ordinates to which give the load at any point. This *load area* we can divide into portions so small that the entire area may be considered as composed of the small trapezoids thus formed. If, for instance, we divide the load area into a number of trapezoids of equal width, as one foot one yard, etc., as the case may be, then the load upon each foot or yard will be given by the area of each of these trapezoids. If the trapezoids are sufficiently numerous, we may consider each as a rectangle whose base is one foot or one yard, etc., as the case may be, and whose height is the mean or centre height. The weight therefore for each trapezoid *acts along its centre line*. We thus obtain a system of parallel forces, each force being proportional to the area of its corresponding trapezoid, and equal by scale to the mean height or some convenient aliquot part of this height. We can then form the *force polygon*; choose a pole; draw lines from the pole to the forces; and then parallels to these lines, thus forming the *string* or *equilibrium* polygon; and so obtain the graphical representation of the moments at every point.

Since, however, the polygon in this case approximates to a curve, that is, is composed of a great number of short lines, the above method is subject to considerable inaccuracy, as errors multiply in going along the polygon.

This difficulty can, however, be easily overcome.

Thus we may divide the load area into *two* portions only, and then draw the force and equilibrium polygon, considering each portion to act at its centre of gravity, and so obtain an equilibrium polygon composed of three lines only. These lines *will be tangents to the equilibrium curve.* (Art. 76.) We thus have three points of the curve, and its direction at these points. In this manner we may determine as many points as may be necessary, without having the sides of the polygon so short or so numerous as to give rise to inaccuracy.

43.—The above will appear more plainly by consideration of a

BEAM UNIFORMLY LOADED.

The curve of load distribution becomes in this case a straight line. The load area is then a rectangle, and hence the load per unit of length is constant. Let us now divide this load area [Fig. 30, Pl. 8], into *four* equal parts, and considering each portion as acting at its centre of gravity, assume a scale of force, and draw the force polygon. Since in this case the reactions at the supports must be equal, we take the pole C, in a perpendicular to the force polygon *at the middle point.* This causes the closing line of the equilibrium polygon to be parallel to the beam itself, which is often convenient. We now draw C 0, C 1, etc., and then form the polygon 0 $a c e g h$. The lines 0 a, $a c$, $c e$, etc., of this polygon, are *tangent to the moment curve* at the points $b, d, f,$ 0 and h, *where the lines of division prolonged meet the sides.* The curve can now be easily constructed, as will appear from the next Art.

Moment Curve a Parabola.—Suppose we had divided the load area into only *two* parts, of the length x and $l - x$ [Fig. 30, Pl. 8]. Then the moment polygon would be $o a k h$, and the horizontal projection of the tangent line $a k$ would be $\frac{1}{2} x + \frac{1}{2}(l-x) = \frac{1}{2} l$.

That is, the horizontal projection of any tangent line to the moment curve is *constant.* But this is a property of the *parabola. The moment curve for a uniform load is therefore a parabola, symmetrical with respect to the vertical through the centre of the beam.*

If, then, we divide o C and C h into equal parts, and join corresponding divisions above and below, we can construct any number of tangents in any position.

[NOTE.—We may prove analytically that the moment curve *is* a parabola,

and hence that the line *a k must be a tangent*. Thus the moment at any point is
$$H y = \tfrac{1}{2} p l x - \tfrac{1}{2} p x^2$$
p being the load per unit of length, l the length, and the reaction at support therefore $\frac{p l}{2}$. Hence $y = \frac{p}{2 H}(lx - x^2)$ for origin O.

When the origin is at d, representing horizontal distances by y' and vertical by x', we have $x = \frac{l}{2} - y'$, and $y = h - x'$, h being the ordinate at middle $= \frac{p l^2}{8 H}$.

Hence by substitution
$$h - x' = \frac{p}{2 H}\left[\frac{l^2}{2} - l y' - \frac{l^2}{4} + l y' - y'^2\right]$$
or reducing
$$y'^2 = \frac{2 H}{p} x'$$
which is the equation of a parabola having its vertex at d.]

We may of course take the pole anywhere, and hence H may have any value. It is in general advantageous in such cases (*i.e.*, for uniform load) to take $H = \frac{p l}{2}$. We have then
$$y^2 = l x,$$
and for $y = \frac{l}{2}$, or for the middle ordinate, we have $x = \frac{l}{4}$.

To draw the moment curve we have then simply to lay off the middle ordinate equal to ¼th the span. The curve can then be constructed in the customary way for a parabola. Any ordinate to this curve multiplied by $H = \frac{p l}{2}$ will then give the moment at that point.

Enough has probably now been said to illustrate the application of our method to the determination of the moment of rotation, bending moment, or moment of rupture. The reader will have no difficulty in applying the above principles to any practical case that may occur.

It will be observed that the customary curve of moments in the graphic methods at present in general use, comes out as a *particular case* of the equilibrium polygon for uniform load.

This polygon has other interesting properties, which we shall notice hereafter. For instance, just as its ordinates [Fig. 30] are proportional to the bending moments or moment of rotation, so also its *area* is proportional to the moments *of* the moments, or the *moment of inertia* of the load area.

As to the *shearing force* at any point of a beam submitted to the action of parallel forces, the reactions at the ends being easily found as above by a line parallel to the closing line in the force polygon, we have only to remember that the shear at any point *is equal to the reaction at one end, minus all the weights between that end and the point in question.*

Thus for a uniformly distributed load we have simply to lay off the reactions which are equal to one-half the load, above and below the ends, and draw a straight line, which thus passes through the centre of the span. The ordinates to this line are evidently then the shearing forces. If we have a series of concentrated loads, we have a broken line similar to A'_1 1' 1'' 2', etc., Fig. 32, Pl 7, where each successive weight as we arrive at it, is subtracted from the preceding shear.

44. Beam continuously Loaded and also Subjected to the Action of Concentrated Loads.—In practice we have to consider not only a continuously distributed load, such as the weight of the truss or beam itself, but also concentrated forces, such as the weight of cars, locomotives, etc., standing upon or passing over the truss.

In Pl. 8, Fig. 31, we have a continuous loading represented by the load area **A** *a b* **B**, and in addition four forces P'_{1-4}. Now, since the total moment about any point is equal to the sum of the several moments, we can treat each method of loading separately and then combine the results. Thus with the force polygon (*b*) we obtain the equilibrium polygon **A**' 1 2 3 .... **B**' for the continuous loading, and with the force polygon (*a*) the equilibrium polygon **A**' 1'' 2'' 3'' .... **B**'' for the concentrated loads. If now in (*b*) we draw **C L** parallel to the closing line **A' B'**, and in (*a*) **C'' L'** parallel to the closing line **A' B''**, we obtain at once the reactions at the supports for each case.

Thus for continuous loading we have **L** 0 for reaction at **A**, and 10 **L** for reaction at **B**; for the concentrated loads, **L'** 0' at **A** and 4' **L'** at **B**. These reactions hold the beam in equilibrium.

For any cross-section *y*, the *shear* to the right is composed of the two components **L** 7 and **L'** 3' (*i.e.*, is equal to the reactions minus the forces between cross-section and support). The moment of **L** 7 is given by the ordinate $o\,y$ to the corresponding polygon, and we may consider **L** 7 as acting at the point of intersection a of the side 7 8 with **A′ B′** (Art. 38). In the same way **L′ 3′** acts at b. We may unite both these reactions and find the point of application of their resultant c, by laying off in force polygon (b) 7 b equal to **L′ 3′**, and then constructing the corresponding equilibrium polygon $e\,a\,d\,c$. The resultant **R** passes through c. This construction remains the same evidently, even when the points a and b fall at different ends of the beam, as may indeed happen. The components will then have opposite directions, and must be subtracted in order to obtain the resultant.

The *total* moment of rotation at y is proportional to the sum of $m\,n$ and $o\,y$. The greatest strain is where this sum is a maximum. In order to perform this summation and ascertain this point of maximum moment it is advantageous to construct another polygon instead of **A′** 1″ 2″, etc., whose closing line shall *coincide* with **A′ B′**. This is easy to do, by drawing in force polygon (a), **L′ C′** parallel to **A′ B′**, and taking a new pole **C′** the same distance out as before, that is, keeping **H** constant, and then constructing the corresponding polygon **A′** 1′ 2′ 3′, etc.

Thus the ordinate $p\,y$ gives the total moment at y. We can make use here also of the principle that the corresponding sides of the two polygons must intersect upon the vertical through **A′** (Art. 26). We have thus the total moment at any point, and can easily determine the point of maximum moment or cross-section of rupture. This point must necessarily lie between the points of maximum moments for the two cases, or coincide with one of them. In the Fig. this point coincides with the point of application of $\mathbf{P'_2}$.

45. Case of Uniform Load.—If the continuous load is *uniformly distributed* we can obtain the above result without being obliged to draw the curve. As in this case we have a very short construction for the determination of the point of greatest moment, it may be well here briefly to notice it.

If we erect ordinates along the length of the beam as an axis of abscissas, equal to the sum of the forces acting beyond any cross-section, the line joining the end points of these ordinates has a greater or less inclination to the axis according as the uniform load is greater or smaller. At the points of application of the concentrated loads this line is evidently *shifted* parallel to itself. Since at the point of maximum strain the sum of the forces either side is zero, this point is given by the intersection of the broken line thus found with the axis.

Thus in Pl. 7, Fig. 32, let **A B** be the beam sustaining a uniform load, and also the concentrated loads $P_1\ P_2\ P_3\ P_4$. The reaction of the uniform load at the supports is equal to half that load. To find the reactions for the concentrated loads we draw the force polygon 0 1 2 3 4, choose a pole **C**, then construct the equilibrium polygon **A'** 1 2 3 4 **B'**, and parallel to **A' B'** draw **C L**. **L** 0 and **L** 4 are the reactions at **A** and **B**. Now through **L** draw A_0 **L** horizontal, make it equal to the length of the beam, and take it as axis of abscissas. [It is of course advantageous here to lay off the forces along the vertical through **B**, as done in the Fig. Then A_0 falls in the vertical through **A** and $1_0\ 2_0\ 3_0\ 4_0$ are directly under the forces themselves.]

The ordinate to be laid off at A_0 is equal to **L** 0 + half the uniform load. Between A_0 and 1_0 the line A'_1 1' is inclined to the axis at an angle depending upon the uniform load. Lay off **L U** equal to this load and draw A_0 **U**. A'_1 1' must be parallel to this line. At 1' the line A'_1 1' is shifted to 1'', so that 1 1'' is the load P_1. Then 1'' 2' is parallel as before to A_0 **U**, and 2' 2'' is the load P_2, and so on. *The intersection 2_0 with A_0 **L** gives the point of maximum moment or cross-section of rupture.* The force P_2 at this point in our Fig. is divided, as shown by **L** in the force polygon, into two portions, one of which is to be added to the forces left, the other to the forces right. The ordinate $y_0\ y'$ at any point gives the *shear* or sum of the forces acting at that point. This force acts up or down according as the ordinate is above or below the axis.

Moreover, the area between the broken line and axis A_0 **L**, limited by this ordinate, gives the *moment of rotation* of the forces beyond the section y, areas below the axis being negative. For a section at z, therefore, we have area $A_0\ A'_1\ 1'\ 1''\ 2'\ 2_0$, minus $2_0\ 2''\ 3'\ 3''\ z'\ z_0$, or what is the same thing, the area $z_0\ z'\ 4'\ 4''\ B'_1\ L$, since the sum of the moments of all the forces is zero.

46. Influence of a Concentrated Load, passing over the Beam.—If in addition to the already existing uniform and concentrated loads, a *new* force operates, we have by (44) simply to construct for this new force its force and equilibrium polygon, and unite the forces and moments thus found with those already existing.

In Pl. 7, Fig. 32, we have assumed a new force P'_1 near the left support. The force polygon is $0'$ $1'$ C', the pole distance being taken the same as before. For any one position of this force we have then the equilibrium polygon A' $1'$ B'', and drawing a parallel C' L' to A' B'' we obtain the reactions $0'$ L' and L' $1'$, which must be added to the reactions already obtained.

If now we take a section y *between* P'_1 and the point of maximum moment 2_0 before found, the sum of the forces either side of this section undergoes the following changes: Upon the side where P'_1 lies, and the point 2_0 does not lie, where therefore the sum was originally an upward force, we have the downward force L' $1'$ (equal to algebraic sum L' $0'$ + $0'$ $1'$). The sum of the forces at the section, or the shearing force, is therefore *diminished*.

The total rotation moment is, however, *increased* by the amount indicated by $m\ n$. Both changes, that of the sum of the forces and the moment of rotation, *increase* as P'_1 approaches y, and are therefore greatest when P'_1 reaches y.

If P'_1 passes y, this point is in the same condition as z with reference to the former position of P'_1; that is, the force and point 2_0 are now both on the same side of the section. For z, then, the original downward force to the left is increased by the force L' $1'$. To the right the upward force is increased by $1'$ L'. In like manner the moment of the forces beyond z is increased by the amount indicated by $o\ p$. This change is greatest when P'_1 reaches z.

Therefore when a load passes over the beam the sum of the shearing forces is *diminished* in all sections between it and the original point of greatest moment, and *increased* in sections beyond this point, while the moment of rotation, or bending moment, for all cross-sections is increased. These changes moreover increase for any section as the load approaches that section. The shear at any point is therefore least, and the moment greatest, when the load reaches that point. As soon, however, as the load passes this point, the shear passes suddenly from its smallest to its greatest opposite value, and then diminishes as the load recedes, together with the moment of rotation. On the other side of the point 2_0 of original greatest moment, the shear and moment increase as the load approaches, and become greatest for any point when the load reaches that point. At the moment of passing, these greatest values pass to their smallest values, and increase afterwards as the load recedes.

Since by the introduction of the load the shear for points upon one side of 2_0 is diminished (between 2_0 and the load), and on the other side increased, and the greatest moment is at the point where the shear is zero, it follows that the point of greatest moment moves in general towards the load. At a certain point, then, both meet. As the load then advances this point accompanies it, passes with it the original position, and follows it up to the point where it would have met the same load coming on from the other side. From this point, as the load continues to recede, it returns, and finally reaches its original position as the load arrives at the further end.

It is evidently of interest to learn the position of these two points, where the load meets and leaves the point of greatest moment, or cross-section of rupture, and this in Fig. 32 we can easily do.

When P'_1 arrives at $1'$, we have evidently the reactions by laying off $L\,E$ equal to P'_1, drawing $A_0\,E$, and through its intersection with the vertical through the weight drawing the horizontal $A'_0\,B'_0$. $L\,B'_0$ is then the increase of reaction at B due to P'_1. The entire reaction is $B'_0\,B'_1$, and the broken line A'_1 $1'\,1''$, etc., holds good still, if we merely change the axis from $A_0\,L$ to $A'_0\,B'_0$. The point of greatest moment, which is still the intersection of the broken line with the new axis, in the present case is not changed by reason of the overpowering influence of P_2. It does not move to meet the load, but awaits it until it reaches P_2, and until, therefore, the new axis takes the position $A''_0\,B''_0$.

If, however, the force P'_1 comes on from the *right*, we have the reactions for any position as z, by laying off $A_0\,E'$ equal to P'_1, drawing $L\,E'$, and then the horizontal $A'''_0\,B'''_0$ through the intersection of $L\,E'$, with the vertical through z. Then $A_0\,A'''_0$ is the reaction at A, due to this position of the load. The intersection x', corresponding to x, shows the point to which the point of greatest moment 2_0 moves to meet the load. As the load passes towards the left, this point moves towards the right, and both come together evidently at the point V_1, corresponding to the new axis $A_0^{iv}\,B_0^{iv}$. The point of greatest moments passes then from 2_0 to V_0, and beyond these two limits it can never pass.

Our construction, then, is simply to lay off the load in opposite directions perpendicularly from each end of the axis $A_0 L$, and join the end points $A_0 E$ and $L E'$. The intersections of these lines with the diagram of shear give the points 2_0 and V_0 required.

47. Load Systems.*—Concentrated loads occur in general in practice in a certain succession, as for instance the forces acting at the points of contact of the wheels of a train of cars passing over the beam, and it is necessary then to investigate the influence of different positions of the train. It evidently amounts to the same thing whether we suppose the weights to move over the beam, or suppose the weights stationary and the beam to move. In either case we obtain every possible position of every weight relatively to the ends of the beam.

The severest load to which we can subject a railway truss, for example, is when the span is filled with locomotives. If we suppose, for illustration, in round numbers, the distance between the three axles of the locomotive 3 ft. 6 in., between the axles of the tender 5 ft. 6 in., between the foremost tender and the back locomotive axle 4 ft., and the entire length of locomotive and tender 34 ft. 6 in., and then suppose the weight upon each locomotive axle 13 tons, and upon each tender axle 8 tons, we have a system of weights in fixed order and at fixed distances, and the truss should be investigated for a series of these systems, as many as can be placed upon the span, passing over it from one end to the other.

In Pl. 9, Fig. 32 (*a*), we assume two such locomotives as shown by P_{1-10}, and construct the force and equilibrium polygons. The forces are symmetrically arranged with respect to a central point, and the pole in the force polygon is therefore taken perpendicular from the middle of the force line.

Now the system of forces being as represented, suppose the span to *shift*. Thus suppose the span of a given length represented by $S_1 S_1$ in the Fig. Then 0 6 is the line closing the polygon for this position of the span, and a parallel to 0 6 in the force polygon, viz., $C L$ gives the reactions at the ends. Let now the span move from $S_1 S_1$ to $s_5 s_5$; we have a new po-

---

* *Elemente der Graphischen Statik*, Bauschinger.

sition for the line closing the polygon and new reactions. As the span continues to shift to the right, the lines closing the polygon revolve, and as their projections are always constant, viz., equal to the span, they are all tangent to a *parabola*, which they therefore envelop.

48. Properties of this Parabola.—This parabola has several important properties which will aid us in the investigation of the case above proposed.* In Pl. 9, Fig. 32 (d), let **XX** be the line along which the span is shifted; a **M** and a **N** the outer sides of the polygon, intersecting at a, along which the closing lines slide as they revolve. For a given position $s\,s$ of the span, $\sigma\,\sigma$ is the corresponding line. $s_0\,s_0$ is the position of the span, for which the centre, c_0, lies in the vertical through a. In this position $\sigma_0\,\sigma_0$ is tangent to the parabola at ω_0, its middle point, and upon this line *lie the centres of all the other lines* (taken of course as reaching from a **N** to a **M**). Now the point of tangency, β, of any other line, as $\sigma\,\sigma$, with the parabola, *is as far from the centre of that line, γ, as the centre of that line is itself from c_0.* We have then only to make $c\,b$ equal to $c\,c_0$, and drop a perpendicular through b to find β. Thus for the position $s_1\,s_1$ and the line $\sigma_1\,\sigma_1$, to find the point of tangency δ_1, make $c_1\,d_1$ equal to $c_1\,c_0$, and draw $d_1\,\delta_1$ perpendicular to intersection with $\sigma_1\,\sigma_1$.

Inversely we may find that position for the span $s\,s$, for which the vertical through a *given* point, b, shall pass through the point of tangency.

We have only to move the span so that its middle point c shall be as far from c_0 as it is already from the given point, or make $c\,c_0$ equal to $c\,b$. (See Art. 75.)

If we shift now the span $s\,s$, and at the same time the point b through an *equal distance*, the intersections of the vertical through b, with the corresponding closing lines of the polygon, will all lie upon the same line $\sigma\,\sigma$.

If therefore b'_1 is such an intersection, b has been moved from b to b'_1, and hence the span from $s\,s$ to $s_1\,s_1$.

49. Different Cases to be Investigated.—We are now ready to investigate the effect of a live load such as represented in Pl. 9, Fig. 32 (a). For the determination of the proportions of the truss the following points are specially important:

---

* See *Elemente der Graphischen Statik*, Bauschinger, pp. 108-114. Also, *Die Graphische Statik*, Culmann, pp. 136-141.

CHAP. V.] MOMENT OF ROTATION—PARALLEL FORCES. 55

1. When a certain number of wheels pass over the truss, but without any passing off, or new ones coming on; what position of the system gives the maximum moment at any given cross-section not covered by the system, and how great is this moment?

2. Under the same supposition as above, what position of the load gives the greatest moment for a given point covered by one of the load systems?

3. Among all the various points of the span, at which is found the greatest maximum moment, for what position of the load does it occur, and how great is it?

4. If the number of wheels is indeterminate, how many must pass on, and what position must they have to give at any point the greatest maximum moment; where is the corresponding cross-section, what position must the load have, and how great is this maximum moment?

The three first questions are easily solved by the aid of the above properties of the parabola, enveloped by the *closing lines* of the equilibrium polygon, corresponding to different positions of the span.

Thus, as regards the first question, let the given cross-section be b, Pl. 9, Fig. 32 (d), and suppose the span $s\,s$ in the position where the vertical through b intersects $\sigma\,\sigma$ at the point of tangency β. When now the span shifts, the intersection of the ordinate through b, with the corresponding tie line, will always lie upon $\sigma\,\sigma$. But this ordinate gives the reduced moments for b (reduced to pole distance **H**.) The greatest of these moments will then be simply the greatest of the ordinates between $\sigma\,\sigma$ and the polygon, and will always be found at an angle of the same. When found, we have at once the position of b, and of course of the span with reference to the given loads. This is always such that a wheel stands over the given section.

Thus in Fig. 32 (a), supposing the four wheels P_6 to P_9 to pass over the span $t_1\,t_1$, we seek the position of the load to give the greatest moment at a point $\frac{1}{8}$ of the span from the left, therefore $\frac{3}{8}$th from the middle.

We lay off the span in such a position, $t_1\,t_1$, that its centre is distant from the intersection a of the outer lines of the polygon by $\frac{3}{8}$th of the span.

The ordinate through the given point now passes through the point of tangency of the tie line and parabola. We draw this tie line t_1 9, and seek the greatest ordinate between it and the polygon. This we find at 7, and directly above 7 the given point must lie, and hence we have the position of the span, viz., $t\ t$. If the scale of tons is ten tons to an inch, of distance 5 ft. to an inch, and the pole distance **H** is assumed $12\frac{1}{2}$ ft. = $2\frac{1}{2}$ inches, the scale of moments will be $10 \times 2.5 \times 5,\ =\ 125$ ft. tons to an inch.

As to the second question; the position of the span required, is that where the vertical through the given point of the system **S** Fig. 32 (*a*), intersects the corresponding tie line at its point of tangency with the parabola; all other tie lines intersect this vertical in a point between the tangent point and the polygon. The middle of the span must then lie midway between the intersection *a* of the outer polygon sides and the point *s*, where the vertical through **S** meets the line **X X**. Thus the span has the position $t_2\ t_2$.

The third question, finally, is easily solved if the parabola enveloped by the tie lines is drawn. The greatest ordinate between this parabola and the polygon gives the greatest moment, and the point and the position of span required, since the middle of the span must be half-way between the point given by this ordinate and *a*.

The greatest moment is always found upon an ordinate through *an angle of the polygon.*

If, however, the parabola is not drawn, we find by trial at several angles, drawing the tie lines and comparing the corresponding ordinates, the ordinate required. Here the following considerations may aid:

When the load is uniformly distributed, the maximum moment is in the middle of the span, and at the same time in the vertical through the intersection *a* of the outer polygon sides. The polygon itself becomes a parabola. The less uniform the load is, the more this point approaches the heaviest loaded side, as also the intersection *a*, though not in the same degree. For loads not *exceedingly* unsymmetrical the point may be sought for, then, in the neighborhood of *a*, *i.e.*, near the resultant of the forces acting upon the truss. Thus in our example we are justified in selecting the corner 7 of the polygon, nearest the point of intersection *a*.

50. Most unfavorable Position of Load upon a Beam of given Span.—The fourth question above requires a somewhat more extended consideration. The most unfavorable position of a system of given concentrated forces is when it causes the greatest moment at the cross-section of rupture. This position is from the preceding, given by taking the centre of the beam midway between the vertical through the point of intersection of the outer sides of the equilibrium polygon and the nearest angle of the same. If with this centre we increase the span, the maximum moment increases until the span has the greatest length possible without more wheels coming on.

Thus for the two wheels P_4 and P_5, Pl. 9, Fig. 32 (a), a is the intersection of the outer polygon sides, and 4 the nearest polygon angle. The almost equally near angle 5 gives at any rate no *greater* moment. In order then that these two weights may cause the greatest maximum moment, the middle of the beam must lie half-way between a_1 and 4; and as the span increases in length this moment increases, and is then greatest when the span reaches to s_1 or P_3.

If now the span still increases so as to *also include* P_3, the point of intersection of the outer polygon sides recedes to a_2, where in our Fig. it coincides almost exactly with the polygon angle 4. Here then, approximately at 4, we must locate the centre of the beam. If we take the same length of span as before, that is, make the half span $a_2 s_2$ equal to the distance from s_1 to the point midway between a_1 and 4, we see by drawing the closing lines for these two positions of the span, that the maximum moments measured upon the vertical through 4 are almost exactly equal in each case. For a *smaller* length of span including the three weights, the maximum moment decreases, and is less therefore than the maximum moment already caused by the two wheels. The span $s_1 s_1$ may then be regarded as the greatest for which the two wheels $P_4 P_5$ give the greatest possible maximum moment. As the space $s_2 s_2$, upon which we have now three wheels, increases, the moment increases, and is greatest when the span, its centre always remaining now at a_2 reaches to s'_2 or to P_2.

If now it still increases so as to *also include* P_2, the intersection of the outer polygon sides retreats to a_3. The nearest polygon angle is still 4, and midway then between a_3 and 4 we must now locate the middle of the beam. If from this centre we lay off the half span equal to $a_2 s'_2$, to s_3, and draw the closing line for this position of the span, we see as before that the moment given by the ordinate at 4 is for either case almost exactly the same. Any less span including the four weights would give a less moment; less, therefore, than the moment already caused by the three weights. The span $s'_2\ s'_2$ then precisely as before, is the extreme limit upon which the three wheels P_3 to P_5 cause the greatest possible maximum moment.

In a precisely similar manner we find that the span $s'_3\ s'_3$ with a centre midway between a_3 and 4 is the limiting span for the four wheels P_2 to P_5.

If now the span still increases so that P_1 comes on, the intersection of the outer polygon sides falls in our Fig. nearly at s_1, and since this point also happens to correspond almost exactly with the angle 3, we take the centre of the beam at s_1. The greater the span now becomes, the greater the maximum moment. The greatest length, however, which the span can have without including P_6, is twice $s_1\ 6$, or twice the distance between s_1 and P_6. If P_6 also comes on, the intersection of the polygon sides is found at a_5, and the nearest polygon angle is 4. Midway then between a_5 and 4 is the new centre of the beam, while before P_6 came on, it was nearly at s_1. But for centre s_1 the half span was $s_1\ 6$, while now it is somewhat less than 4 6; therefore considerably smaller. Since, however, we wish to follow the span as it continues increasing, we must compare those two spans which are equal before and after the coming on of P_6. The right-hand ends of these spans, viz., s'_4 and s_5 must evidently be distant each side of 6, by the half distance of their centres s_1 and 4, or a_2 (more accurately the point halfway between a_3 and 4, but a_3 and 4 lie in our Fig. so nearly together that the centre cannot be indicated more exactly). We make then $s_1\ s'_4 = a_2\ s_5 = M_1\ 6$, provided that M_1 is taken halfway between the centre s_1 and a_2.

An exact construction shows that the maximum moments for these two spans, the one given by the ordinate through 3, the other by the ordinate through 4, are almost exactly equal, and moreover, that the maximum moment for the span $S_1\ S_1$ of equal length whose centre is at M_1 is also almost exactly equal, when measured upon the vertical through M_1. We can therefore take $S_1\ S_1$ as the limit of those spans for which the five wheels P_1 to P_5 cause the greatest maximum moment.

Taking on now the seventh wheel, the intersection of the

outer polygon sides is at a_6 and the nearest polygon angle is 5. Half-way between a_6 and 5 we must then take the centre, while before it lay at a_2 (nearly). If we take then M_2 half-way between a_2 and this new centre, we find precisely as before the span $S_2 S_2$ with centre M_2, and right end at P_7, as the limiting span for the six wheels P_1 to P_6. The same holds good for the span $S_3 S_3$ with centre M_3, for the seven wheels P_1 to P_7, and so on. If, according to supposition, P_3 P_4 P_5 are 3 ft. 6 in. apart, P_2 and P_3 4 ft., and P_1 and P_2 5 ft. 6 in. apart; then for spans up to s_1 s_1 = say 8 ft., the two wheels P_4 P_5 will give the greatest maximum moment, and their place upon the beam is given by the position of the centre (half-way between a_1 and 4). From about 8 ft. to 15 ft. span, or s_2 s_2 the three wheels P_3 to P_5 give the greatest maximum moment, and the centre of the span is located at a_2. For spans from 15 ft. to 19 ft. span, or s'_3 s'_1, the four wheels P_2 to P_5 give the maximum moment, and the centre is at s_1; and so on. Thus for a span of any given length we have at once the weights and their position, in order to cause the greatest maximum moment, as also the place of this moment, viz., the point vertically over that angle of the equilibrium polygon nearest the centre of the span. The ordinate through this point included by the equilibrium polygon, and the closing line for the given span, taken to the moment scale gives this moment at once; or this ordinate taken to the scale of force must be multiplied by the previously assumed *pole distance*.

51. Greatest Moment of Rupture caused by a System of Moving Loads at a given Cross-Section of a Beam of given Span.—For beams or trusses of long span, which are as a rule caused to vary in cross-section, it is not sufficient merely to find the greatest maximum moment which a given system of concentrated forces can cause; we must also know for a number of individual cross-sections, the maximum moments which can ever occur.

For this purpose the force and equilibrium polygons being first constructed, we shift as above the given span along a horizontal line, and draw for each successive position of the span the corresponding closing line in the equilibrium polygon, marking the point where each closing line is intersected by a vertical through the given cross-section, which of course moves with the span, keeping always the same position with reference to the ends. The points thus obtained form a curve, and the greatest ordinate between this curve and the polygon gives the greatest moment which can act at the given cross-section. This greatest ordinate will always be found at an angle of the polygon, and hence a weight must always rest upon the cross-section. Since the cross-section itself must lie upon this ordinate, we have directly the position of the span with reference to the given forces. The closing line for this position being then drawn, a parallel to it in the force polygon gives the reactions for this position.

The reader will do well to make the construction indicated for an assumed span and system of weights, to convenient scales, checking the results by computation.*

The above method applies more particularly to *solid* or "*plate*" girders, beams, or trusses. It may of course be applied to *framed* structures also, such as those illustrated in chapter first. Thus the moment at any point, divided by the depth of truss at that point, gives the strain in flanges. The more preferable, as perhaps also the simplest method of determining the strains in such cases, however, is to find the reactions due to each individual weight. Each reaction can then be followed through the structure, as explained in that chapter, and the strains in every member for every weight in every position can thus be obtained and tabulated. An inspection of the table will then give at once the strains due to the united action of any desired number of these weights.

We have thus *two* methods for the solution of such cases; first, by the composition and resolution of forces, and, second, by the equilibrium polygon and moments of rupture, and may, if we choose, check the results obtained by one method by the other. In most practical cases involving *framed structures*, however, the first method is preferable as being simpler, quicker of application, and of superior accuracy.

For solid-built beams or "plate girders," etc., the second method comes more especially into play. The determination of the strains in a structure of this kind from the known moment of rupture at any point, requires a knowledge of the *moment of inertia* of the cross-section at that point, and this may also be found by the Graphical method.

---

* This construction is given in Art. 15, Fig. VIII., of the Appendix.

CHAPTER VI.

MOMENT OF INERTIA.

52. Thus far we have seen that by the graphic method we can in any practical case determine the moment of the exterior forces acting upon a piece at any cross-section of that piece. But the exterior forces give rise to and are resisted by molecular or *interior* forces. Now the moment of the exterior forces being found, the cross-section of the piece at any point being known, and one of the dimensions of this cross-section being assumed, it is required to find the other dimension, so that the strain per unit of area of cross-section shall be less than the recognized safe strain of the material as found by experiment.

The moment of the exterior forces at any cross-section we call the moment of rupture; and designate it by **M**. Let $d =$ the depth of cross-section.*

$y =$ the variable distance of any fibre above or below the neutral axis.

$\beta =$ the breadth of the section at the distance y from the neutral axis, and consequently a variable, except in the case of rectangular sections.

$s =$ the horizontal unit strain exerted by fibres in the cross-section at a given distance c from the neutral axis.

Then since the fibres exert forces which are proportional to their distance from the neutral axis or to their change of length, the unit strain in any fibre at a distance y from the neutral axis will be $\frac{s\,y}{c}$. Let the depth of this fibre be $d\,y$, then, since the breadth of section is β, the total horizontal force exerted by the fibres in the breadth β, will be $\frac{s}{c}\beta\,y\,d\,y$. The moment of this force about the neutral axis will be $\frac{s}{c}\beta\,y^2\,d\,y$, and the

---

* *Theory of Strains*, Stoney, p. 43, Art. 67.

integral of this quantity will be the sum of the moments of all the horizontal elastic forces in the cross-section round the neutral axis, that is, equal to the *moment of rupture* of the section in question. We have therefore

$$\mathbf{M} = \frac{s}{c}\int \beta y^2 \, dy.$$

For a rectangular cross-section, for instance, β is constant and equal to the breadth b. Representing the depth by d we have $\mathbf{M} = \dfrac{b\,d^3\,s}{12\,c}$, or if we make c the distance of the extreme fibres $= \dfrac{d}{2}$

$$\mathbf{M} = \frac{s\,b\,d^2}{6}$$

from which $\mathbf{M}$ being known, as also s, if we assume b we can find d or the reverse.

The integral $\int \beta y^2 \, dy$ is the *moment of inertia* of the cross-section, and may be defined as *the sum of the products obtained by multiplying the mass of each elementary particle by the square of its distance from the axis.* [See *Supplement to Chapter VII.*, Art. 10.]

From the above, we see its importance in determining the strain at any distance from the neutral axis, or in proportioning the cross-section, so that the resulting strain shall be less than a given quantity at any point. We see also that for a rectangular cross-section the moment of inertia is $\dfrac{b\,d^3}{12}$, where b is the breadth and d the depth.

53. Graphical Determination.—We have already seen that the moment of a force, as $\mathbf{P}_1$ (Pl. 6, Fig. 20) with reference to any point, as o, is given by the ordinate $n\,m$ multiplied by the constant $\mathbf{H}$ (Art. 38). The ordinate $n\,m$ then represents the product of $\mathbf{P}_1$ multiplied by the horizontal distance of b from n. But the *area* of the triangle $b\,n\,m$ is $m\,n \times \frac{1}{2}\,b\,n =$ $\mathbf{P}_1 \times \frac{1}{2}\,b\,n^2$, that is, *the area of the triangle $b\,n\,m$ represents one-half the moment of inertia of $\mathbf{P}_1$ with respect to o.* Just as the exterior ordinates of the equilibrium polygon have been shown to have a certain significance, and to represent the moments of the forces, so the exterior *areas* of the equilibrium polygon represent the moments of the moments, or the moments of inertia. Thus in Pl. 8, Fig. 30, the exterior parabolic area $o\,C\,h$ should be one-half the moment of inertia of the rectangle or load area $o\,p\,r\,h$, with reference to the resultant of the area forces as an axis.

Let us see if this is so. The area of the triangle $o\,h\,C$ is $\frac{1}{2}$ $o\,h\,\times$ the ordinate $S\,C$. This ordinate $S\,C$ gives, as we have seen, the moment, with respect to S, of the reaction. We can therefore find its value. Thus if p is the load per unit of length, and l is the length, $\frac{p\,l}{2}$ is the reaction, and $\frac{p\,l^2}{4}$ this moment. The area of the triangle $o\,C\,h$ is therefore $\frac{l}{2} \times \frac{p\,l^2}{4} = \frac{p\,l^3}{8}$.

The parabolic area $o\,d\,h$ is $\frac{2}{3}$ of the circumscribing rectangle. This rectangle is $l \times S\,d$. The ordinate $S\,d$ is equal to $SC - dC$. We have already found SC and dC is the sum of the moments of P_1 and P_2, or $\frac{p\,l}{2} \times \frac{l}{4} = \frac{p\,l^2}{8}$. Hence $S\,d = \frac{p\,l^2}{4} - \frac{p\,l^2}{8} = \frac{p\,l^2}{8}$. The area of the circumscribing rectangle is then $\frac{p\,l^3}{8}$. Two-thirds of this is $\frac{2\,p\,l^3}{24}$, which subtracted from $\frac{p\,l^3}{8}$ gives for half the moment of inertia $\frac{1}{24}\,p\,l^3$. Hence the moment of inertia is $\frac{1}{12}\,p\,l^3$, as should be.

54. We see therefore the significance of the *area of the equilibrium polygon*.

If, when a number of forces are given, we form the force polygon, and then the equilibrium polygon, the ordinates to this last give the *moments* to the assumed pole distance. If now we *take these moments themselves as forces* applied at the same points, form a new force polygon with new pole distance, and new equilibrium polygon, the ordinates to this new polygon to the new pole distance will give the moments *of* the moments or the *moments of inertia* of the forces. The same method is applicable to moments of a higher order, but in practice we have only to do with those of the second order alone.

55. Radius of Gyration.—The moment of inertia of a system of parallel forces $P_1\,F_2$ etc., in a plane, with reference to an axis from which the points of application are distant q_1 q_2, etc., is then $\Sigma\, P\, q^2$. This is the product of three quantities, one of which is measured by the scale of force, and the other two by the scale of length. We can therefore regard it as the product of the square of a certain length by the sum of the given forces, or $\Sigma\, P\, q^2 = k^2\, \Sigma\, P$. We call k the *radius of gyration*.

In order to find the moment of inertia of a system of parallel forces then, we must by the preceding Art. construct *two* force and equilibrium polygons. If the pole distances are H and H', and the segments into which the axis is divided by the produced sides of the polygons are P'_1 P'_2 and P''_1 P''_2 etc., respectively, then

$$\Sigma\, P\, q^2 = H\, H'\, \Sigma\, P''$$

and the radius of gyration is given by

$$H\, H'\, \Sigma\, P'' = k^2\, \Sigma\, P$$

or,
$$k = \sqrt{\frac{H\, H'\, \Sigma\, P''}{\Sigma\, P}}$$

This expression is easy to construct. Thus for example in Pl. 11, Fig 33, let $o\, n\, C$ be the first force polygon, $o\, n$ the force line, containing the forces P; C the pole, and H the pole distance. Make $o\, b$ equal to the second pole distance $H,'$ and draw $b\, c$ parallel to $n\, c$ and $c\, t$ parallel to H. Then

$$c\, t = h = \frac{H\, H'}{\Sigma\, P}$$

whence
$$k = \sqrt{h\, \Sigma\, P''}$$

If, therefore, in Fig. 33 (b), $m''_0\, m''_n$ is the segment of the axis cut off by the outer sides of the second equilibrium polygon, that is, if $m''_0\, m''_n = \Sigma\, P''$, we have only to prolong $m''_0\, m''_n$ to L, making $m''\, L = h$, and describe a semicircle upon $m''_0\, L$, and erect the perpendicular $m''\, k$, which will be equal to k. In general, the pole distance H and H' can be taken arbitrarily, but it is often advantageous to take H (sometimes H' also) equal to $\Sigma\, P$. Then

$$k = \sqrt{H'\, \Sigma\, P''}$$

We should then have in Fig. 33 simply to increase $m''_0\, m''_n$ by the second pole distance H', and then proceed as above to find k.

CHAP. VI.]   MOMENT OF INERTIA.   65

It is to be remembered that q_1 q_2, etc., the distances of the points of application of the forces from the axis, may be *measured in any direction*, and H is parallel to this direction, and is *not* therefore necessarily perpendicular to $o\,n$.

The above will be rendered plain by reference to Fig. 34, Pl. 10. We suppose four forces applied at the points A_1 $_2$ A_3 A_4 respectively, and acting parallel to XX. Required the moment of inertia of these forces and the radius of gyration, the distances q_1 q_2, etc., being measured parallel to YY. First we form the force polygon by laying off along XX, 0 1, 1 2, 2 3, 3 4, parallel, and in the direction of action of the forces, choosing a pole C, and drawing C 0, C 1, C 2, etc. We now construct the corresponding *equilibrium polygon*, CI, III, IIIII, IIIIV, etc. The segments 0 1', 1' 2', 2' 3', etc., represent the *statical moments* of the forces with reference to XX. That is, these segments to the scale of force multiplied by the pole distance C y *parallel to* YY to the scale of distance, give the statical moments of the forces. Now we take these segments themselves as forces, and suppose them acting at the former points of application. With the same pole as before we draw C 0, C 1', C 2', etc., and form the corresponding equilibrium polygon CI, III', IIIII', etc. The sum of the segments of XX cut off by the outer lines of this polygon, or $o\,y$, to the scale of force multiplied by H H' or $\overline{C\,y^1}$ *gives the moment of inertia of the forces* with respect to XX.

This moment then is $M = \overline{0\,y} \times \overline{C\,y^1}$
where $\overline{0\,y} = \Sigma P''$ and $\overline{C\,y^1} = H H'$.

The radius of gyration k is, as we have seen, given by

$$k = \sqrt{\frac{H H' \Sigma P''}{\Sigma P}},\quad \Sigma P \text{ being equal to } 0\,4 \text{ in the Fig. Hence}$$

$$k = \sqrt{\frac{\overline{C\,y^1} \times \overline{0\,y}}{0\,4}}.$$

If, then, we lay off $0\,d = 0\,4$, and make $0\,c = C\,y$, and make the angle $d\,c\,e$ a right angle, we shall find a point e to the right and $0\,e$ will be equal to $\dfrac{C\,y^1}{0\,4} = \dfrac{H H'}{\Sigma P}$. Upon $e\,y$ now describe a semi-circle, the point of intersection b' with the perpendicular through 0 will give (Art. 55)

$$0\,b' = \sqrt{\frac{\overline{0\,y} \times \overline{C\,y^1}}{0\,4}} = \sqrt{\frac{H H' \Sigma P''}{\Sigma P}} = k = \text{radius of gyration.}$$

The square of this line, then, multiplied by ΣP or $0\,4$, will give

5

at once the moment of inertia of the given four forces with reference to **X X** and **Y Y** as axes. If we were to suppose the same forces with the same points of application to act parallel to **Y Y** instead of **X X**, the distances $q_1 q_2$ being measured parallel to **X X** instead of **Y Y**, we should have the force polygon C_1 O 1_1 2_1 3_1 4_1 instead of C 0 1 2 3 4, and a precisely similar construction would give us $o\,x$ multiplied by pole distance for the moment of inertia, and 0 a' for the radius of gyration. We recommend the reader to follow through the construction as shown by Fig. 34.

56. Curve of Inertia—Ellipse and Hyperbola of Inertia.—If having found the radius of gyration as above, we lay it off from the axis on either side, in a direction parallel to the directions in which $q_1\,q_2$, etc., are supposed measured, and through the points thus determined draw two parallels to the axis **M'** and **M''** on either side, and then suppose the axis to revolve in the plane of the forces about any point as **O** situated in the axis; the lines **M'** and **M''** also revolve and enclose a curve of the second degree, whose centre coincides with **O**. Thus, if in Pl. 10, Fig. 34, we lay off **O** b along **Y Y** on both sides of **X X** equal to $o\,b' = k$ already found, and then let **X X** revolve about **O**, **K J** and **J K** will also revolve, and enclose either an ellipse or hyperbola.

In order to prove this, take **O** as an origin of co-ordinates. Let the co-ordinates of the points of application of the forces $\mathbf{A}_1$ $\mathbf{A}_{11}$, etc., be $x_1 y_1$, $x_{11} y_{11}$, etc. From each of these points **A** draw parallels to the axis of y, intersecting the axis of x in the points **C**. Then **O C** $= x$, **A C** $= y$. Now pass through the point **O** an axis of moments **M** in any direction, and project for each point **O C A** parallel to this axis upon the line q, which measures the distance of each point from the axis of moment (*not* necessarily perpendicular distance). This projection is evidently equal to q. Denote by a and β the ratios by which distances along **X** and **Y** must be multiplied, in order to obtain their projections upon q, by lines parallel to **M**. Then
$$q = a\,x + \beta\,y$$
for each point of application, and hence
$$\Sigma\,\mathbf{P}\,q^2 = \Sigma\,\mathbf{P}\,(a\,x + \beta\,y)^2$$
or since for one and the same axis **M**, and direction q, a and β are constant,
$$\Sigma\,\mathbf{P}\,q^2 = a^2\,\Sigma\,\mathbf{P}\,x^2 + \beta^2\,\Sigma\,\mathbf{P}\,y^2 + 2\,a\,\beta\,\Sigma\,\mathbf{P}\,x\,y.$$

In this expression a and β will vary with the position of **M** and the direction of q, but $\Sigma P x^2$, $\Sigma P y^2$ remain unchanged. These last expressions are, however, nothing more than the moments of the second order (moments of inertia) of the given force system with reference to the co-ordinate axis, the distances of the points of application being measured in the direction of the axis. They are known if the force system is given and the co-ordinate system assumed.

If we put $\Sigma P x^2 = a^2 \Sigma P$, $\Sigma P y^2 = b^2 \Sigma P$, $\Sigma P x y = f^2 \Sigma P$, b and a, etc., are the radii of gyration of the moments of inertia with reference to x and y, and the above equation becomes

$$\Sigma P q^2 = \Sigma P [a^2 a^2 + \beta^2 b^2 + 2 a \beta f^2]$$

If we conceive for the assumed position of **M**, the radius of gyration k to be found, and **M′** and **M″** drawn on either side at a distance $\pm k$, measured parallel to q, and indicate the distances cut off by these lines from the co-ordinate axes by $\pm x_e$ $\pm y_e$, and then project these distances parallel to **M** upon the direction of q or k, we have $k = a\, x_e = \beta\, y_e$, whence

$$a = \frac{k}{x_e} \quad \beta = \frac{k}{y_e}$$

and these values substituted in the above equation give

$$\Sigma P q^2 = k^2 \Sigma P \left[\frac{a^2}{x_e} + \frac{b^2}{y_e} + \frac{2 f^2}{x_e y_e}\right] = \pm k^2 \Sigma P$$

where k^2 is essentially positive in the second term.

Hence,

$$\frac{a^2}{x_e^2} + \frac{b^2}{y_e^2} + \frac{2 f^2}{x_e y_e} = \pm 1 \dots\dots\dots\dots\dots\dots (1)$$

If we suppose the axis **M** to change its position revolving about O, the segments x_e y_e cut off from the axes of x and y by **M′** and **M″** alone will change in this equation. It is therefore the equation of the curve enclosed by **M′ M″**. If this curve is known for a given force system, then the moment of inertia for *any* axis passing through its centre is easily found. We have only to draw parallel to the axis two tangents to this curve, one on either side, and measure their distance from **M**, in the direction in which the distances q of the points of application from the axis are taken. This distance is the radius of gyration, and the moment of inertia is simply the product of its square by the algebraic sum of the forces.

We call the curve represented by the above equation therefore, the *curve of inertia*. If we refer the curve to co-ordinate axes which coincide with the conjugate diameters, the equation becomes

$$\frac{A^2}{x^2} + \frac{B^2}{y^2} = \pm 1$$

where x and y are the new ordinates, and A, B, the conjugate semi-axes of the curve. A and B are therefore the radii of gyration of the force system, measured in the direction of the co-ordinate axes, and hence

$$A^2 = \frac{\Sigma P X^2}{\Sigma P}, \quad B^2 = \frac{\Sigma P Y^2}{\Sigma P}$$

where x and y are the co-ordinates of the points of application of the given forces.

Since $\Sigma P q^2 = k^2 \Sigma P$ if the sign of $\Sigma P q^2$ is the same as ΣP, k^2 is positive. When, on the other hand, these signs are different, k^2 is negative. That is, when all the forces act in the same direction k^2 is positive, and we have

$$\frac{A^2}{x^2} + \frac{B^2}{y^2} = 1$$

which is the equation of an *ellipse*.

If, however, the parallel forces act in different directions, k_2 may be positive or negative. For cases where k^2 is negative, either A^2 or B^2 will be negative, and we shall have

$$\frac{A^2}{x^2} - \frac{B^2}{y^2} = \pm 1$$

or,

$$-\frac{A^2}{x^2} + \frac{B^2}{y^2} = \mp 1.$$

Both cases coincide. The double curve consists of two hyperbolas with common assymptotes, common centre, and equal semi-axes. For every axis M passing through the common centre O, we have a pair of parallel tangents *either* to one *or* the other hyperbola. The corresponding k^2 is positive for the one, negative for the other.

If, then, in the method of construction to which we shall presently refer, the square of the semi-axis B, which lies in the axis of Y, is negative, that hyperbola whose imaginary axis lies in Y gives k^2 positive, the other gives k^2 negative, and reversely for the other case. If the axis of moments M coincides with one of the common assymptotes, the radius of gyration and moment of inertia with respect to it of the given force system is zero.

57. Construction of the Curve of Inertia.—The curve of inertia for a given system of parallel forces and given centre O, is determined by the direction of any two conjugate diameters, since as we have seen in Art. 55, Pl. 10, Fig. 34, these directions being assumed we can find the radii of gyration with respect to **X X** and **Y Y**, and can thus determine O a and O b, the semi-diameters. We have then to develop a principle by means of which these directions may be determined.

If we denote the distances of the points of application of the forces from the axis of **M** measured in any direction by y, then the statical moments of the forces, **P** y, are indeed dependent upon the direction in which y is measured, but their *relative* values remain the same. If then being found for any direction of y, these statical forces are considered as being themselves parallel forces acting at the points of application, and their *centre of action* is found (for gravity—centre of *gravity*) for some other value of y, this centre of action remains unchanged. For any axis passing through this centre of action the sum of the moments of the forces is zero. If therefore we take a point O in the axis **M** as origin of a system of co-ordinates, whose axis **O X** may lie at will in the plane of the forces, while **O Y** passes through the centre of action; the sum of the moments of the statical moments **P** y, considered as forces acting at the points of application, with reference to **O Y**, will be zero. These moments however, provided that the distances of the points of application are measured along the co-ordinate axes, are the *moments of inertia*, viz., Σ **P** y x. If these are zero we see that the general equation of the curve of inertia (1) Art. 56, becomes that of a hyperbola referred to its conjugate diameters as axes. With the centre O therefore, *the line joining* O *with the centre of action, gives the direction of the conjugate diameter of the curve.*

This is the principle required. By means of it we can find the conjugate diameters of the inertia curve, for a given centre O, and thus construct it.

58. Construction of the Curve of Inertia for four parallel forces in a Plane. Example.—As an example let us take the four parallel forces in Pl. 10, Fig. 34, supposed to act in different directions, parallel to **X X** at the points **A₁ A₂**, etc.

As before we have the force polygon **C 0 1 2 3 4** for an arbitrary axis as **X X**, and from the corresponding equilibrium polygon, we determine the statical moments with reference to **X X**, 01' 1'2', etc., *to the basis* **C 0**. These moments we again consider as parallel forces acting at **A₁ A₂**, etc., for which we have **C 0 1' 2' 3' 4'** and corresponding equilibrium polygon **C I II' III'**, etc. We then determine the centre of action **S**, by a second polygon 0" **I II" III"**, etc., the sides of which are respectively perpendicular to the first, according to the process for finding the centre of gravity, Art. 30. *The line joining* **O** *with* **S** *gives the direction of* **Y Y**, the diameter of the curve conjugate to **X X**. To find the *length* of the semi-diameters **O** b and **O** a, we must find the moments of inertia of the forces with reference to **X X** and **Y Y**, taking the distances of the points of application as measured *parallel* to these lines.

Therefore instead of **C 0**, we must take **C** y *as basis* or pole distance, and then find the radii of gyration as already indicated in Art. 55, viz., **O** b' and **O** a'. These distances laid off along **Y Y** and **X X** give the semi-conjugate diameters of the curve of inertia.

From the Fig. we see that the force **P₁** whose direction from left to right we shall always consider positive, and $\Sigma P = 0\ 4$ have the same sign. On the other hand the total moment of inertia 0 y and the moment of inertia of **P₁**, viz., 0 1" have different signs. The square of radius of gyration $k^2 = \dfrac{\overline{Cy^2} \times \overline{0y}}{\Sigma P}$ is therefore negative, the radius itself or the semi-diameter **O** b is imaginary.

In similar manner, we see that **O** a the radius of gyration for **Y Y** is real, since the total moment of inertia **O** x and $\Sigma P = 0\ 4_1$, have the same signs. The curve is then a double hyperbola with the conjugate semi-diameters **O** a and **O** b.

It is then easy to find the assymptotes **K K** and **J J**, and by bisecting the angle which they make, the principal axes **A A** and **B B**. In order to find the length of these axes, we have the well-known principle that for any point as a, the product of ak and k**O** (ak being parallel to the assymptote **J J**) is equal to $\dfrac{1}{4}$ the sum of the squares of the semi-axes $\left(\dfrac{A^2+B^2}{4}\right)$. If then

CHAP. VI.] MOMENT OF INERTIA. 71

we find kl, the mean proportional of Ok and ka, and lay it off *twice* from O to D along the assymptote O K, O D is the diagonal of a rectangle whose sides are the principal axes. We thus find the vertices A, A, B, B.

We can thus construct the curves. Then for *any* position of the axis X X as it revolves about O, we can find the corresponding radius of gyration and consequently the moment of inertia, by simply drawing tangents to the curve above and below the new position of X X and *parallel* to it. The radius of gyration thus obtained measured to the scale of length and multiplied by the algebraic sum of the forces, or 0 4 to the scale of force, will give the moment of inertia required for the assumed position of the axis.

59. Central Curve. Central Ellipse.—If the point O about which the axis turns *coincides with the centre of action* (or gravity) of the forces, we call the curve enclosed by the parallels M′ M″ at the distance k on either side, the *central curve*. When the parallel forces all act in the same direction this curve is always an ellipse.

For the central curve the principle proved in Art. 57 and the method of construction given in Art. 58, are no longer applicable, for the algebraic sum of the statical moments of the given forces is zero for every axis through the centre of gravity. We cannot therefore find the centre of gravity of the moments of the forces, when considered as forces themselves and applied at the given points of application.

If we divide, however, these moments considered as forces into two portions or groups, and find the centre of gravity of each group, the *line joining these two points* has an important property, viz., that for every moment axis parallel to it, *the algebraic sum of the moments of the statical moments considered as forces*, that is, the algebraic sum of the moments of inertia of the forces, *is zero*. In other words, $\Sigma P e e'$ is zero, e being the distances of the points of application from the first axis, which passes through the centre of gravity of the forces, and e' the distances from the axis parallel to the line joining the two centres of gravity of the two groups of statical moments considered as forces. If we draw then through the centre of gravity of the forces themselves the moment axis X X, and take it as the axis of abscissas of a co-ordinate system whose Y axis passes also through the centre of gravity of the forces and is parallel to the line joining the two centres of gravity of the statical moments considered as forces, then the moments of inertia $\Sigma P y x$ are zero, and hence as in the preceding Art. *this axis of* **Y** *is conjugate to* **X X**.

This holds good not only for the central curve, but also for every inertia curve, whose centre **O** instead of coinciding with the centre of gravity of the forces, lies in the axis passing through that centre. In this case also the axis through the centre **O** parallel to the line of union above, is a conjugate to **X X**. Still more, the half length of this conjugate diameter is in both cases the radius of gyration of the force system for the axis **X X** and the direction of **Y**.

Hence *in every inertia curve of a system of parallel forces, whose centre lies in an axis passing through the centre of gravity of the forces, the diameters conjugate to this axis are parallel and equal.* All these inertia curves are therefore touched by two lines parallel to this axis and equally distant on either side. This distance is the radius of gyration for this axis.

For any such inertia curve, whose centre **O** is distant i from the centre of gravity **S** of the forces, we call **E** and $\mathfrak{E}$ the parallel conjugate axes to **S O** for this curve, and the central curve respectively; **q** and q the distances from them of any point of application, these distances measured parallel to **S O**, and considered positive when the point of application lies on the same side of **E** or $\mathfrak{E}$ respectively as the centre of gravity **S** from **E**. Then i, the distance apart of **E** and $\mathfrak{E}$ is essentially positive, and if we indicate by a and a the lengths of the semi-conjugate diameters for the inertia and central curve respectively, we have

$$a^2 \Sigma P = \Sigma P q^2 \text{ and } a^2 \Sigma P = \Sigma P q^2$$

where **q** and q stand in the simple relation

$$q = q + i.$$

Hence

$$\Sigma P q^2 = \Sigma P (q+i)^2 = \Sigma P q^2 + 2 i \Sigma P q + i^2 \Sigma P.$$

Since $\mathfrak{E}$ passes through the centre of gravity $\Sigma P q = o$, and therefore

$$\Sigma P q^2 = \Sigma P q^2 + i^2 \Sigma P = a^2 \Sigma P.$$

Hence

$$a^2 = a^2 + i^2,$$

an equation which gives the relation between the lengths of the semi-conjugate diameters of the central and any inertia curve, whose centre lies upon an axis through the centre of gravity of the forces, at a distance i from this centre.

Any two curves at equal distances either side of the centre of gravity are therefore equal. If the semi-diameter of the central curve a is real, and therefore a^2 positive, a^2 is also positive and greater than a^2. All the inertia curves are therefore of the same kind as the central curve, and enclose the centre of gravity. If, however, a^2 is negative, and the central curve therefore an hyperbola; all those inertia curves whose centres are distant from the centre of gravity by a distance i less than a are hyperbolas also. For a distance i equal to a, the curves reduce to straight lines equal and parallel to the conjugate diameter of the central curve. For i greater than a, the curves become ellipses.

60. Centre of Action of the Statical Moments of the Forces.*—We again suppose, through the centre of action of the forces S [Fig. 35, Pl. 11] a line NN drawn which cuts the central curve at A and A'. Two such points we have in every case, except when the curve is an hyperbola, and NN coincides with an assymptote.

Let $\mathfrak{E}$ be the conjugate axis to NN in the central curve, E a parallel to it through any point o distant i from S, and also conjugate to NN in the inertia curve whose centre is o. Then since the statical moments of the forces with reference to NN is zero, the centre of action of the statical moments with respect to E, considered as forces acting at the points of application, will be somewhere upon NN. It is required to find where.

We call q the distance of any point of application from E, measured parallel to NN, and positive when upon the same side of E as S, then i is essentially positive.

As before, q is the distance of the points of application from $\mathfrak{E}$, also measured parallel to NN, and positive in the same direction as q.

Then we have always
$$\mathsf{q} = q + i.$$
and for the moments of inertia of the forces with respect to E and $\mathfrak{E}$ $\quad \Sigma P \mathsf{q}^2 = \Sigma P (q+i)^2 = \Sigma P q^2 + i^2 \Sigma P$
or when a is the semi-diameter of the central curve, $SA = SA'$ and $\quad \Sigma P \mathsf{q}^2 = (a^2 + i^2) \Sigma P$

---

* See *Supplement to Chap. VII., Art.* 10, *latter part.*

Let now m be the distance of the centre of gravity or action, of the moments of the forces with respect to **E**, from **E**, and 𝔪 its distance from $\mathfrak{E}$, positive the same as q and q. Then
$$m = \mathfrak{m} + i$$
and since the sum of the moments is equal to the moment of the resultant:
$$\Sigma\,\mathbf{P}\,\mathbf{q}^2 = m\,\Sigma\,\mathbf{P}\,\mathbf{q}.$$
But the sum of the moments **P q** of the forces with reference to **E**, is equal to the product of the sum of the forces into the distance i of the centre of gravity of the forces from **E**. Hence
$$\Sigma\,\mathbf{P}\,\mathbf{q} = i\,\Sigma\,\mathbf{P},$$
and therefore
$$m\,i\,\Sigma\,\mathbf{P} = \Sigma\,\mathbf{P}\,\mathbf{q}^2 = (a^2 + i^2)\,\Sigma\,\mathbf{P},$$
or,
$$m\,i = a^2 + i^2.$$
Introducing the value for m
$$(\mathfrak{m} + i)\,i = a^2 + i^2$$
or
$$\mathfrak{m}\,i = a^2.$$

If now a^2 is positive, which is always the case for an ellipse as central curve, 𝔪 is also positive, and is therefore to be laid off from S along **N N** on the opposite side of $\mathfrak{E}$ from o. If then we conceive an axis **E'** drawn parallel to **E**, and symmetrical with reference to S, which axis we shall call for convenience the *symmetrical axis to* **E**, we see from the above relation that **M** *is the pole of this axis in the central curve*.

If, however, a^2 is negative, therefore a imaginary, 𝔪 is negative, and must be laid off from S towards o, and the point **M** thus found is therefore the pole of the axis **E** itself, or in the case of an hyperbola is the pole of **E'** in that hyperbola which is *not* cut by **N N**, and for which therefore **A A'** is imaginary.

Hence we have the principle—

If we consider the statical moments of the forces with reference to any axis as **E** *as themselves forces acting at the given points of application, the centre of gravity of these moment forces does not coincide with the centre of gravity of the original forces, but is the pole* * *in the central curve of an axis* **E'** *parallel and symmetrical to* **E**.

In those cases where the central curve becomes an hyper-

---

* POLAR LINE OF A POINT, in the plane of a conic section, is a line such, that if from any point of it two straight lines be drawn tangent to the conic section, the straight line joining the points of contact will pass through the given point, which is called a *pole*.

bola, we must observe whether the diameter conjugate to the moment axis is real or imaginary. In either case the centre of gravity is the pole of the line symmetrical to the moment axis in that hyperbola for which that diameter is real or imaginary.

The construction is given in Pl. 11, Fig. 35.

Upon $\mathbf{S}o' = \mathbf{S}o$ we describe a semi-circle. With $\mathbf{S}$ as centre, and $\mathbf{S}\mathbf{A}' = a =$ semi-diameter of the central curve, describe an arc, and from the intersection with the semi-circle drop a perpendicular upon $\mathbf{S}o'$. The point $\mathbf{M}$ thus found is the centre of gravity of the moments. For: $a^2 = \overline{a\,\mathbf{M}^2} + \mathrm{m}^2$ and $\overline{a\,\mathbf{M}^2} = \mathrm{m}\,(i - \mathrm{m})$ hence $a^2 = \mathrm{m}^2 + \mathrm{m}\,i - \mathrm{m}^2 = \mathrm{m}\,i$. The central curve being known as also the distance i, the point $\mathbf{M}$ can be readily found.

61. Cases where the Direction of the Conjugate Axis of the Inertia Curve can be at once Determined.—There are certain special and practical cases in which the conjugate directions or axis of the inertia curve can be at sight determined, so that only the length of the semi-diameters remains to be found. The most important of such cases are as follows:

· (1.) When in a system of parallel forces, these forces can be so grouped in pairs, that the lines joining the points of application of each pair are all parallel, and the centres of gravity of each pair all lie in the same straight line. Then for the central curve and all inertia curves whose centres lie upon this straight line, the direction of the axis conjugate to this line is the same as that of the lines joining the points of application of each pair.

This is easy to prove. For, for each pair, the sum of the moments with respect to the line joining their centres of gravity, is zero. These moments regarded as forces and applied at the points of application, give therefore for each pair two parallel opposite and equal forces, the sum of the moments of which for any line parallel to the line joining the points of application, is zero. This is the case for all the pairs, and therefore the direction of the lines joining the points of application is that of the axis conjugate to the line joining the centres of gravity, for the central curve as also all inertia curves whose centres lie upon this last line.

(2.) When the forces can be so grouped that the points of application of each group lie in parallel lines, and the centres of gravity of the groups lie in the same straight line. Then this straight line gives the direction for the central curve and every inertia curve whose centre lies upon it, of the diameter conjugate to an axis passing through the centre and parallel to the lines joining the points of application.

For if we take any such axis, the points of application of the forces in each group are equally distant. The statical moments for each group are then proportional to these distances. If, therefore, they are considered as forces, their centre of gravity coincides with that of the forces themselves, and lies therefore in the line joining the centres of gravity of the groups. The centre of gravity of the whole force system lies then in this line, which is therefore the direction of the axis conjugate to the line parallel to the lines joining the points of application, in the central curve, and also all curves whose centres lie upon this line.

(3.) When the forces can be so grouped that the centres of the central curves of each group lie in the same straight line, and the diameters in each curve conjugate to this line, are parallel. Then in the central curve of the entire system, the diameter conjugate to this line is also parallel to these diameters. For, for any axis parallel to these diameters, the centres of gravity of the moments of the forces in each group lie upon the line joining the centres of the curves. The centre of gravity of the moments for the entire system lies then also upon this line, which is therefore the direction of the axis conjugate to an axis parallel to the diameters of the curves, for any inertia curve whose centre lies upon this line.

In all these cases, if the directions thus found are perpendicular, we have to do with the principal axes.

62. Practical Applications.—We can now apply the above principles to practical cases, and as in the determination of the moment of inertia of irregular figures, we have to deal with triangles, parallelograms and trapezoids, we have first to consider these three cases.

1st. The Parallelogram. Pl. 11, Fig. 36.

The moment of inertia of a parallelogram is, as is well known,

$\mathbf{M} = \frac{1}{12} a b^3$,* a being the breadth and b the depth.

---

* $\int_{-\frac{1}{2}b}^{+\frac{1}{2}b} a x^2 \, dx = \frac{1}{12} a b^3$

Hence $k^2 = \dfrac{1}{12} b^2 =$ radius of gyration, or $k = \sqrt{\dfrac{1}{2}b \times \dfrac{1}{6}b}$.
That is, the radius of gyration is a mean proportional between $\dfrac{1}{2}b$ and $\dfrac{1}{6}b$.

The centre of gravity of the parallelogram is at **O** the intersection of the diagonals, and this is therefore the centre of the central curve.

If we suppose the parallelogram divided into laminæ parallel to **D C**, and suppose each lamina divided by **G H** parallel to **B C**, the centres of gravity of each will lie upon **G H**. Right and left of **G H** we then have a group of forces whose points of application lie in lines parallel to **G H**, and the lines joining any pair, one on each side of **G H**, are parallel. By (1) of the preceding Art., therefore, **G H** and **E F** are conjugate axes of the central curve. For the lengths of the half diameters, we find the mean proportional between $\dfrac{1}{2}b$ and $\dfrac{1}{6}b$, $\dfrac{1}{2}a$ and $\dfrac{1}{6}a$, respectively, by the half circles **B F** and **B H**. We thus find k and k', and can then construct the central ellipse directly, or find the principal axes, and then construct it. The centre of action of the moments of the parallelogram, with reference to any axis parallel to **A B**, is as we have seen, Art. 60, the pole of a line parallel and equally distant from **O** on the other side. If we draw this line then, as **D C**, then from **G** draw two tangents to the central ellipse, and unite the points of tangency by a line; the intersection of this line with **O G** is the centre of gravity of the moments of the forces themselves considered as forces, or area of the parallelogram, with reference to **A B**.

2d. Triangle. Pl. 11, Fig. 37.

The moment of inertia of a triangle for the axis **B C** is $\dfrac{1}{12} a h^3$ * whence $k^2 = \dfrac{1}{6} h^2$, and for an axis **E F** distant $i = \dfrac{1}{3} h$, which passes through the centre of gravity,

$$a^2 = k^2 - i^2 = \dfrac{1}{18} h^2. \quad \text{(Art. 59.)}$$

---

* $\int_0^h a \dfrac{h-x}{h} x^2 \, dx = \dfrac{1}{12} a h^3$, h being the line **A D**, $a = $ **B C**.

The conjugate axes of the central curve are by principle 1 or 2 of the preceding Art. **E F** and **A D**.

The above value of a is then the length of the semi-diameter along **A D**, or $a = \sqrt{\frac{1}{6}h \times \frac{1}{3}h}$. That is, a is a mean proportional between $\frac{1}{6}h$ and $\frac{1}{3}h$. This is found by the semi-circle **O D** Fig. 37.

The moment of inertia of the triangle with respect to **A D** is $\frac{1}{6}h\left(\frac{1}{2}a\right)^3$. The radius of gyration then is $\sqrt{\frac{1}{6}\left(\frac{1}{2}a\right)^2} = \sqrt{\frac{1}{2}\left(\frac{1}{2}a\right) \times \frac{1}{3}\left(\frac{1}{2}a\right)}$ or a mean proportional between $\frac{1}{2}$ and $\frac{1}{3}$ of $\frac{a}{2}$ or **D C**.

This is given by the semi-circle on $\mathbf{D\,G} = \frac{1}{2}\mathbf{D\,C}$, and we thus have the four points 1 2 3 4 of the central ellipse, and the semi-diameters 0 1 and 0 3, and can therefore construct it. From the central ellipse as before, we can find the centre of gravity of the moments considered as forces for any axis parallel to **B C** or **A D**, as also in either case, the radius of gyration and therefore moment of inertia, for any axis passing through **O**.

3d. *Trapezoid.* Pl. 11, Fig. 38.

Here the lines **E F** joining the centres of the parallel sides, and **G H** parallel to these sides, and passing through the centre of gravity 0, are the conjugate axes of the central ellipse.

For the axis **A B** and direction **E F**, the moment of inertia is

$$\frac{1}{12}(a+3b)h^3,\ *$$

a and b being **A B** and **C D**, and $h = \mathbf{E\,F}$. The square of radius of gyration is then

$$k^2 = \frac{\frac{1}{12}(a+3b)h^3}{\frac{1}{2}(a+b)h} = \frac{1}{6}\frac{a+3b}{a+b}h^2$$

$*\int_0^h \left[a - (a-b)\frac{x}{h}\right] x^2\, dx = \frac{1}{12}(a+3b)h^3.$

For the radius of gyration for **G H**, at a distance $i = \frac{1}{3} h$ $\frac{a+2b}{a+b}$ we have

$$a^2 = k^2 - i^2 = \frac{1}{6}\frac{a+3b}{a+b} h^2 - \frac{1}{9} h^2 \left(\frac{a+2b}{a+b}\right)^2 = \left[\frac{1}{18} + \frac{1}{9}\right.$$

$$\left.\frac{ab}{(a+b)^2}\right] h^2.$$

This radius a is half the diameter along **E F**.

To construct it, put $(3\,a)^2 = \frac{1}{2} h^2 + \frac{ab}{(a+b)^2} h^2$.

Describe a semi-circle upon **E F**, and at the centre o_1, and at the intersection of the diagonals k, erect perpendiculars o_1 **J** and **K L**. Then $\overline{FJ}^2 = \frac{1}{2} h^2$ and $\overline{KL}^2 = \frac{ab}{(a+b)^2} h^2$, since **E K** $= \frac{b}{a+b} h$ and **K F** $= -\frac{a}{a+b} h$. If therefore we lay off **K L** equal to **J M** from **J**, we have

$$\mathbf{F\,M} = \sqrt{\frac{1}{2} h^2 + \frac{ab}{(a+b)^2} h^2} = 3\,a,$$

and hence the half diameter sought is one-third **F M**. We thus find 0 1 and 0 2.

To find the other semi-diameter we have the moment of inertia for **E F** and direction **G H**, $\frac{h}{48}(a^3 + a^2 b + a\,b^2 + b^3)$*, hence the square of the radius of gyration is

$$\frac{\frac{h}{48}(a^3 + a^2 b + a\,b^2 + b^3)}{\frac{1}{2}(a+b)\,h} = \frac{1}{24}\left(a^2 + b^2\right) = \frac{1}{6}\left[\left(\frac{1}{2}a\right)^2 + \left(\frac{1}{2}b\right)^2\right]$$

This last expression is easily constructed. In the right-angled triangle **F B N**, the hypothenuse **F N** $= \sqrt{\left(\frac{1}{2}a\right)^2 + \left(\frac{1}{2}b\right)^2}$, **B N** being made equal to **C E**. If we describe then a semi-

---

*$\left[\int_0^b h y^2\,dy + \int_b^{\frac{a}{2}} h \frac{a-2y}{a-b} y^2\,dy\right]$

circle upon $\mathbf{F\,U} = \frac{1}{2}\,\mathbf{F\,N}$, and make $\mathbf{F\,W} = \frac{1}{3}\,\mathbf{F\,N}$, $\mathbf{F\,V}$ is the semi-diameter sought. We thus find 0 3 and 0 4, and can now construct the central ellipse. This being constructed we can find the centre of gravity of the moments with reference to any axis parallel to $\mathbf{A\,B}$ or $\mathbf{E\,F}$, according to Art. 60, or the moment of inertia for any axis through 0, by drawing a parallel tangent to the ellipse. The distance from 0 to the point of tangency gives then the radius of gyration for that axis.

4th. Segment of Parabola. Pl. 11, Fig. 39.

Let the segment be limited by $\mathbf{B\,C} = 2\,h$, and $\mathbf{A\,D} = l$. Then it is evident that these two axes are conjugate (Art. 61), and the centre of the central curve is 0, the ratio of $\mathbf{A}\,0$ to $0\,\mathbf{D}$ being as 3 to 2. Hence $\mathbf{A\,D}$ and $\mathbf{E'\,F'}$, parallel to $\mathbf{C\,D}$ through 0, are conjugate axes of the central curve. To find the length of the semi-diameters along these axes we find first the moment of inertia of the segment with reference to an axis $\mathbf{Y\,Y}$ parallel to $\mathbf{E'\,F'}$ and tangent to the parabola at $\mathbf{A}$. We have then for this moment of inertia

$$\int_0^l 2\sqrt{2\,p\,x} \times x^2\,dx = \frac{4}{7}\,l^3\,h$$

where p is the parameter of the parabola, and $l = \mathbf{A\,D}$. Since the area of the segment is $\frac{4}{3}\,h\,l$, we have for the square of the radius of gyration

$$k^2 = \frac{3}{7}\,l^2.$$

The square of the radius of gyration then for $\mathbf{E'\,F'}$ whose distance from $\mathbf{A}$ is $i = \frac{3}{5}\,l$ is

$$a^2 = k^2 - e^2 = \frac{3}{7}\,l^2 - \frac{9}{25}\,l^2 = \frac{12}{175}\,l^2,$$

a being the semi-diameter along $\mathbf{A\,D}$. It is easier here to compute a, viz., $a = 0.26186\,l$, and lay it off from $\mathbf{O}$, thus finding 3 and 4.

For the other semi-diameter we find the moment of inertia for $\mathbf{A\,D}$ and the direction $\mathbf{E'\,F'}$. Thus

$$\int_{-h}^{+h}(l-x)\,y^2\,dy = \int_{-h}^{+h}\left(l - \frac{y^2}{2\,p}\right)y^2\,dy = \frac{4}{15}\,l\,h^3.$$

The radius of gyration squared is, therefore,

$$\frac{\frac{4}{15} l h^3}{\frac{4}{3} l h} = \frac{1}{5} h^2,$$

and hence the radius of gyration is $\beta = 0.44721\ h$. Laying this off from 0, we obtain 1 and 2, and can therefore now draw the central ellipse.

63. Compound or Irregular Cross-Sections.—Every cross-section may be divided up into trapezoids, triangles, parallelograms and parabolic segments, and the above cases will aid us, therefore, in the application of the graphic method to compound or irregular cross-sections. The engineer is often called upon to determine the moment of inertia of such sections as the **T**, double **T**, or different combinations of these in proportioning the different pieces of bridges, such as chords, struts, floor-beams, etc., as also in many other constructions. The calculation for such cross-sections is sometimes very laborious. As an example of the application of the graphical method best illustrating the above principles, we take the cross-section shown in Fig. 40, Pl. 12.

First we divide the cross-section into a series of trapezoids. The first segment, bounded by a curve, we may consider a parabolic area. These trapezoids we reduce to equivalent rectangles of common base a [Art. 32], and take the corresponding heights as forces. These forces we lay off in the force polygon and choose a pole **C** at distance **H** from force line, drawing **C** 0, **C** 1, **C** 2, etc. Parallel to these lines we have the first equilibrium polygon **I II III** ... **VIII**, the intersection of the two outer sides of which gives the point of application of the resultant. The intersection **S** of the resultant with the axis of symmetry gives the *centre of gravity* of the cross-section [Art. 30]. The segments $o\ 1'$, $1'2'$, $2'3'$, etc., cut off from o **S**, give the statical moments of the forces with reference to o **S** to the basis **H**. We now choose another pole **C′** at distance **H′**, and form another force polygon, considering these moments as forces, and applied at the centres of action of the moments of the separate areas into which the whole cross-section has been divided. These centres of action can be determined by forming the *central curve* for each area according to Art. 62, and then applying the principle of Art. 60. A little consideration will show that these centres of gravity will coincide approximately with the centres of gravity of the areas themselves, except for areas (3) (4) (5) and (6). Finding then for these areas the centres of action of the moments considered as forces, we construct the equilibrium polygon O' I' II' .... $VIII'$. The distance $0''$ $8''$ cut off by the first and last sides of this polygon gives the moment of inertia to the pole distances H and H' and the reduction base a. Thus $0''$ $8''$ measured to scale of force and multiplied by $a\,H\,H'$ is the moment of inertia of the cross-section with reference to $o\,S$.

The radius of gyration is then $k = \sqrt{\dfrac{a\,\overline{H\,H'}\,\overline{0''8''}}{a\,\overline{0\,8}}}$.

The division will be performed if we take $H' = \overline{0\,8} = \Sigma P$. This we can easily do now without drawing a new polygon, since what is required is the intersection of the outer sides only. Thus take a new pole o_1' distant from $o\,S$, $H' = \overline{0\,8}$. Now we know that each side of the new polygon for this pole distance will intersect the corresponding side of the first in a line parallel to $o\,C_1'$ [Art. 27]. Since the new polygon may start from any point, we may take the first side to coincide with $O\,VIII'$. Then the line of intersection of any two sides is $O\,VIII'$ $8''$. Produce any side as $IV'\,V'$ to intersection e with this line; from e draw $e\,a_1'$ parallel to $C_1'\,4'$.

Through a' the intersection of $o'\,I'$ and $V'\,IV'$, the resultant of (1) (2) (3) and (4), must pass. The change of pole cannot affect this resultant, which must therefore pass through a_1', the intersection of $e\,a_1'$ with the vertical through a' parallel to $o\,S$. Hence $o_1'\,a_1'$ is the direction of the last side of the new polygon, and $8''0_1''$ is the moment of inertia for the new pole distance $o\,C_1' = 0\,8$. The radius of gyration then is $k = \sqrt{H\,\overline{0_1''8''}}$. In other words, k is a mean proportional between H and $0_1''8''$. The construction of k is given by the semi-circle described upon $0_1''8'' + H$. The ordinate to this semi-circle through $0_1''$ perpendicular to $o\,S$ gives k. We thus find the semi-diameter $S\,a = S\,a'$ of the central ellipse.

In order to find the other semi-diameter $S\,b = S\,b'$, we might divide the cross-section into areas by lines parallel to $S\,X$, and then proceed as above. This is, however, unnecessary. With the same areas as before, we can find the central curve for that area on each side of $X\,X$, and then the centre of application of the moment of each of these areas with respect to **X X** itself, considered as a force. The method of procedure is then precisely as before. We draw a polygon the sides of which are respectively perpendicular to those of the first polygon, and thus find the statical moments $0'''\ 1'''\ 1'''\ 2'''$, etc., to basis **H**.

Choosing then a pole **C'''** at distance **H'''** and drawing the corresponding polygon, we have $\overline{0\,8}^{\text{IV}}$ for the moment of inertia. The radius of gyration is then $k = \sqrt{\dfrac{a\,\mathbf{H}\,\mathbf{H'''}\,\overline{0\,8}^{\text{IV}}}{a.\,0\,8}}$.

We have taken $\mathbf{H'''} = \dfrac{1}{2}\overline{0\,8}$, hence $k = \sqrt{\dfrac{1}{2}\mathbf{H}\,\overline{0\,8}^{\text{IV}}}$. Hence k is a mean proportional between $\dfrac{1}{2}\overline{0\,8}^{\text{IV}}$ and **H**. The construction is given in the Fig. by a semi-circle upon $\mathbf{H} + \dfrac{1}{2}\overline{0\,8}^{\text{IV}}$. We thus find the semi-axis $\mathbf{S}\,b' = \mathbf{S}\,b$, and can now construct the central ellipse. We have thus found graphically not only the moments of inertia of the cross-section with respect to **X X** and **Y Y**, but, by means of the central ellipse, for any other axis in the plane of the Fig. passing through **S**.

64.—The above method of procedure holds good generally for any cross-section, except that, when there is no axis of symmetry, the centre of gravity must be found by a second equilibrium polygon whose sides are respectively perpendicular to those of the first. When the moment of inertia with reference to a single axis only is required, the above method becomes quite short and simple, as well as accurate. In our Fig. the scale used as also the number of divisions taken make the process appear more complicated than it really is.

With this we shall close our discussion of moment of inertia, merely observing, that all the principles deduced in this chapter for forces acting in a plane hold equally good for forces in *space*. The central curve then becomes an area, we have a moment plane instead of moment axis **M**, and the ellipse and hyperbola of inertia become ellipsoid and hyperboloid respectively.

For a much fuller discussion of the subject than is possible here, we refer the reader to *Culmann's Graphische Statik*, pp. 160–206; also *Bauschinger's Elemente der Graphischen Statik*, pp. 116–168. To the latter we are largely indebted in the preparation of the present chapter; Plates 10 and 12 are, with slight alteration, reproduced from that work.

PART II.

APPLICATION TO BRIDGES.

65.—Under the head of *Parallel Forces* we have already given the general application of the graphical method to the determination of the moments and shearing forces in beams resting upon two supports only. We shall now take the subject up more in detail, and show the methods of determining the maximum strains for all the possible conditions of loading which may occur in Bridge Girders. In the following we shall adhere closely to the development of the subject as given by *Winkler*. [*Der Brückenbau*, Wien, 1872.]

66. Forces which act upon a Bridge.—The forces which act upon a bridge may be enumerated as follows:

1st. *The weight of the bridge itself.*—This, previous to the calculation of the strains, is *unknown*, since it depends upon the intensity of the strains themselves. It is customary to *assume* the weight to begin with, by comparison with existing structures of similar character, and then to find the resulting strains. The weight answering to these strains can then be easily ascertained; the strength of the materials used being known, and compared with the assumed weight. According as it is less or greater, the weight was then assumed too great or the reverse. A second approximation to the true weight may then be made, and the strains proportionally diminished or increased. As rules for estimating the weight of bridge girders under 200 feet span, we have, for weight of girder **G**,

$$G = \frac{W\, l^2}{12\, f\, d},$$

where **W** = the assumed approximate total distributed load in tons, including the weight of girder;

l = length in feet;
d = depth in feet;

$f=$ the working strain in tons per *sq. foot* of cross-section. (See *Stoney, Theory of Strains,* vol. ii., p. 441.)

We have also the rule: "Multiply the distributed load in tons by 4; the product is the weight of the main girders, end-pillars and cross-bracing in pounds per running foot." Iron is taken at 5 tons per sq. inch tension, and 4 tons per sq. inch compression.

2*d. The moving or live load;* which is determined by the purpose of the bridge. This load can take various positions upon the bridge, and may even be divided into several portions. It is therefore an important problem to determine that distribution which shall cause the maximum strains.

The live load is, as the term implies, in *motion,* so that, in combination with the deflection, there is a centrifugal force, or increase of pressure. This is, however, in practice disregarded, while such a coefficient of safety is chosen in proportioning the parts, that the increase of strain due to this cause is fully covered.

3*d. Horizontal forces,* caused by the wind and the passage of loads.

4*th. Pressures at the supports.* The known forces cause reactions at the supports, which evidently must also be considered as forces acting upon the bridge girder. For straight girders, these reactions are vertical, while in suspension and arch systems they are inclined.

67. Bridge Loading.—The heaviest load to which a railway bridge can be subjected is when it is covered from end to end with locomotives. "The standard locomotive is assumed to be 24 feet long, and to have six wheels with a 12-foot base; to have half its weight resting on the middle wheels, and one-fourth on the leading and trailing pairs respectively, which are supposed to be at equal distances on either side of the middle wheels." (See Stoney, vol. ii., p. 405.) The standard engine is assumed to *weigh* 24 tons, 30 tons and 32 tons, according to the construction. This makes the standard load 1 ton, $1\frac{1}{4}$ ton, or $1\frac{1}{3}$ ton per foot of single line. Short bridges of less than 40 feet span must be considered as subject to concentrated loads from single engines.

The maximum load for public bridges is recommended by Stoney at 100 lbs. per sq. ft.

68. In the Straight Truss all the Outer Forces act in a

Vertical Direction.—The strain in any cross-section depends upon, first, the resultant of all the outer forces acting either side of the cross-section; and second, the statical moment of these forces with reference to the cross-section. The first, or the algebraic sum of all the forces acting between the cross-section and either end, we call the *shearing force* for this cross-section, and indicate it by **S**. It is also designated as *vertical force*, or *transverse* force. The moment of the resultant, or the algebraic sum of the moments of all the exterior forces, with reference to any cross-section, we call the *moment* for this cross-section, and indicate it by **M**. It is also called *bending moment*, or *moment of rupture*. For example, in a lattice girder with horizontal flanges the strains in the web are proportional to the shearing forces, those in the flanges to the bending moments.

The shearing force is considered positive when it acts on the left side upwards, or on the right side downwards. The moment **M** is positive, when on the left side the tendency of rotation is to the left, on the right side to the right, or *when it tends to make the girder convex upwards*, that is, causes compression in the lower fibre or flange.

CHAPTER VII.

SIMPLE GIRDERS.

69. Action of Concentrated Loads—Invariable in Position.—By "simple girder" we understand a girder resting upon two supports only, in opposition to a continuous girder which rests upon more than two.

Suppose a number of forces $P_1 \ldots P_5$ acting at various points. [Fig. 41, Pl. 13.] We form the force polygon by laying off the forces to scale one after another; then choose a pole **O**, and draw **O** 0, **O** 1, **O** 2, etc., to the points of division. Parallel to these lines we draw the lines of the equilibrium polygon between the corresponding force lines prolonged. If now we *close* the polygon thus formed by the line **A B**, and draw through **O** the parallel **O L** to **A B**, the segments 0 **L** and **L** 5 of the force line give the reactions V_1 and V_2. Further, the shearing force between **A** and P_1 is $S_1 = V_1 = $ **L** 0 ; between P_1 and P_2, $S_2 = V_1 - P_1$; at P_3, $S_3 = V_1 - P_1 - P_2$, etc. That is, *the shearing forces are the distances of the points of the force polygon from* **L**. It is easy, then, to construct them, as shown in the lower shaded area of the Fig. (See also Art. 46.)

If in the equilibrium polygon we let fall at any point a vertical as **I K**, and from **K** draw **K L** perpendicular to **A B**, and indicate by **H** the horizontal pull, by **L** the strain in **A B**, and by **M** the sum of the moments of all forces left of **I K**, then, for equilibrium about **K**, we have $M = L \times \overline{K L} = L \times I K \cos I K L$, or, since the angle $I K L = L O H$ in force polygon, $L \times \cos I K L = H$, and hence $M = H \times I K$, or representing the variable ordinate **I K** by y:

$$M = H y.$$

But **H** is the distance of the pole **O** from the force line ; *the moment at any point is therefore proportional to the vertical height of the equilibrium polygon*. (See also Art. 38.) If we take **H** equal to the unit of force, we have

$$M = y,$$

so that in this case *the moment at any point is directly given by the ordinate of the polygon at that point.* It is this important property of the equilibrium polygon which renders it especially serviceable in the graphical solution of this and similar problems.

70. Concentrated Load—Variable Position—Shearing Force.—If the load lies to the right of any given cross-section, then the shearing force at this cross-section will be $S = V_1$, or, since we regard a force to the left acting up as positive, S is positive. As the load P moves towards the left, V_1 or S increases. When the load is to the left of the cross-section, the shearing force at the cross-section is $S = V_1 - P$, and since P is always greater than V_1, S is negative. The nearer P approaches the cross-section, the smaller is S—algebraically.

Hence: *a concentrated load causes a positive or negative shear, according as it is to the right or left of the cross-section considered, and the shearing force is greater the nearer the load is to the cross-section.*

Moments.—If the load lies to the right of the cross-section, the moment is $M = -V_1 x$, x being the distance of the cross-section from the left support. M is therefore negative and increases with V_1; that is, as the load approaches the cross-section. If the load is on the left of the cross-section, $M = -V_2 (l - x)$, V_2 being the reaction at the right support. Here also M is negative and increases with V_2; that is, as the load approaches the cross-section.

Hence: *a concentrated load wherever it lies causes in every cross-section a negative moment, which for any cross-section is a maximum, when the load is applied at that cross-section.*

71. Position of a given System of Concentrated Loads causing Maximum Shearing Force.—If P_1 is the *sum* of all the loads to the left of any cross-section, the shear at that cross-section is $S = V_1 - P_1$. As the system moves to the left without any load passing off the girder or any load passing the cross-section, V_1 and therefore S increases as long as S is positive, or as long as $V_1 > P_1$. If a load passes off the girder, then for the remaining loads S increases anew as the system moves to the left, until a load of the system passes the cross-section in question. The same holds good for a system moving to the right, where S is negative.

Hence: *the shearing force is a maximum for any point,*

when there is a load of the system at that point, and the maximum is positive or negative, according as the load lies just to the right or left of the point.

Since for a single load (Art. 70) **S** is positive or negative, according as the load is to the right or left, **S** will be in general a positive or negative maximum when all the loads lie to the right or left, and the heaviest nearest the cross-section. Only in cases where a small load precedes, can **S** be greatest when the second load lies upon the point in question.

If **P** is the resultant of all the loads and β its distance from the *right* support,

$$\mathbf{V}_1 = \mathbf{P}\,\frac{\beta}{l}, \text{ and therefore } \mathbf{S} = \mathbf{P}\,\frac{\beta}{l} - \mathbf{P}_1.$$

Therefore **S** will vary as the first power of x, the distance of the cross-section from the left support, provided that no wheel passes beyond the support. *Therefore, between any two cross-sections for which the load on the girder remains the same, the shear* **S** *is represented by the ordinates to a straight line.*

72. Construction of the Maximum Shearing Forces.— Construct the force polygon with the given loads; choose a pole **O** [Pl. 13, Fig. 42 (*a*)] and draw the corresponding equilibrium polygon. It is required to determine the shear **S** at a cross-section distant x from the left support, under the supposition that the first load **P**$_1$ of the system, moving towards the left, acts at this cross-section.

Determine upon the outer side **P**$_1$ **A** of the polygon passing through the point **P**$_1$, a point **A** distant from **P**$_1$ by the distance x, and then find the point **B** upon the polygon distant from **A** by l, the length of span, and draw **A B**. Parallel to **A B** draw **O L** in the force polygon, then $\mathbf{A L} = \mathbf{V}_1 = \mathbf{S}$, the shear at **P**$_1$. Drop a vertical through **B** intersecting **P**$_1$ **A** produced, in **M**; then the triangles **O A L** and **A M B** are similar, and therefore $\mathbf{S} = \mathbf{A L} = \mathbf{B M}\,\frac{a}{l}$, when a is the pole distance. If we choose $a = l$, then $\mathbf{S} = \mathbf{B M}$.

Hence: *the maximum shearing forces are proportional to the vertical segments between the equilibrium polygon and the prolongation of the outer side taken at the end of the system, or are equal to these segments if the pole distance is taken equal to the span;* provided that the last load is at the cross-section.

We have, therefore, the simple construction given in Pl. 13, Fig. 42 (*b*). The positive and negative values of **S** equally distant from the right and left supports are equal, so that it is only necessary to construct **S** for one value. [See also Art. 78.]

If the second load is to be at the cross-section, and if *e* is the distance between the first and second, we draw first a line whose equation is $y = P_1 \dfrac{e-x}{l}$, and construct, as above, a polygon, for which the second load lies on the right support **B**, and whose second side (between second and third loads) coincides with the above line. The ordinates to this line above the axis of abscissas will give maximum of + **S**.

73. Maximum Moments.—Since, according to Art. 70, a concentrated load causes a negative moment at any point, wherever it may lie, we must have evidently loads upon both sides of any point, in order that the moment may be a maximum. Since a single load causes a greater moment at any point the nearer it lies to that point, the greatest load must lie nearest the cross-section in question. The method of loading, causing maximum moments, can be best determined for a *distributed* load (not necessarily *uniform*). In this case the equilibrium polygon becomes a curve [Pl. 13, Fig. 43]. If in this curve we draw **A B**, and take **C** so that **A C** : **C B** :: $x : l-x$, then **C D** = **M** for x. Suppose **A B** to take the position **A′ B′**, the horizontal protection of **C C′** being indefinitely small, then **C′ D′** = **M** + d **M**. In order now that **M** may be a maximum, **C′ D′** must be equal to **C D** or **C C′** parallel to **D D′**. If in the force polygon **O A₁** is parallel to **A A′**, **O B₁** to **B B′**, and **O D₁** to **D D′**, then **A₁ D₁** and **D₁ B₁** are the loads upon **A C** and **B C**.

Draw through **C** a vertical, and through **A, A′, B, B′**, parallels to **C C′** or **D D′** intersecting this vertical in **E, E′, F, F′**.

Then   **C E** : **C F** :: **A C** : **B C** :: $x : l-x$,
    C E′ : **C F′** :: **A′ C′** : **B′ C′** :: $x : l-x$;

therefore

   C E : **C F** :: **C E′** : **C F′**, or **C E′** : **C E** :: **C F′** : **C F**;
also   **C E′** − **C E** : **C F′** − **C F** :: **C E** : **C F**,
that is,   **E E′** : **F F′** :: $x : l-x$.

If now we draw through **A′** and **B′** parallels to **C C′**, or **D D′** to intersections H and I, we have $AH = EE'$, $IB' = FF'$. Since the triangle $AA'H$ is similar to OA_1D_1 and $BB'I$ to OB_1D_1, and since $A'H = BI$, we have

$A_1D_1 : B_1D_1 :: AH : B'I :: EE' : FF' :: x : l-x$.

Since A_1D_1 equals the load P_1 on AC, and B_1D_1 the load P_2 on BC, we have $P_1 : P_2 :: x : l-x$.

The same will hold true approximately for concentrated loads. Hence, *in order that the moment at any point may be a maximum, the system of loads must have such a position that the loads either side of this point are to each other as the portions into which the span is divided.*

In Pl. 13, Fig. 44, let CD give the moment at C. If the line AB moves so that the horizontal projections of AC and BC remain equal to x and $l-x$, then as long as the ends A and B move on the same straight lines, the point C will also move in a straight line. The point C describes, therefore, a broken line. The verticals between this line and the polygon correspond to the moments for various positions of the load and a given value of x. Evidently the greatest ordinate will be over an angle of the equilibrium polygon which is not under an angle of the line described by C—that is, for M maximum, a load must lie upon the cross-section.

For any cross-section, then, the moment is a maximum when a load is applied at this cross-section. Which of the loads must be so applied is determined by the preceding rule.

74. Construction of Maximum Moments. — After the equilibrium polygon has been constructed, in order to find M for a point C (Pl. 13, Fig. 45), we determine two points F and G upon the polygon which are distant horizontally from the load on the given cross-section corresponding to the angle E by distances AC, BC. Then draw FG, and the vertical KE is equal to M when the pole distance is unity. We make $CI = EK$. In this way we can construct the moments for different loads of the load system at the given cross-section, and thus determine that position of the load which gives the maximum moment at the cross-section.

Generally when $KE = y$, and the pole distance is a, we have $M = ay$. The pole distance a is measured to the scale of force, and then y is given by the scale of length. The unit for M, in order that M may be equal to y, is evidently $\dfrac{1}{a}$th part of the unit of length (when the pole distance is a force units), or, what is the same thing, one unit of length is equal to a moment units. The same equilibrium polygon can be used for any number of girders of various spans, hence the method is of very rapid application.

75. Absolute Maximum of Moments.—Since for any cross-section **M** is a maximum when a load lies at that section, a load must also lie upon the cross-section for which **M** is an absolute maximum.

If the line **A B** slides upon the equilibrium polygon, altering its length so that its horizontal projection is constant and equal to l, it will envelop a portion of a parabola so long as its ends move in the same sides of the polygon. [Pl. 13, Fig. 46.] The curve thus produced is therefore composed of portions of a parabola. Let the ordinate **D C** correspond to the moment at the point of application of the load **P**. **D C** will be evidently greatest when **A B** is tangent to the curve at **C**, so that *the maximum of the moments occurring at* **D** *is given by the distance* **C D** *between the polygon and curve enveloped by* **A B**.

Let the prolongation of the sides upon which **A B** slides meet in **E**, and **F G** be the tangent to the parabola at the point **H** in the vertical through **E**, so that **F H = H G**, and let **I** be the intersection of **A B** and **F G**. Draw through **A** a parallel to **E B**, intersecting **F G** in **K**. Then the horizontal projections of **A F** and **A K** are equal, since those of **E F** and **E G** are equal.

Since, however, the projections of **F G** and **A B** as also of **A F** and **G B** are equal, **A K** must be equal to **G B**. Hence **A I = B I**. In a parabola the distances of the three diameters passing through two points and the point of intersection of the corresponding tangents are equal, hence the projections of **H I** and **C I** are equal.

The middle point **I** *of the tangent* **A B** *lies, then, half way between the angle* **D** *vertically below the point of tangency and the intersection* **E** *of the sides upon which it slides.*

Since the projection of **A B** is l, its construction is easy. The construction must, of course, be repeated for each angle, in order to determine that for which **M** is an absolute maximum.

The above principle may, then, be thus expressed: *The moment at any load is a maximum, when this load and the resultant of all the loads are equally distant from the centre of the girder.* (See also Art. 48.)

CHAP. VII.] SIMPLE GIRDERS. 93

76.—In Arts. 46 to 50 the above principles have been already deduced so far as relates to the moments alone, and a reference to Art. 49 will show their application to the investigation of the effect of a system of loads moving over the girder. We pass on, therefore, to

CONTINUOUSLY DISTRIBUTED LOADING.

Suppose the load p per unit of length laid off as ordinate. The area thus obtained we call the *load area*. Pl. 13, Fig. 46 (b).

The equilibrium polygon becomes here a curve, for which the same law holds good. If we draw tangents to the curve at the points D' and E' corresponding to D and E, intersecting in C', then the resultant of the load upon DE passes vertically through C', or C' is vertically under the centre of gravity of the area $D D'' E'' E$.

If we consider the load area divided into a number of parts, the resultant for each will pass through the intersection of the tangents at the points vertically under the lines of division. Since these tangents are parallel to the lines in the force polygon corresponding to these lines of division, they form the equilibrium polygon for the concentrated loads, or resultants of the portions into which the load area is divided.

Hence: *if we divide the load area into portions, and replace each by a single force, the sides of the corresponding polygon are tangent to the equilibrium curve at the points corresponding to the lines of division.* (Art. 42.)

77. Total Uniform Load.—In this case the reactions at the supports are $V_1 = V_2 = \frac{1}{2} p l$. Hence, for any cross-section distant x from the left support, the shearing force is

$$S = V_1 - p x = \frac{1}{2} p (l - 2 x).$$

For $x = \frac{1}{2} l$; $S = 0$. S is greatest for $x = 0$ and for $x = l$; that is, maximum $S = + \frac{1}{2} p l$, and $S = - \frac{1}{2} p l$.

The moment at any cross-section is

$$M = - V x + \frac{1}{2} p x^2 = - \frac{1}{2} p x (l - x).$$

M will be greatest for $x = \frac{1}{2} l$, and

$$\text{Max. } \mathbf{M} = -\frac{1}{8} p\, l^2.$$

The shearing forces are, then, given by a straight line intersecting the span in the middle, the ordinate at either end being $\frac{1}{2} p\, l$. [Pl. 14, Fig. 47.]

The moments, as we have already seen [Art. 44, Fig. 30], are given by a parabola whose vertex is in the centre of the span and whose middle ordinate is $\frac{1}{8} p\, l^2$. Since we have seen [Art. 70] that a load at any point causes at every point a negative moment, the *maximum moment at any point will be when the whole span is loaded.*

78. Method of Loading causing Maximum Shearing Force.—We have seen [Art. 70] that a single load causes at any point a positive or negative shear, according as it lies upon the right or left side of the cross-section at that point. Hence, for a uniform load,

The shearing force will be a positive or negative maximum according as the load reaches from the right or left support to the cross-section in question. For the positive maximum we have $\mathbf{V}_1 = p\,(l-x)\,\dfrac{(l-x)}{2\,l} = \dfrac{1}{2} p \dfrac{(l-x)^2}{l}$. Therefore, max. $+\mathbf{S} = \dfrac{1}{2} p \dfrac{(l-x)^2}{l}$.

For the graphical determination we can apply the method given in Art. 72, Fig. 42, by which we have for max. $+\mathbf{S}$ and max. $-\mathbf{S}$ two *parabolas* whose vertices are at the ends of the span, and whose ordinates at these points are $+\dfrac{p\,l}{2}$ and $-\dfrac{p\,l}{2}$. Since, however, each point is found thus from the preceding, the construction is not very exact. We may deduce a better construction as follows. [Pl. 14, Fig. 48.] Through any point F of the curve drop a vertical intersecting A B in C and the line B K parallel to the tangent at F in G. Let the tangent at F intersect A B in H. Then $CH = BH$; hence, $CF = \dfrac{1}{2} CG$.

We have, then, $A E = \dfrac{1}{2} A D = \dfrac{1}{2} p\, l$. Since $C F = \dfrac{1}{2} C G$, we have also $A I = \dfrac{1}{2} A K$; therefore, $A I : A E :: A K : A D$.

CHAP. VII.]   SIMPLE GIRDERS.   95

Hence the following construction:

$$\text{Make } \mathbf{A\,E} = \frac{1}{2} p\, l.$$

Divide **A E** and **A B** into an equal number of equal parts, and draw lines from **B** to the points of division of **A E**, and verticals through the points of division of **A B**. The curve passes through the points of intersection of corresponding lines.

79. Live and Dead Loads.—Let p be the load per unit of length for dead, and m for live load. The maximum moment for any point will be as before.

$$\mathbf{M} = -\frac{1}{2}(p+m)\, x\,(l-x);\quad \text{that is, will be}$$

given by a parabola whose middle ordinate is $-\frac{1}{8}(p+m)\, l^2$.

For the shearing force, we have

$$\text{Max.} + \mathbf{S} = \frac{1}{2} p\,(l-2x) + \frac{1}{2} m\, \frac{(l-x)^2}{l};$$

$$\text{Max.} - \mathbf{S} = \frac{1}{2} p\,(l-2x) - \frac{1}{2} m\, \frac{x^2}{l}.$$

Indicate **A C** [Fig. 49, Pl. 14] by x_1, for which max.$-\mathbf{S} = 0$. then

$$0 = p\, l\,(l-2x_1) - m\, x_1^2,$$

or

$$x_1^2 + 2\frac{p}{m} l\, x_1 - \frac{p}{m} l^2 = 0;$$

hence

$$\frac{x_1}{l} = -\frac{p}{m} + \sqrt{\frac{p^2}{m^2} + \frac{p}{m}}.$$

For the point **D** for which max.$+\mathbf{S} = 0$, $\mathbf{B\,D} = x_1$. The shearing force within **A C** is positive, within **B D** negative, while within **C D** it is both positive and negative.

For	$l = 5,$	10,	20,	50,	75,	100,	150 metres.
$\frac{p}{m} =$	0.12	0.19	0.31	0.64	1.05	1.55	3.12
$\mathbf{AC} = \mathbf{BD} =$	0.24	0.29	0.33	0.38	0.42	0.44	$0.46\,l$
$\mathbf{CD} =$	0.52	0.42	0.34	0.24	0.16	0.12	$0.08\,l$;

that is, **C D** *diminishes with increasing span.*

Recapitulation.—For girders of a length of about 100 feet or more, then, we may consider the live load as distributed per unit of length. The maximum shearing force can then be easily found according to the preceding Art., while the maximum moments will be given by the ordinates to the parabola for full live and dead load [Fig. 30, Art. 44]. For a *framed* structure, we have simply to multiply the shear at any point by the secant of the angle which the brace at that point makes with the vertical, in order to find the strain in that brace. The moment, divided by the depth of truss at the point in question, gives the strain in the flanges. For a *plate* girder, the moment being found as above, and one dimension as the depth given, we can, from Art. 52, so proportion the other dimension as that the strain in the outer fibre shall not exceed the amount allowable in practice. The preceding Art. as also Arts. 78 and 44 and 52 are all that we need to refer to for all practical cases of parallel flange girders of large span.

The preceding will complete our discussion of the simple girder. We have only to remark here that the strains due to rolling load will, in general, be most satisfactorily found by the method of resolution of forces, as illustrated in Art. 12. By this method we first find the reactions at the supports for a *single apex load*, either graphically or by a simple calculation $\left[V_1 = \frac{P(l-x)}{l} \right]$, and then follow this reaction through the girder, and find the resulting strains. We can thus find and tabulate the strains in every piece due to a weight at each and every apex. The maximum strains can, then, be easily taken from the table thus formed. When the live load is supposed thus concentrated at each apex, it is, as we have seen in Art. 12, unnecessary to follow through every reaction. The reactions due to the first and last weights are sufficient to fill out the table. For solid-built beams or plate girders, the principles of the present Chap., therefore, come more especially into play. (See also remarks at close of Chap. V.)

The preceding principles will, it is hoped, be found sufficient to enable the reader to find the maximum moments and shear at each and every cross-section of a beam of given span resting simply upon two supports, and acted upon by any given forces or system of forces in any given position. The reader will do well to take examples of simple trusses, and check the results obtained by the method given in Chap. I. by the above principles. The method of tabulation of single apex loads upon which we lay so much stress is fully given by Stoney ["Theory of Strains," vol. i.], and the examples there given will be found of service.

Finally, then, the strains in upper and lower chords *are greatest for full load over whole span*. We have, therefore, only to erect upon the given span a parabola whose centre ordinate is $\frac{(p+m) l^2}{8}$, where p is the load per unit of length for dead, and m for live load [Art. 44]. The ordinates to this parabola at any point give at once the maximum moment at that point. The depth of truss at this point, if a framed structure, or the moment of inertia of the cross-section at this point, if it is a solid beam [Art. 52], being known, the strain in the flanges or outer fibres may be easily determined. The strain in the web is given by the maximum shear. For dead load alone this is given by the ordinates to a straight line passing through the centre of span, whose extreme ordinates are $\frac{p\,l}{2}$ [Art. 77]. The maximum shear due to live load alone ($m\,l$) will be given by the ordinates to two semi-parabolas, convex to the span, having their vertices at each end, and the extreme ordinates $\frac{m\,l}{2}$ [Art. 78]. At any point, the *greatest* of the two ordinates to these parabolas is to be taken. For live and dead loads together, Art. 79 may also be useful. The shear being known, the strain in any diagonal is equal to the shear multiplied by the secant of the angle made by the diagonal with the vertical [Art. 10 of Appendix] for *parallel flanges*. For flanges *not* parallel, we must find the *resultant* shear as given in Art. 16 (4) of Appendix, or, better still, the flanges once known, the diagonals can be diagrammed according to the principles of Chap. I.

For the investigation of load systems, the principles of Arts. 70–75 will be found sufficient, and the application of these principles we have already sufficiently illustrated in Arts. 49–51.

SUPPLEMENT TO CHAPTER VII.

CHAPTER I.

METHODS OF CALCULATION.

1.—In Chapter I. of the text we have already obtained a method of diagram which will be found both simple and general, and by which we can readily determine the strains for any given loading in *any* framed structure, no matter how irregular in its shape or dimensions, *provided only that all the outer forces are known.*

In Chap. VII. we have also been put in possession of another method of diagram, by which we may for any structure of the above class, *framed or not*, determine the moment at any point, and can then properly proportion the cross-section.

Thus far, indeed, we are unable to apply these methods to the continuous girder or braced arch, as in these cases there are not only upward reactions but also end moments, and in the latter case a thrust also, which must first be determined. The determination of these requires that the elasticity of the material and cross-section of the structure be taken into account. But with these exceptions, and they are of rare occurrence in practice, we can already solve any case which may present itself.

In the Appendix, if he has attended to our numerous references to it, the reader will have already become familiar with two corresponding methods of *calculation*, viz., that by resolution of forces and that by moments.

It is, however, in many cases desirable to know not only the strains in every piece of a structure, but also the *deflection* of the structure, and this also requires a knowledge of the theory of flexure or of elasticity. For the sake of completeness, therefore, aiming as we do to put the reader in possession of methods of calculation as well as of graphic determination, we shall devote a few pages here to a brief notice of these two above-mentioned methods of calculation, and then pass on to the theory of elasticity itself. This latter has been too generally considered by those unacquainted with the methods of the calculus as difficult and abstruse. It is true that the calculus must be called into requisition; but so simple are the processes for beams of single span—and it is with these only we have at present to do—that we indulge the hope that by going back to first principles we may enable even those at present unacquainted with the calculus to follow our demonstrations intelligently, and to comprehend perfectly and even apply readily the method for themselves.

We cannot, indeed, make the reader familiar with *all* the principles of the calculus, but all these principles are by no means needed. Its fundamental idea, a few of its terms and applications, are all that he need be familiar with in order to perform the simple integrations we shall encounter, as readily as the most skilled mathematician. This portion of the present Supplement may, perhaps, be considered by many as unnecessary and superfluous. We are, indeed, justified in assuming such knowledge. But as we believe our plan practicable, we cannot resist the desire of making our development intelligible to *all*, and thus rendering our treatment of the simple girder at least complete.

The practical man as well as the mathematician may thus have at his disposal the powerful aid of the calculus, so far at least as his purposes require it, and be able to deduce for himself the formulæ which hitherto he has accepted "upon faith." It may also not be improbable that here and there one may be found who, pleased with the simplicity of the principles and the fruitfulness of their application, may be led to further prosecute the study for his own satisfaction.

We shall first, then, notice briefly the two methods of calculation above referred to; then devote a few pages to the development of those principles and rules of the calculus of which we shall make use, and finally apply these principles to the discussion of the curve of deflection of loaded beams.

2. Ritter's Method.—This method is referred to in Art. 14. It rests simply upon the principle of the lever, or the law of *statical moments;* requires no previous knowledge, and converts the most difficult cases of strain determination into the most elementary problems of mechanics. Ritter, in his "Theorie eiserner Dach- und Brücken-Constructionen," has applied this simple principle in such detail and fullness, and so clearly set forth its elegance and simplicity, that it very generally, and justly, goes by his name.

"Its results are clear and sharp as the results of Geometry, and of direct practical application. There is hardly another branch of engineering mechanics which, for such a small amount of previous study, offers such satisfactory results, and which is so suited to engage the interest of the beginner."

We have given in the Appendix to Chap. I. (Arts. 6, 9, 10) detailed examples of its application. Throughout this work similar illustrations of its use will be met with, so that it is only necessary here to state more fully than in the text its general principle.

If any structure holds in equilibrium outer forces, it does so by virtue of the *strains* or inner forces which these outer forces produce. Now the outer forces being always given, we wish to find the interior forces or strains. If, then, the structure is framed, and we conceive it cut entirely through, the strains in the pieces thus cut must hold in equilibrium all the outer forces acting between the section and either end. Thus, in Fig. 6, Pl. 2, a section cutting **D**, 7 and **H** completely severs the truss. Then the strains in these three pieces must hold in equilibrium the reaction at **A** and all the forces between **A** and the section.

Now the principle of statical moments is simply that, when any number of forces in a plane are in equilibrium, the algebraic sum of their moments with respect to any point in that plane must be zero.

The application of this principle is simply so to choose this point of moments as to *get rid of all the unknown strains in the pieces cut, except one only;* and then the other forces being known in intensity, position, and direction of action, we can easily find this one; since, when multiplied by its known lever arm, it must be equal and opposite to the sum of the moments of the known forces.

In a properly constructed frame it will, in general, always be possible to pass a section cutting only three pieces. Then, by taking as a centre of moments the intersection of any *two*, we can easily find the strain in the third.

Even if *any* number of pieces are thus cut, if all but one meet at a common point, the strain in this one can be determined.

Thus, in Fig. IV., Pl. 1 of the Appendix, a section may be made cutting 2 3, $d\,h$, $h\,c$ and $c\,\text{Y}$. But all these pieces, except the last, meet in 2, and the strain in this last piece may, therefore, be easily determined.

The above is all that is necessary to be said as to this method. The examples already referred to will make all points of application and detail plain as we proceed. We see no reason why the reader who has mastered Chapter I. and diligently followed out the examples as given in the Appendix, should not now be able to both calculate and diagram the strains in any framed structure all of whose outer forces are known.

3. Method by Resolution of Forces.—We have also yet another method of calculation, based upon the principle that, if any number of forces in a plane are in equilibrium, the sum of their vertical and horizontal components are respectively zero. In structures all the forces acting upon which are vertical, and such are all bridge and roof trusses, etc., of single span, we have only to regard the *vertical* components.

In this connection we have to call attention to the following terms and considerations. The *shear* or shearing force at any point is the algebraic sum of all the outer forces acting between that point and one end. These outer forces are the weights and reactions at the ends. At any apex of a framed structure, where several pieces meet, the horizontal components of the strains in these pieces must balance, or the structure would move; and for the same reason, the algebraic sum of the vertical components must be equal and opposite to the *shear*. The shear being known, if the strains in all the pieces but one are also known, that one can be easily found. Thus the algebraic sum of all the vertical components of the strains in the other pieces being found, and added or subtracted from the shear, as the case may be, the *resultant* shear, multiplied by the secant of the angle made by the piece in question with the vertical, gives at once its strain.

This method is also fully explained in the Appendix, Art. 16 (4), and a practical rule is there given for properly adding the vertical components and determining whether the result is to be added to or subtracted from the shear. This rule we owe to *Humber*.* We have thus *two* methods of calculation, which, for the sake of convenience, we may speak of as *Ritter's* and *Humber's*. Corresponding to Humber's method we have also a graphic solution, based upon the same principles precisely. This we have set forth in Chapter I., and may call *Prof. Maxwell's* method. In Chapter II. and the following we have also become acquainted with the graphic solution corresponding to Ritter's method, or the method of moments, which we may speak of as *Culmann's*. It is to this method, based upon the properties of the *equilibrium polygon*, that the graphical statics properly owes its value and fruitfulness, and to it is due whatever pretensions it can claim as a system. It will be seen hereafter that it alone can furnish a general method applicable to *all* structures, whether framed or not; whether all the outer forces are known or not. By the same general method we are enabled to find the centre of gravity and moment of inertia of areas, and to solve thus a great variety of practical problems —through which, however different, runs one universal method, one simple routine of construction.

---

* *Strains in Girders, calculated by Formulas and Diagrams.*—Van Nostrand, New York.

CHAPTER II.

PRINCIPLES OF THE CALCULUS NEEDED IN OUR DISCUSSION.

4. Differentiation and Integration.—We need but a very few simple ideas and conclusions in order to have at our disposal the whole theory of flexure for beams of single span. Those to whom these ideas are not familiar already may find them indeed new, but will not find them difficult or even abstruse, and with attention to the following will, we venture to think, make a valuable acquisition.

The sign $\int$ is called the "sign of integration," and integration means simply *summation*. It arises merely from the lengthening of the original letter S, first used by Leibnitz for the purpose. The letter d is called the "sign of differentiation;" in combination with a letter, as $d\,x$, it reads "differential of x," and signifies simply the *increment* which has been given to the variable x. So much for terms.

Now suppose we have the equation
$$y = 5\,x^2, \quad \ldots \ldots \ldots \quad (1)$$
in which x and y, although varying in value, must always vary in such a way that the above equation holds always true. This being the case, let us give to y an *increment*—that is, supposing it to have some definite value for which, of course, x is also definite in value, *increase* this value by $d\,y$.

Then x will be increased by *its* corresponding amount $d\,x$, and as the above relation must always hold true, we have
$$y + d\,y = 5\,(x + d\,x)^2 \quad \ldots \ldots \quad (2)$$
or
$$y + d\,y = 5\,(x^2 + 2\,x\,d\,x + d\,x^2).$$

Inserting in this the value of y from (1), we have
$$d\,y = 5\,(2\,x\,d\,x + d\,x^2), \quad \ldots \ldots \quad (3)$$
which is the value of the increment of y or $d\,y$, in terms of x and the increment of x or $d\,x$. That is, the increments are *not* connected by the same law as the variables. The variable y is always 5 times the square of the variable x, but the increment of y is *greater* than 5 times the square of the increment of x by an amount indicated by $5 \times 2\,x\,d\,x$. From (3) we have
$$\frac{d\,y}{d\,x} = 5\,(2\,x + d\,x), \quad \ldots \ldots \quad (4)$$
which gives the value of the *ratio* of the two increments. Now, if we assume a certain value for x, we find easily from (1) the corresponding value of y. If we increase this value of x by a certain assumed increment, $d\,x$, we find easily from (3) the *corresponding* increment of y, or $d\,y$. Then (4) would give us the ratio of these two increments.

Now we see at once from (4) that the smaller we consider dx to be, the nearer this ratio approaches the limiting value $5 \times 2\, x$. We may suppose dx as small as we please, and then this ratio will differ as little as we please from $5 \times 2\, x$. This value, $5 \times 2\, x$, *forms, then, the limit towards which the value of the ratio* $\dfrac{dy}{dx}$ *approaches* as dx diminishes, but which limit evidently it can never *actually* reach or *exactly* equal. Because, in order that this should be the case, dx must be zero. But if dx is zero, that is, if x is not increased, y also is not increased; dy is, therefore, zero, and there is *no ratio at all.*

Now, just here comes in what we may regard as the central principle of the calculus.

If two varying quantities are always equal and always approaching certain limits, then those limits must themselves be equal.

The principle is too obvious to need demonstration. "Two quantities always equal present but one value, and it seems useless to demonstrate that one variable value cannot tend at the same time towards two constant quantities different from one another. Let us suppose, indeed, that two variables always equal have different limits, **A** and **B**; **A** being, for example, the greatest, and surpassing **B** by a determinate quantity $\triangle$.

The first variable having **A** for a limit will end by remaining constantly comprised between two values, one greater, the other less than **A**, and having as little difference from **A** as you please; let us suppose this difference, for instance, less than $\dfrac{1}{2}\triangle$. Likewise the second variable will end by remaining at a distance from **B** less than $\dfrac{1}{2}\triangle$. Now it is evident that, then, the two values could no longer be equal, which they ought to be according to the data of the question. These data are then incompatible with the existence of any difference whatever between the limits of the variables. Then these limits are equal." *

Now let us apply this principle to equation (4). In this equation $\dfrac{dy}{dx}$ is a variable *always* equal to $5\,(2\,x+dx)$. But $5\,(2\,x+dx)$, as we diminish dx, approaches constantly the limit $5 \times 2\, x$; and as $\dfrac{dy}{dx}$ is always equal to $5\,(2\,x+dx)$, *it also* constantly approaches the same limit. These limits, then, are equal, and the *limit* of $\dfrac{dy}{dx} = 5 \times 2\, x$.

Now, if we conceive, and such a conception is certainly possible, dx to be the difference between x and its *consecutive* or *very next* value, such that between these two values *there is no intermediate value of* dx; then dy will be the difference between two *consecutive* values of y; and regarding, then, dx and dy in this light, $\dfrac{dy}{dx}$ will be the *limit* of the ratio of the in-

---

* *The Philosophy of Mathematics.* Bledsoe.

crements, since the increments are then limiting increments, and can be no smaller without disappearing.

We have thus
$$\frac{dy}{dx} = 5 \times 2\, x,$$
which is an *exact* relation between the increments upon this supposition. From this we have $dy = 5 \times 2\, x\, dx$.

If now we *sum up* all the increments dy, then by virtue of the supposition we have made, $\int dy$ must equal y. We thus suppose y to *flow*, as it were, unbrokenly along by the consecutive increments dy, just as the side of a triangle moving always parallel to itself, and limited always by the sides, describes the area of that triangle, while the change dy of its length is the difference between two immediately contiguous positions. Upon this supposition, we repeat, $\frac{dy}{dx}$ is the *limit* of the ratio of the increments, which limit is, as we see from (4), equal *exactly* to $5 \times 2x$. We do not reject or throw away dx from the right of that equation "because of its small size with reference to $2x$," nor, thus rejecting it upon one side of the equation, do we retain it upon the other "in order to retain a trace of the letter x"!; but simply pass to the limit, and then, according to our fundamental principle above, equate those limits themselves. But if $\int dy = y$, then the integral of $5 \times 2x\,dx$, or $\int 5 \times 2x\,dx = y = 5\,x^2$. By "differentiating," as we say, equation (1) we get (5), and by "integrating" (5) we obtain (1).

Hence we see the appropriateness of the term "*fluent*" given by Newton to the quantity dy or $2x\,dx$. So also we see the appropriateness of the term "*ultimate ratio*" * for $\frac{dy}{dx}$ itself.

---

* Liebnitz undoubtedly discovered the calculus independently of Newton, but he considered dx as a quanity so "infinitely" small that in comparison with a finite quantity it could be disregarded "as a grain of sand in comparison with the sea." We see, indeed, from eq. (4) that if dx *upon one side* be zero, we get the same value for $\frac{dx}{dy}$ as before. But if dx is zero on one side, it should be zero *on the other side also*. No matter how small we suppose dx to be, we have no right to get rid of it by *disregarding* it. That Liebnitz recognized this cannot be doubted, and he was therefore inclined to consider his method as approximate only. But to his surprise he found his results *exact*, differing from the true by not even so much as a "grain of sand." There was to him ever in his method this mystery, nor could he conceive what these quantities could be which, though disregarded, gave true results. Bishop Berkeley challenged the logic of the method, and adduced it as an evidence of "how error may bring forth truth, though it cannot bring forth science." Strange to say, even the disciples of Newton were unable to answer Berkeley without taking refuge in the undoubted truth of their results. And yet Newton in his *Principia* lays it down as the corner-stone of his method, that "*quantities which during any finite time constantly approach each other, and*

The whole of the calculus is but the deduction of rules for finding from given equations as (1) their "differential equations" as (5), or inversely of finding from the differential equation by "integration," or summation, the equation between the variables themselves.

Such of these rules as we *need for our purpose* we can now deduce.

5. Differentiation and integration of powers of a single variable.—We have already seen that the $\int dy = y$ and $\int 2x\,dx = x^2$, hence $d(x^2) = 2x\,dx$.

If we should take $y = x^3$, we should have, in like manner, as before,
$$y + dy = (x + dx)^3 = x^3 + 3x^2\,dx + 3x\,dx^2 + dx^3,$$
or
$$dy = 3x^2\,dx + 3x\,dx^2 + dx^3,$$
or
$$\frac{dy}{dx} = 3x^2 + 3x\,dx + dx^2,$$
and passing to the limits, as before,

$\frac{dy}{dx} = 3x^2$, or $dy = 3x^2\,dx$. Hence the differential of x^3 or $d(x^3) = 3x^2\,dx$, and reversely, the integral of $3x^2\,dx$ or $\int 3x^2\,dx = x^3$. In similar manner, we might find
$$d(x^5) = 5x^4\,dx \text{ and } \int 5x^4\,dx = x^5.$$

Comparing these expressions, we may easily deduce general rules which will enable us at once *upon sight* to "differentiate," that is, find the relation connecting the increments; and "integrate" or sum up the successive consecutive values of the variable; for any expression containing the power of a single variable.

These rules are as follows:

To differentiate:

"*Diminish the exponent of the power of the variable by unity, and then multiply by the primitive exponent and by the increment of the variable.*"

Thus, $d(x^2) = 2x\,dx$, $d(x^3) = 3x^2\,dx$, $d(x^7) = 7x^6\,dx$, $d(x^{\frac{3}{2}}) = \frac{3}{2}x^{\frac{1}{2}}\,dx$, $d(x^n) = n\,x^{n-1}\,dx$, etc.

To integrate:

"*Multiply the variable with its primitive exponent increased by unity, by the constant factor, if there is any, and divide the result by the new exponent.*"

---

before the end of that time approach nearer than any given difference, are equal." There can be little doubt that Newton saw clearly that although the quantities might never be able to actually reach their limits, yet that those limits themselves were equal, and *hence* the increment could be left out in the equation, but *not* because by any means it was of insignificant size. His terms "*ultimate ratio*" and "*fluent*" are alone sufficient to indicate that he understood the true logic of the method he discovered; while Liebnitz seems to have stood gazing with wonder at the workings of the *machine* he had found, but whose mechanism he did not understand. [See *Philosophy of Mathematics*. Bledsoe. Lippincott & Co., 1868.]

Thus $\int 2x\,dx = \frac{2x^2}{2} = x^2$ $\int 3x^2\,dx = \frac{3x^3}{3} = x^3$ $\int x^6\,dx = \frac{x^7}{7}$

$\int x^{\frac{1}{2}}\,dx = \frac{x^{\frac{3}{2}}}{\frac{3}{2}} = \frac{2}{3}x^{\frac{3}{2}}$ $\int n\,x^{n-1}\,dx = \frac{n\,x^n}{n} = x^n$, etc.

It is of this latter rule that we shall make especial use in what follows.

6. Other Principles—Integration between limits, etc.— We may observe from (1) and (4) that a *constant factor may be put outside the sign of integration*. Thus $\int 5 \times 2x\,dx = 5\int 2x\,dx = 5\,x^2$.

It is also evident without demonstration that the *integral of the sum of any number of differential expressions is equal to the sum of the several integrals*.

Thus $\int \left[x\,dx + x^2\,dz + y^4\,dy + x^4\,dx \right]$

is the same as $\int x\,dx + \int x^2\,dz + \int y^4\,dy$, etc.

If in (1) we had

$$y = 5\,x^2 + a,$$

where a is a *constant*, we should have

$$y + dy = 5\,(x + dx)^2 + a = 5\,(x^2 + 2x\,dx + dx^2) + a,$$

or $dy = 5\,(2x\,dx + dx^2)$, or $\frac{dy}{dx} = 5\,(2x + dx)$;

whence

$$\frac{dy}{dx} = 5 \times 2x, \text{ or } dy = 5 \times 2x\,dx, \text{ or}$$

just the same as before.

The integral of this will then be $y = 5\,x^2$ as before, whereas it *should* be $y = 5\,x^2 + a$.

If two differential equations, then, are equal, *it does not necessarily follow that the quantities from which they were derived are equal*.

We should, then, *never forget when we integrate to annex a constant*. The *value* of this constant will in any given case be determined by the limits between which the integration is to be performed.

We indicate these limits by placing them above and below the integral sign. Thus the integral of $x^2\,dx$ between the limits of $x = +h$ and $x = -h$ is $\int_{-h}^{+h} x^2\,dx$. If we integrate $x^2\,dx$, we have, then, $\int x^2\,dx = \frac{x^3}{3} + C$, where C is a constant whose value must be determined by the conditions of the special case considered. If we introduce the value of $x = h$ for one limit, we have $\frac{h^3}{3} + C$. For $x = 2h$ for another limit, we have $\frac{8\,h^3}{3} + C$.

We have, then, two equations, viz.:

when $x = h$, $\int_{x=h} x^2\,dx = \frac{h^3}{3} + C,$

and when $x = 2h$ $\quad\displaystyle\int^{x=2h} x^2\, d\, x = \frac{8\, h^3}{3} + C$;

and by *subtracting* one from the other, we have for the integral between the limits $x = 2h$ and $x = h$, $\displaystyle\int_h^{2h} x^2\, d\, x = \frac{7}{3} h^3$, and C thus disappears.

We have, then, only to substitute in succession the values of the variable which indicate the limits, and *subtract the results*.

If also there is but one limit, we could determine C if there were also a condition, such, for instance, as that $\int x^2\, d\, x$ should equal h when $x = 2h$.

The ratio $\dfrac{d\, y}{d\, x}$ is called the "*first* differential coefficient;" if it were to be differentiated again, the next ratio, viz., that of the differential *of* the differential of y to differential of x^2, or $\dfrac{d^2\, y}{d\, x^2}$, is the "*second* differential coefficient," and so on.

Thus, $y = x^5$; $d y = d(x^5) = 5 x^4 d x$, or $\dfrac{d y}{d x} = 5 x^4$; differentiating again, $\dfrac{d^2 y}{d x} = 20 x^3\, d\, x$, or $\dfrac{d^2 y}{d x^2} = 20 x^3$, and so on to *third* differential coefficient, etc.

7. Example.—As an example of the application of our principles, let it be required to determine the area of a triangle. Let the base be b and the height h. Take the base as an axis, and at a distance of x above the base draw a line parallel to b, and at a very small distance $d\, x$ above this line draw another, thus cutting out a very small strip. (Let the reader draw the Fig.) Now for the base y of this strip we have the proportion $h - x : y :: h : b$, or $y = b - \dfrac{b\, x}{h}$, hence the area of the strip is $b\, d x - \dfrac{b\, x\, d\, x}{h}$. But the area of this rectangular slip is not equal to the area of that portion of it comprised within the triangle. It projects over at each end, and is, therefore, somewhat greater. Thus for the small trapezoid actually within the triangle we have for the upper side y', $h - (x + d\, x) : y' :: h : b$, or $y' = b - \dfrac{b}{h}(x + d\, x)$. Hence $y - y' = \dfrac{b\, d\, x}{h}$, and the area of the projecting portion of the rectangle, that is, its *excess* over the trapezoid, is then $(y - y')\, d\, x$, or $\dfrac{b\, d\, x^2}{h}$. Therefore, $b\, d\, x - \dfrac{b\, x\, d\, x}{h} - \dfrac{b\, d\, x^2}{h} = d\, a$, or $\dfrac{d\, a}{d\, x} = b - \dfrac{b\, x}{h} - \dfrac{b\, d x}{h}$, where $d\, a$ is the area of the small trapezoid itself. Now these latter two quantities are always equal for *any* value of $d\, x$. But as $d\, x$ decreases, one side of the equation approaches the limit $b - \dfrac{b\, x}{h}$, and $\dfrac{d\, a}{d\, x}$, therefore, approaches this same limit. The rectangle itself is, then, the limit of the ratio of the area of the small trapezoid to its height, and we can then *equate the limits themselves*, remembering that in this case $d\, a$ is the area passed over by the side y in passing from one position to the consecutive or *very next*. We have, then, $da = b\,dx - \dfrac{b\,x\,dx}{h}$, and if we integrate this expression, that is, sum up all the da's, we have the area of the triangle. Therefore,

$$\mathbf{A} = \int b\,dx - \dfrac{b\,x\,dx}{h} = b\,x - \dfrac{b\,x^2}{2\,h} + \mathbf{C},$$

where $\mathbf{C}$ is the constant of integration, which we must *never* forget to annex. Now, in the present case we wish to sum up all the areas da, or "integrate," between the limits $x = o$ and $x = h$. But for $x = o$, $\mathbf{A}$ must be zero, and hence we have $\mathbf{C} = o$ for the condition that x starts from the base. If in addition to this condition we make $x = h$, we have the sum of all the areas between $x = o$ and $x = h$.

$$\mathbf{A} = b\,h - \dfrac{b\,h}{2} = \dfrac{b\,h}{2}, \text{ as should be.}$$

The above reasoning is somewhat prolix.

If we thoroughly appreciate that dx is the difference between two *consecutive* values of x, we see at once that we obtain the limiting value of the rectangle directly by multiplying its base by dx. The sum of all these must be the area. This conception of dx enables us to curtail much of our reasoning.

Let us take the same problem again, but this time take the axis through the centre of gravity of the triangle; that is, at $\tfrac{1}{3}h$ above the base. Then for the base y at any distance x above this axis, we have

$$\tfrac{2}{3}h - x : y :: h : b, \text{ or } y = \tfrac{2}{3}b - \dfrac{b\,x}{h}.$$

Multiply this by dx upon the above conception of dx, and we have at once *not* for the rectangle upon y, but for its *limiting value*, that is, for the area of that portion of the rectangle included within the triangle,

$$da = y\,dx = \tfrac{2}{3}b\,dx - \dfrac{b\,x\,dx}{h}.$$

Integrating this, then, we have

$$\mathbf{A} = \int \tfrac{2}{3}b\,dx - \dfrac{b\,x\,dx}{h} = \tfrac{2}{3}b\,x - \dfrac{b\,x^2}{2\,h} + \mathbf{C},$$

where $\mathbf{C}$ is a constant to be determined by the limits as before. For one limit, $x = -\dfrac{1}{3}h$, and hence we have

$$\mathbf{A}' = -\dfrac{5}{18} b\,h + \mathbf{C}.$$

For the other limit, $x = +\dfrac{2}{3}h$, and hence we have

$$\mathbf{A}'' = \dfrac{4}{18} b\,h + \mathbf{C}.$$

If we *subtract* the first from the second, $\mathbf{C}$ disappears, and we have $\mathbf{A} = \mathbf{A}'' - \mathbf{A}' = \dfrac{9}{18} b\,h = \dfrac{1}{2}b\,h$, as before.

We might also have integrated first between the limits $x = 0$ and $x = \dfrac{2}{3}h$

For $x = 0$, $C = 0$, and the area *above* the axis is then $\frac{4}{18} b h$. For $x = 0$ and $x = -\frac{1}{3} h$, we have for the area *below* the axis $-\frac{5}{18} b h$. This area has a different sign because below. If we give it the same sign as the other, and then add it, we have the total area. If it also had been above, the total area would have been the difference. Generally, then, we *subtract* according to our rule.

8. Significance of the first differential coefficient.—Any equation between two variables of the first degree is the equation of a straight line. If of the second degree, it represents one of the conic sections, an ellipse, circle, parabola, or hyperbola. Of a higher degree, a *curve* generally. If, then, we take the axis of x horizontal and y vertical, and if $d y$ and $d x$ are the *consecutive* increments of y and x, that is, the difference between any value and the *very next*, the ratio $\frac{d y}{d x}$ is evidently *the tangent of the angle which a tangent to the curve at any point makes with the horizontal.*

If, then, we make $\frac{d y}{d x} = 0$, and find the value of the variable x corresponding to this condition, we find evidently the value of x for which the tangent to the curve is horizontal. If now the curve is *concave* towards the axis, this value of x, substituted in the original equation, will give the *maximum* or greatest value of the ordinate y; because for the point just one side of this the tangent slopes one way, and for the point just the other side it slopes the other. The point where the tangent is horizontal must then be the highest.

If the curve is, on the other hand, *convex* to the axis, the value of x, which makes $\frac{d y}{d x} = 0$, substituted in the original equation, will give y a *minimum* value for similar reasons. By setting the first differential coefficient, then, equal to zero, we may find that value of x which corresponds to the maximum or minimum value of the ordinate, as the case may be. In the case of the deflection of simple beams upon two supports, the curve is always *concave* to the axis, and hence we obtain by this process always the *maximum* deflection.

The above comprises all the principles of which we shall make use in the discussion of the theory of flexure. With a little study, we believe that any one familiar with analytical operations, even although he may never have studied the differential or integral calculus, can follow us intelligently in what follows. Whatever points may still be a little obscure will clear up as he sees more plainly than now their application.

CHAPTER III.

THEORY OF FLEXURE.

9. Coefficient of Elasticity.—Let us now take up the theory of flexure, and see if it is not possible so to present the subject that, in the light of the preceding principles, we may be able to solve all such problems as present themselves.

If a weight P acts upon a piece of area of cross-section A, and elongates or compresses it by a small amount l, we know from experiment that, within certain limits, twice, three times, or four times that weight will produce a displacement of $2\,l$, $3\,l$, $4\,l$, etc. These limits are the limits of elasticity. Within them practically, then, the displacement *is directly as the force*. If we assume this law as *strictly* true for all values of the displacement, and if we denote the original length by L, then, since the force per unit of area is $\frac{P}{A}$, and since this unit force causes a displacement l, in order to cause a displacement L equal to the original length, this unit force must be $\frac{L}{l}$ *times* as great, or equal to $\frac{P}{A}\frac{L}{l}$. This force we call the *modulus* or *coefficient of elasticity*. It is always denoted by E. Hence

$$E = \frac{PL}{Al} \quad \dots \dots \dots \quad (6)$$

The coefficient of elasticity, then, *is the unit force which would elongate a perfectly elastic body* BY ITS OWN LENGTH. It is a theoretical force then; but as the law of perfect elasticity upon which its value is based is true practically within certain limits, by experiments made within those limits, knowing P, A, and L, and measuring l, we can find what the force *would have to be* if the law were always true. Such experiments have been made, and the values of E for different materials are to be found in any text-book upon the strength of materials.

From (6) we have for the unit force of displacement

$$\frac{P}{A} = \frac{El}{L} \quad \dots \dots \dots \quad (7)$$

These expressions will be found useful as enabling us to replace often expressions containing an unknown displacement by a definite or experimentally known value.

10. Moment of Inertia.—This is also a convenient abbreviation, and enables us to replace unknown expressions by a, in any given case, perfectly determinate value.

The moment of inertia, with respect to any axis, is the algebraic sum of the products obtained by multiplying the mass of every element of a given cross-section by the square of its distance from that axis.

If a parallelogram stand on end, and then its support be suddenly pulled away from under it, it will fall over backwards. But to knock it over thus requires force. The force which in this case overturns it is that due to *inertia*. At every point of the surface there is, then, a force acting, depending upon the mass of this point. But not *alone* upon the mass. A force at the top acts evidently with more effect to turn the body over than one at the bottom, which merely tends to make it slide. The *moment* of each element of the area is, then, a measure of the force which at each point causes rotation, and the *sum* of these moments is, then, the measure of the overturning action of the whole force of inertia upon the surface. The *moment of this latter force*, or the sum of the moments *of* the moments, is, then, the *moment of inertia* of the cross-section. Each element of the surface must then be multiplied by the *square* of its lever arm, and the sum of all the results thus obtained taken. In other words, the moment of each element is *itself* considered as a force, and then *its* moment again taken. The sum is denoted by I. For any given dimensions and axis it is a perfectly definite quantity, and may thus often replace expressions containing unknown quantities.

The principles of the calculus just developed will enable us to determine it in some cases, at least, very readily. Its value for various forms of cross-section, in terms of the given dimensions, is given in every text-book upon the strength of materials.

Let us suppose a rectangular cross-section of breadth b and height h, and take the bottom as axis. The area of any elementary strip is, then, $b\,dx$. If its distance from the bottom is x, we have for its *moment* $b\,x\,dx$, and for its *moment of inertia*, then, $b\,x^2\,dx$. Integrating this expression, we have

$$\int b\,x^2\,dx = \frac{b\,x^3}{3} + C.$$

This integral is to be taken between the limits $x = 0$ and $x = h$. For $x = 0$, $b\,x^2\,dx = 0$, and hence $C = 0$. For $x = h$, then, we have $\frac{b\,h^3}{3}$. If the axis had been taken through the centre of gravity, we should have the above integral between the limits $+\frac{h}{2}$ and $-\frac{h}{2}$. For $+\frac{h}{2}$ we have $\frac{b\,h^3}{24} + C$. For $-\frac{h}{2}$, $-\frac{b\,h^3}{24} + C$. Subtracting one from the other (Art. 6), we have $\frac{b\,h^3}{12}$ for the moment of inertia. For a *triangle* of height h and base b, we have for axis through centre of gravity, from Art. 7, for the area of the very small strip at distance x, $\frac{2}{3}b\,dx - \frac{b}{h}x\,dx$. Multiplying this by x^2, we have for its moment of inertia $\frac{2}{3}b\,x^2\,dx - \frac{b}{h}x^3\,dx$. The integral of this is $\frac{2}{9}b\,x^3 - \frac{b\,x^4}{4\,h} + C.$

For $x = \frac{2}{3}h$, this becomes $\frac{4}{243}b\,h^3 + C.$

For $x = -\frac{1}{3}h$, we have $-\frac{11}{972} b h^3 + C$.

Subtracting one from the other (Art. 6), we have $\frac{27}{972} b h^3$, or $\frac{1}{36} b h^3$ for the moment of inertia. The moment of inertia of the rectangle $I = \frac{h^3}{12}$ may be written $\frac{I}{2} = \frac{b h}{2} \times \frac{1}{4} h \times \frac{2}{3} \frac{h}{2}$, or the moment of inertia of the half parallelogram is equal to its area, into the distance of its centre of gravity multiplied by $\frac{2}{3}$ds. of its height. We see at once that when we consider, then, the statical moments as themselves forces, *the centre of action of these moment forces does not coincide with the centre of gravity of the area*. This principle we have already noticed in Chap. VI., Art. 60.

We can also put $\frac{I}{A} = \frac{h^2}{12} = \left(\frac{1}{2\sqrt{3}} h\right)^2$. This value $\frac{1}{2\sqrt{3}} h$ is called the *radius of gyration*. It is evidently the distance from the axis to that point at which, if the mass were concentrated or sum of all the forces were considered as acting, their moment of inertia would be that of the cross-section itself. The value of $\frac{I}{A}$ is, in general then, the *square* of the radius of gyration. We have already shown in Chap. VI. how to find it graphically for various cross-sections.

We are now ready to take up the case of a deflected beam, and to find the differential equation of its curve of deflection.

11. Change of Shape of the Axis.—In the Fig. given in the Supplement to Chap. XIV., we have represented a beam deflected from its original straight line by outer forces. Let the two sections $A\,C$, $B\,D$ be *consecutive* sections, parallel before flexure, and remaining plane after. Let the length of the axis $m\,a$ be s, then $n\,a = d\,s$, and let $d\,\phi$ be the very small angle between the sections after flexure.

If the deflection is small, s will be approximately equal to x, and $d\,s$ to $d\,x$. The elongation of any fibre at a distance v from the centre is, then, $v\,d\,\phi$. The unit force corresponding to this elongation is from (7) $T = E \frac{d\phi}{d x} v$. If $d\,a$ is the cross-section of any fibre as $d\,c$, then the whole force of extension is

$$\frac{E\,v\,d\,a\,d\,\phi}{d\,x}.$$

The *moment* of this force is, then, $\frac{E\,v^2\,d\,a\,d\,\phi}{d\,x}$. The integral of this between the limits $+\frac{h}{2}$ and $-\frac{h}{2}$ will give the entire moment of rupture. But this is equal and opposite to the moment M of all the outer forces; hence

$$M = \frac{E\,d\phi}{d\,x} \int_{-\frac{h}{2}}^{+\frac{h}{2}} v^2\,d\,a.$$

But, as we have just seen, this integral is the *moment of inertia* **I** of the cross-section with reference to the axis through the centre. Hence,

$\mathbf{M} = \dfrac{\mathbf{E\,I}\,d\,\phi}{d\,x}$. Since ϕ is a very small angle, it may be taken equal to its tangent, or equal to $\dfrac{d\,y}{d\,x}$; hence $\dfrac{d\,\phi}{d\,x}$ $\dfrac{d^2 y}{d\,x^2}$ and $\mathbf{M} = \mathbf{E\,I}\,\dfrac{d^2 y}{d\,x^2}$.

But $v\,d\,\phi : v :: d\,x : r$, where r is the radius of curvature;

hence $\qquad \dfrac{v\,d\,\phi}{v} = \dfrac{d\,x}{r} \qquad$ or $\qquad \dfrac{d\,\phi}{d\,x} = \dfrac{1}{r}$.

Therefore, $\qquad \mathbf{M} = \mathbf{E\,I}\,\dfrac{1}{r} = \mathbf{E\,I}\,\dfrac{d^2 y}{d\,x^2} = \dfrac{\mathbf{T\,I}}{v}$ . . . . . (8)

and $\qquad\qquad\qquad\qquad \mathbf{T} = \dfrac{\mathbf{E}\,v}{r}$ . . . . (9)

Equation (8) is our fundamental equation.

In any given case we have only to write down the expression **M** for the moment of the outer forces at any point, and equate it with $\mathbf{E\,I}\,\dfrac{d^2 y}{d\,x^2}$. Integrating once we shall then have for **I** constant, of course, $\mathbf{E\,I}\,\dfrac{d\,y}{d\,x}$ and, integrating again, **E I** y in terms of x, or the equation of the deflection curve itself. Making $\mathbf{E\,I}\,\dfrac{d\,y}{d\,x} = 0$, we can then find the point of maximum deflection, and inserting in the value for **E I** y the value of x thus found, can find the maximum deflection itself. The discussion of any case reduces thus to a simple routine, and every case is in many respects but a repetition of the same processes.

12. Beam fixed at one end and loaded at the other— Constant cross-section.—We shall always consider a moment *positive* when it causes compression in the lower fibre; *negative* when it causes tension in that fibre. Distances to the right of the origin are *always* positive, to the left negative. Hence on the *left* of any section an upward force is negative, a downward force positive; while on the right of the section the upward force is positive and the downward one negative. The reader should always draw the Fig. for each case discussed, and in the beginning, at least, review these conventions each time.

Now let a beam of length l have the weight **P** at the free end, and let it be fixed horizontally or "walled in" at the right end. Then the moment at any point distant x from the left or free end is $\mathbf{M} = + \mathbf{P}\,x$.

(a) *Change of shape.*

From (8) we have now

$$\mathbf{E\,I}\,\dfrac{d^2 y}{d\,x^2} = +\mathbf{P}\,x.$$

Integrating once (Art. 5) we have

$$E I \frac{dy}{dx} = \frac{P x^2}{2} + C,$$

where C is the constant of integration to be determined (Art. 6) by the given conditions. Now by the condition in this case, when $x = l$, $\frac{dy}{dx}$ must be zero, because the end is fixed, and the tangent there must therefore be horizontal (Art. 8). Hence $C = -\frac{P l^2}{2}$, and

$$E I \frac{dy}{dx} = \frac{P x^2}{2} - \frac{P l^2}{2}.$$

We have thus introduced the condition that x cannot be *greater* than l. Integrating again (Art. 5)

$$E I y = \frac{P x^3}{6} - \frac{P l^2 x}{2} + C.$$

Here again we have a constant to be determined, and here again we have the condition that for $x = l$, y must be zero, since at the fixed end there can be no deflection. Therefore, $C = \frac{P l^3}{3}$ and

$$E I y = \frac{P}{6}\left(2 l^3 - 3 l^2 x + x^3\right) = \frac{P}{6}(2 l + x)(l - x)^2.$$

The deflection will evidently be greatest at the free end, and here, therefore, for $x = 0$, we have

$$y = \Delta = \frac{P l^3}{3 E I}.$$

If the cross-section is rectangular, $I = \frac{1}{12} b h^3$ (Art. 10), and the maximum deflection $\Delta = \frac{4 P l^3}{E b h^3}$.

(b) *Breaking weight.*

We have also from equation (8) $M = \frac{T I}{v}$, where T is the tensile strain in any fibre distant v from the centre. For $v = \frac{h}{2}$, T is the *tensile* strain in the outer fibre, and $M = \frac{2 T I}{h}$. For $v = -\frac{h}{2}$ we have the compressive strain in the outer fibre upon the other side, or $M = \frac{2 C I}{h}$. Theoretically the two should be equal. Practically they are not. In fact, if we put for M its value, we have $P x = \frac{2 T I}{h}$, or for a rectangular cross-section $P x = \frac{1}{6} \cdot T b h^2$. This is greatest for $x = l$, hence the breaking weight $P = \frac{T b h^2}{6 l}$. From this we have $T = \frac{6 P l}{b h^2}$. Now experimenting with beams of various materials, known dimensions and given weights, we may find experimentally T. It would seem that this value thus found should equal either the tenacity or crushing strength of the material, but the results of experiment show that it never equals either, but is always intermediate between T and C. Calling this intermediate value R, we have

$$P = \frac{2\,R\,I}{h\,l} \quad \ldots \ldots \ldots \quad (10)$$

The formula is based upon the condition of *perfect elasticity*, while R is determined by experiments made at the breaking point when the condition of perfect elasticity is no longer fulfilled. In the following table the tabulated values of R are correct for solid rectangular beams, and sufficiently exact for those which do not depart largely from that form. If instead of we use the values of T or C, whichever is the smaller, we shall *always be on the safe side*, since R is invariably intermediate between these.

In general we shall refer to the equation

$$M = \frac{2\,T\,I}{h} \quad \ldots \ldots \ldots \quad (11)$$

when we have occasion to find the breaking strength. But it must be always remembered that in any practical example we should replace T by R for rectangular beams, or by T or C, whichever is the smaller, for others. We give also the values of the coefficient of elasticity E. (Wood's Resist. of Materials.)

	T	C	R	E
Cast-iron	16,000	96,000	36,000	17,000,000
Wrought-iron	58,200	30,000	33,000	25,000,000
English Oak	17,000	9,500	10,000	1,451,200
Ash	17,000	9,000	10,000	1,645,000
Pine	7,800	5,400	9,000	1,700,000

All in pounds per square inch.

2. Beam of uniform strength.

Suppose the cross-section or I is *not* constant, but varies so that at every point the strain T is constant. From (11) we have

$$M = P\,x = \frac{2\,T\,I}{h} \text{ for the outer fibre, whence }$$

$T = \dfrac{P\,h\,x}{2\,I}$. For a rectangular cross-section $T = \dfrac{6\,P\,x}{b\,h^2}$. Now suppose the breadth and height at the *fixed end* are b_1 and h_1. Then at this end $T = \dfrac{6\,P\,l}{b_1\,h_1^2}$. But this must be equal to T at any other point; hence

$$\frac{6\,P\,x}{b\,h^2} = \frac{6\,P\,l}{b_1\,h_1^2} \text{ or } \frac{b\,h^2}{b_1\,h_1^2} = \frac{x}{l}.$$

If we suppose the height constant, we have for the varying breadth at any point $b = b_1\,\dfrac{x}{l}$. That is, the breadth must vary as the ordinates to a straight line, and the plan of the beam is a triangle with the weight P at the apex.

If the breadth is constant, $h = h_1 \sqrt{\dfrac{x}{l}}$, or the elevation of the beam is a parabola with the weight at apex. If the cross-section is always *similar*, that is, if $\dfrac{b_1}{h_1} = \dfrac{b}{h}$, we have $b = \dfrac{b_1 h}{h_1}$, and substituting in the equation above $h = h_1 \sqrt[3]{\dfrac{x}{l}}$, which is a paraboloid of revolution.

(a) *Change of shape*

From (8) we have

$$\frac{d^2 y}{d x^2} = \frac{P x}{E I} = \frac{P x}{E \times \frac{1}{12} b h^3},$$

where b and h are variable. If we suppose the height h constant and always equal to h_1, then, as we have seen, $b = b_1 \dfrac{x}{l}$; hence for rectangular cross-section

$$\frac{d^2 y}{d x^2} = \frac{12 P l}{E h_1^3 b_1}.$$

Integrating, since for $x = l$, $\dfrac{dy}{dx} = 0$, we have

$$\frac{d y}{d x} = \frac{12 P l x}{E h_1^3 b_1} - \frac{12 P l^2}{E h_1^3 b_1}.$$

Integrating again, since for $x = l$, $y = 0$, we have

$$y = \frac{6 P l x^2}{E h_1^3 b_1} - \frac{12 P l^2 x}{E h_1^3 b_1} + \frac{6 P l^3}{E h_1^3 b_1}.$$

For the maximum deflection $x = 0$, and

$$\Delta = \frac{6 P l^3}{E h_1^3 b_1}.$$

The above value of y can be written

$$y = \frac{6 P l}{E b_1 h_1^3} (l^2 - 2 l x + x^2) = \frac{6 P l}{E b_1 h_1^3} (l - x)^2 = \frac{6 P l^3}{E b_1 h_1^3} \left(1 - \frac{x}{l}\right)^2;$$

but $\dfrac{6 P l^3}{E b_1 h_1^3}$ is $\dfrac{3}{2}$, the deflection of a beam of constant cross-section $b_1 h_1$, as already found. Calling this deflection Δ_0, we have

$$y = \frac{3}{2} \Delta_0 \left(1 - \frac{x}{l}\right)^2$$

for the deflection at any point, or $\Delta = \dfrac{3}{2} \Delta_0$ for the maximum deflection.

In a similar manner, for constant *breadth*, we have

$$y = 2 \Delta_0 \left[1 - 3 \frac{x}{l} + 2 \sqrt{\left(\frac{x}{l}\right)^3} \right], \Delta = 2 \Delta_0 = \frac{8 P l^3}{E b_1 h_1^3}.$$

For similar cross-sections, we have
$$y = \frac{9}{5}\Delta_0\left[1 - \frac{5}{2}\frac{x}{l} + \frac{3}{2}\sqrt[3]{\left(\frac{x}{l}\right)^5}\right], \quad \Delta = \frac{9}{5}\Delta_0 = \frac{36}{5}\frac{P\, l^3}{E\, b_1\, h_1^3}.$$

If we call the *volume* of the beam of constant cross-section **V**, then in the first case the volume $\mathbf{V}_1 = \frac{1}{2}\mathbf{V}$; in the second, $\mathbf{V}_2 = \frac{2}{3}\mathbf{V}$; in the third, $\mathbf{V}_3 = \frac{3}{5}\mathbf{V}$; or

$$\mathbf{V} : \mathbf{V}_2 : \mathbf{V}_3 : \mathbf{V}_1 = 30 : 20 : 18 : 15.$$

The maximum deflections, as we see above, are as

$$2\Delta_0, \frac{9}{5}\Delta_0, \frac{3}{2}\Delta_0, \text{ or as } 20, 18, \text{ and } 15.$$

That is, the deflections at the ends for a beam of uniform strength in the three cases *are as the volumes*.

13. Beam as before fixed at one end—Uniform load—Constant cross-section.—If p is the load per unit of length, we have for the moment at any point distant x from the free end,

$$\mathbf{M} = p\, x \times \frac{x}{2} = \frac{p\, x^2}{2}, \text{ and hence } \frac{p\, x^2}{2\, E\, I} = \frac{d^2 y}{d\, x^2}.$$

This moment is greatest for $x = l$, and hence Max. $\mathbf{M} = \dfrac{p\, l^2}{2}$.

For the *breaking weight*, then, from (11)

$$\frac{p\, l^2}{2} = \frac{2\, T\, I}{h} \text{ or } p\, l = \frac{4\, T\, I}{h\, l},$$

or *twice* as great as for an equal weight at the end.

For the *change of shape*, we integrate twice, precisely as before, the expression $\dfrac{d^2 y}{d\, x^2} = \dfrac{p\, x^2}{2\, E\, I}$, and obtain thus

$$y = \frac{p}{24\, E\, I}\left(3\, l^4 - 4\, l^3\, x + x^4\right).$$

The maximum deflection, then, is

$$\Delta = \frac{p\, l^4}{8\, E\, I},$$

or only $\frac{3}{5}$ths as great as for an equal load at the end.

2. Constant strength.

We have, as before, from (11) $\mathbf{M} = \dfrac{p\, x^2}{2} = \dfrac{2\, T\, I}{h}$, whence $\mathbf{T} = \dfrac{p\, h\, x^2}{4\, I}$; or for rectangular cross-section, $\mathbf{I} = \dfrac{1}{12}\, b\, h^3$ and $\mathbf{T} = \dfrac{3\, p\, x^2}{b\, h^2}$. If b_1, h_1 are the breadth and heighth of the fixed end section, then, since **T** must be always constant,

$$\frac{3\, p\, x^2}{b\, h^2} = \frac{3\, p\, l^2}{b_1\, h_1^2} \text{ or } \frac{b\, h^2}{b_1\, h_1^2} = \frac{x^2}{l^2}.$$

For height constant, $b = b_1\left(\dfrac{x}{l}\right)^2$

For breadth constant, $h = h_1 \frac{x}{l}$.

For similar cross-sections, $h = h_1 \sqrt[3]{\left(\frac{x}{l}\right)^2}$

The first is in plan a parabola; the second, in elevation a triangle; the third, a paraboloid of revolution.

For the *change of shape*, we have, by proceeding in the same manner as in Art. 12, $\Delta = 2 \Delta_0$, $\Delta = 4 \Delta_0$, and $\Delta = 3 \Delta_0$ in the three cases, where Δ_0 is the deflection of a similar beam of constant cross-section $b_1 h_1$.

14. Beam supported at both the ends—Constant cross-section—Concentrated load.—Let the weight P be distant from the left end by a distance l_1 and from the right end by l_2. Let the distance of any point from the left end be x. For the upward reaction at the left end,

$$V_1 \times l = P l_2 \quad \text{or} \quad V_1 = P \frac{l_2}{l}.$$

The moment, then, at any point between the left end and F, for x less than l_1, is $M - \frac{P l_2 x}{l}$. For any point to the right of P, or x greater than l_1, $M' = -\frac{P l_2 x}{l} + P(x - l_1)$. Instead of this, however, we may take the reaction at the other end, $V_2 = P \frac{l_1}{l}$; and then for x greater than l_1,

$$M' = -V_2(l - x) = -\frac{P l_1 (l - x)}{l}.$$

The moment is evidently greatest at the point of application of the load, or for $x = l_1$. Hence the maximum moment is $-\frac{P l_1 l_2}{l}$.

(a) *Breaking weight.*

From (11) $M = \frac{2 T I}{h} = -\frac{P l_1 l_2}{l}$, or, for the breaking weight, $P = \frac{2 T I l}{h l_1 l_2}$. For rectangular cross-section, $I = \frac{1}{12} b h^3$ and $P = \frac{T b h^2 l}{6 l_1 l_2}$.

For a load in the middle, $l_1 = l_2 = \frac{1}{2} l$ and Max. $M = -\frac{1}{4} P l$, and $P = \frac{8 T I}{h l}$, or 4 times as great as for a beam of same length fixed at one end and free at the other.

(b) *Change of shape.*

We have, then, from (8), for x less than l_1,

$$\frac{d^2 y}{d x^2} = -\frac{P l_2 x}{E I l} \quad \text{and} \quad \frac{d^2 y'}{d x^2} = -\frac{P l_1 (l-x)}{E I l} \text{ for } x \text{ greater than } l_1.$$

Integrating, we have

$$\frac{d y}{d x} = -\frac{P l_2 x^2}{2 E I l} + C, \quad \frac{d y'}{d x} = -\frac{P l_1}{E I l}\left[l x - \frac{x^2}{2}\right] + C'.$$

For $x = l_1$, these two values of $\frac{d y}{d x}$ are equal, and hence, since $l_2 = l - l_1$, we have $C' = C + \dfrac{P\, l_1{}^2}{2\, E\, I}$. We have then the two equations

$$\dfrac{dy}{dx} = -\dfrac{P\, l_1\, x^2}{2\, E\, I\, l} + C \quad \text{and} \quad \dfrac{dy'}{dx} = -\dfrac{P\, l_1}{E\, I\, l}\left[l\, x - \dfrac{x^2}{2}\right] + \dfrac{P\, l_1{}^2}{2\, E\, I} + C,$$

containing both the same constant C.

Integrating these, we have

$$y = -\dfrac{P\, l_1\, x^3}{6\, E\, I\, l} + C\, x + C_1, \quad y' = -\dfrac{P\, l_1}{E\, I\, l}\left[\dfrac{l\, x^2}{2} - \dfrac{x^3}{6}\right] + \dfrac{P\, l_1{}^2\, x}{2\, E\, I} + C\, x + C_2.$$

In the first of these, for $x = 0$, $y = 0$; hence $C_1 = 0$.

For $x = l_1$, $y = y'$; and hence $C_2 = -\dfrac{P\, l_1{}^3}{6\, E\, I}$.

For $x = l$, $y' = 0$; and hence, finally, $C = \dfrac{P\, l_1\, l_2}{6\, E\, I\, l}(2\, l - l_1)$.

We have, therefore, by substitution of these constants,

$$y = \dfrac{P\, l_2\, x}{6\, E\, I\, l}(2\, l\, l_1 - l_1{}^2 - x^2) \quad y' = \dfrac{P\, l_1}{6\, E\, I\, l}(l-x)(-l_1{}^2 + 2\, l\, x - x^2).$$

For $x = l_1$, we have the deflection at the load $y = \dfrac{P\, l_1{}^2\, l_2{}^2}{3\, E\, I\, l}$.

Inserting the value of C in the value for $\dfrac{dy}{dx}$ above, and placing the value of $\dfrac{dy}{dx}$ equal to 0, we have for the value of x, which makes y a maximum, $x = \sqrt{\dfrac{1}{3}(2\, l - l_1)\, l_1}$, an expression holding good only for x *less* than l_1. Inserting this in the value for y, we have for the maximum deflection itself

$$\Delta = \dfrac{1}{27}\dfrac{P\, l_1\, l_2}{E\, I\, l}(2\, l - l_1)\sqrt{3\, l_1\, (2\, l - l_1)}.$$

If the load is in the middle, we have for the curve of deflection

$$y = \dfrac{1}{48}\dfrac{P}{E\, I}x\,(3\, l^2 - 4\, x^2),$$

and for the deflection itself $\quad \Delta = \dfrac{P\, l^3}{48\, E\, I}.$

The greatest deflection is *not*, then, at the weight, except when the load is in the middle. When this is the case, the deflection is only $\tfrac{1}{16}$th of the deflection for the same length of beam fixed at one end and loaded at the other or free end.

15. Beam as before supported at the ends—Uniform load.—For a load p per unit of length, the entire load is $p\, l$. The reactions at each end are $\dfrac{p\, l}{2}$, and the moment at any point is

$$M = -\dfrac{p\, l}{2}x + \dfrac{p\, x^2}{2} = -\dfrac{p\, x}{2}(l - x).$$

M is evidently greatest at the centre, and hence

$$\text{Max. } M = -\dfrac{p\, l^2}{8}.$$

For the *breaking weight*, then, from (8)

$$\frac{p\,l^2}{8} = \frac{2\,T\,I}{h}, \text{ or } p\,l = \frac{16\,T\,I}{h\,l},$$

or 4 times as much as for a beam of same length loaded uniformly and fixed at one end.

For the *change of shape*, we have

$$\frac{d^2 y}{d x^2} = -\frac{p\,x\,(l-x)}{2\,E\,I}.$$

The constants of integration are determined by the conditions that, for $x = \frac{l}{2}$, $\frac{d y}{d x} = 0$; $x = 0$, $y = 0$; and $x = l$, $y = 0$. Integrating, then, twice under these conditions, we have

$$y = \frac{p}{24\,E\,I}\,(l^3 - 2\,l\,x^2 + x^3)\,x.$$

This is greatest at the centre, or for $x = \frac{l}{2}$; hence the maximum deflection is

$$\Delta = \frac{5}{384}\cdot\frac{p\,l^4}{E\,I},$$

or only ⅛ths of a beam of the same length fixed at one end and uniformly loaded.

16. Beam supported at one end and fixed at the other —Constant cross-section— Concentrated load.—Let the left end be fixed horizontally so that the tangent to the deflected curve at that point is always horizontal, and therefore $\frac{d y}{d x} = 0$.

Let the distance of the weight **P** from left be a, and the distance of any point x.

Then, for x *less* than a, we have

$$\mathbf{M} = -\mathbf{V}\,(l-x) + \mathbf{P}\,(a-x);$$

for x greater than a,

$$\mathbf{M}' = -\mathbf{V}\,(l-x),$$

where **V** is the reaction at the free end, and is so far unknown.

If we put $\mathbf{M} = \frac{d^2 y}{d x^2}$ and $\mathbf{M}' = \frac{d^2 y'}{d x^2}$, and integrate as usual, and remember that for $x = 0$, $\frac{d y}{d x} = 0$, and for $x = a$, $\frac{d y'}{d x} = \frac{d y}{d x}$, we have

$$\frac{d y}{d x} = -\frac{x}{2\,E\,I}\left[\mathbf{V}\,(2\,l-x) - \mathbf{P}\,(2\,a-x)\right]$$

$$\frac{d y'}{d x} = -\frac{1}{2\,E\,I}\left[\mathbf{V}\,x\,(2\,l-x) - \mathbf{P}\,a^2\right].$$

Integrating again and determining the constants by the conditions that, for $x = 0$, $y = 0$, and for $x = a$, $y = y'$, we have

$$y = -\frac{x^2}{6\,E\,I}\left[\mathbf{V}\,(3\,l-x) - \mathbf{P}\,(3\,a-x)\right]$$

$$y' = -\frac{1}{6\,E\,I}\left[\mathbf{V}\,x^2\,(3\,l-x) - \mathbf{P}\,(3\,x-a)\,a^2\right].$$

CHAP. III.]   SUPPLEMENT TO CHAP. VII.   121

Now, for $x = l$, $y' = 0$; hence $\mathbf{V} = \mathbf{P}\,\dfrac{a^2\,(3\,l-a)}{2\,l^3}$. If the load is in the middle, $\mathbf{V} = \dfrac{5}{16}\,\mathbf{P}$.

$\mathbf{V}$, or the reaction at the free end, is now known, and substituting it in the value of y' above, we have the equation of the deflection curve between the weight and the free end.

$$y' = -\frac{\mathbf{P}}{6\,\mathbf{E}\,\mathbf{I}}\left[\frac{a^2\,(3\,l-a)\,(3\,l-x)\,x^2}{2\,l^3} - (3\,x-a)\,a^2\right].$$

Substituting it also in the value of $\dfrac{d\,y'}{d\,x}$ above, and placing then $\dfrac{d\,y'}{d\,x}$ equal to zero, we find for the value of x, which makes the deflection a maximum, when x is greater than a, $x = l - l\sqrt{\dfrac{l-a}{3\,l-a}}$.

Substituting this value of x in the value of y' above, we have for the maximum deflection itself

$$\Delta = \frac{\mathbf{P}\,a^2}{6\,\mathbf{E}\,\mathbf{I}}\,(l-a)\,\sqrt{\frac{l-a}{3\,l-a}}$$

When the weight is at the middle, this becomes $\Delta = \dfrac{\mathbf{P}\,l^3}{48\,\mathbf{E}\,\mathbf{I}} \times \dfrac{1}{\sqrt{5}}$, or only $\dfrac{1}{16\,\sqrt{5}}$, as much as for a beam of same length fixed at end and with load at other end, and only $\dfrac{1}{\sqrt{5}}$ as much as for same beam simply supported at ends.

Breaking weight.

Having now $\mathbf{V}$, we know $\mathbf{M}$ and $\mathbf{M}'$. Rupture will occur where the moment is greatest, that is, either at the fixed end or at the weight. Now the moment at $\mathbf{P}$ is $-\mathbf{V}\,(l-a) = -\mathbf{V}\,l + \mathbf{V}\,a$. The moment at the fixed end is $-\mathbf{V}\,l + \mathbf{P}\,a$. Now, as $\mathbf{V}$ is always less than $\mathbf{P}$, we see at once that for any value of a less than l, the moment at the weight is greatest.

We have for the moment at the weight from (8)

$$\mathbf{M} = -\mathbf{V}\,(l-a) = -\mathbf{P}\,\frac{a^2\,(3\,l-a)}{2\,l^3}\,(l-a) = \frac{2\,\mathbf{T}\,\mathbf{I}}{h},\text{ and hence}$$

for the breaking weight $\mathbf{P} = \dfrac{4\,\mathbf{T}\,\mathbf{I}\,l^3}{h\,a^2\,(3\,l-a)\,(l-a)}$.

If the weight is in the middle, $\mathbf{P} = \dfrac{64}{5}\,\dfrac{\mathbf{T}\,\mathbf{I}}{h\,l}$,

or $\frac{8}{5}$ths as much as for the same beam supported at the ends.

17. Beam as before fixed at one end and supported at the other—Uniform load.—In this case the moment at any point is $\mathbf{M} = -\mathbf{V}\,(l-x) + \dfrac{1}{2}\,p\,(l-x)^2 = \mathbf{E}\,\mathbf{I}\,\dfrac{d^2\,y}{d\,x^2}$. Integrating twice and determining the constants by the conditions that, for $x = 0$, $\dfrac{d\,y}{d\,x} = 0$ and $y = 0$, we easily obtain

$$y = \frac{x^2}{24\,EI}\left[4\,V\,(3\,l-x) - p\,(6\,l^2 - 4\,l\,x + x^2)\right].$$

For $x = l$, $y = 0$, and hence $V = \frac{3}{8}p\,l$.

Substituting this value of V

$$y = \frac{p\,x^2}{48\,EI}(l-x)(3\,l - 2\,x) \text{ and}$$

$$\frac{dy}{dx} = \frac{p\,x}{48\,EI}(6\,l^2 - 15\,l\,x + 8\,x^2).$$

Putting this last equal to zero, we find $x = \dfrac{15 - \sqrt{33}}{16}\,l$, or $x = 0.5785\,l$, for the value of x, which makes the deflection a maximum, and this inserted in the value of y gives for this maximum deflection itself,

$$\Delta = \frac{39 + 55\,\sqrt{33}}{10^4}\,\frac{p\,l^4}{EI} = 0.0054\,\frac{p\,l^4}{EI}.$$

For the *breaking weight*, we have, since the greatest moment is at the fixed end and equal to $\frac{1}{8}p\,l^2$, $M = \frac{1}{8}p\,l^2 = \frac{2\,T\,I}{h}$; hence $p\,l = \dfrac{16\,T\,I}{h\,l}$.

The strength is, then, $\frac{3}{2}$ times as great as for the same load in the middle, but no greater than for a beam of same length and load supported at both ends.

18. Beam fixed at both ends—Constant cross-section—Concentrated load.—Taking our notation as before (Art. 12), we have in this case not only a reaction at the right end, but also a positive *moment* there as well, both of which must be found. If l_1 be the distance from left end to weight, and l_2 from weight to right end, and if V_1 and V_2 are reactions, M_1 and M_2 the moments at left and right ends respectively, then for equilibrium we must have $-V_1 + V_2 = P$, $M_1 + V_1\,l_1 = M_2 - V_2\,l_2$.

For x less than l_1 we have $M = M_1 + V_1\,x = EI\,\dfrac{d^2 y}{dx^2}$. Integrating once, since the constant is zero, because, for $x = 0$, $\dfrac{dy}{dx} = 0$, we have

$$\frac{dy}{dx} = \frac{1}{2\,EI}\left[2\,M_1 + V_1\,x\right]x.$$

Integrating again, since, for $x = 0$, $y = 0$, and the constant is zero,

$$y = \frac{1}{6\,EI}\left[3\,M_1 + V_1\,x\right]x^2.$$

For the distance from the right end to the weight we may obtain similar expressions, if we take that end as the origin, only we should have $-V_2$ and M_2 in place of V_1 and M_1. At the weight itself $\dfrac{dy}{dx}$ and y must in each case be equal, but $\dfrac{dy}{dx}$ of opposite sign. Therefore we have the equations

$$(2\,M_1 + V_1\,l_1)\,l_1 = -(2\,M_2 - V_2\,l_2)\,l_2,$$
$$(3\,M_1 + V_1\,l_1)\,l_1^2 = (3\,M_2 - V_2\,l_2)\,l_2^2.$$

From these two equations, and the two equations above, viz., $-\mathbf{V}_1 + \mathbf{V}_2 = \mathbf{P}$ and $\mathbf{M}_1 + \mathbf{V}_1\, l_1 = \mathbf{M}_2 - \mathbf{V}_2\, l_2$, we can determine $\mathbf{V}_1$, $\mathbf{V}_2$, $\mathbf{M}_1$ and $\mathbf{M}_2$.

Thus from the last two we have

$$\mathbf{V}_1\, l_1 + \mathbf{V}_2\, l_2 = \mathbf{M}_2 - \mathbf{M}_1 = \mathbf{V}_1\, l_1 + \mathbf{P}\, l_2 + \mathbf{V}_1\, l_2 = \mathbf{V}_1\, l + \mathbf{P}\, l_2,$$

or $\quad \mathbf{V}_1\, l = \mathbf{M}_2 - \mathbf{M}_1 - \mathbf{P}\, l_2.$

So also $\mathbf{V}_2\, l = \mathbf{M}_2 - \mathbf{M}_1 + \mathbf{P}\, l_1$, and substituting these in the equations above, we have

$$(\mathbf{M}_1 + \mathbf{M}_2)\, l = \mathbf{P}\, l_1\, l_2,$$
$$\mathbf{M}_1\, l\,(2\, l_1 - l_2) - \mathbf{M}_2\, l\,(2\, l_2 - l_1) = \mathbf{P}\, l_1\, l_2\, (l_1 - l_2);$$

and from these we have, finally,

$$\mathbf{M}_1 = \mathbf{P}\, \frac{l_1\, l_2^2}{l^2}, \quad \mathbf{M}_2 = \mathbf{P}\, \frac{l_1^2\, l_2}{l^2};$$

and then from the values of $\mathbf{V}_1\, l$ and $\mathbf{V}_2\, l$ above

$$\mathbf{V}_1 = -\mathbf{P}\, \frac{l_2^2\,(3\, l_1 + l_2)}{l^3}, \quad \mathbf{V}_2 = +\mathbf{P}\, \frac{l_1^2\,(l_1 + 3\, l_2)}{l^3}.$$

Change of shape.

Substituting these values, we can now find

$$\frac{d\, y}{d\, x} = \frac{\mathbf{P}\, l_2^2\, x}{2\, \mathbf{E}\, \mathbf{I}\, l^3}\left[2\, l\, l_1 - (3\, l_1 + l_2)\, x\right]$$

$$y = \frac{\mathbf{P}\, l_2^2\, x^2}{6\, \mathbf{E}\, \mathbf{I}\, l^3}\left[3\, l\, l_1 - (3\, l_1 + l_2)\, x\right].$$

Hence y is a maximum for $x = \dfrac{2\, l\, l_1}{3\, l_1 + l_2}$, and the maximum deflection itself is

$$\Delta = \frac{2\, \mathbf{P}\, l_1^3\, l_2^2}{3\, \mathbf{E}\, \mathbf{I}\, (3\, l_1 + l_2)^2}.$$

This expression will be itself a maximum for $l_1 = l_2$ or $l_1 = \tfrac{1}{2}\, l$, that is, the maximum deflection for a weight in the middle is at the weight and equal to

$$\Delta = \frac{\mathbf{P}\, l^3}{192\, \mathbf{E}\, \mathbf{I}}.$$

This deflection is greater than the maximum deflection for any other position of the weight, which in general is not found at the weight itself, but at some other point between the weight and farthest end.

We see above that the deflection in this case for load in middle is only one-fourth as much as for same beam and load when supported at the ends.

Breaking weight.

For the greatest moment, which we easily find to be at the end, we have

$$\mathbf{M} = \frac{\mathbf{P}\, l_1\, l_2^2}{l^2} = \frac{\mathbf{P}\, l_1\, (l - l_1)^2}{l^2}.$$

This is a maximum for $l_1 = \tfrac{1}{3}\, l$. That is, *the greatest moment at the end occurs when the load is distant one-third of the length from that end.* The value of this greatest moment is $\frac{4}{27} \mathbf{P} l$. Hence we have from (11) $\frac{4}{27} \mathbf{P} l$ $= \frac{2\,\mathbf{T}\,\mathbf{I}}{h}$, or $\mathbf{P} = \frac{27\,\mathbf{T}\,\mathbf{I}}{2\,h\,l}$, or $\frac{27}{16}$ as great as for the same beam supported at the ends only. If the weight is in the middle, however, we have $\frac{\mathbf{P}\,l}{8} = \frac{2\,\mathbf{T}\,\mathbf{I}}{h}$ or $\mathbf{P} = \frac{16\,\mathbf{T}\,\mathbf{I}}{h\,l}$, or twice as much as the same beam supported at the ends.

19. The above is sufficient to introduce the reader to the theory of flexure. He can now discuss for himself the above case for *uniform load*, and prove that the maximum deflection is at the centre and equal to $\frac{p\,l^4}{384\,\mathbf{E}\,\mathbf{I}}$. That the greatest moment is at the end and equal to $\frac{1}{12}\,p\,l^2$, and that the breaking weight is $p\,l = \frac{24\,\mathbf{T}\,\mathbf{I}}{h\,l}$. We may also observe that both in the beam fixed at one end and supported at the other, and fixed at both ends, the moment at the fixed end is positive. From this end it decreases towards the weight, and finally reaches a point where the moment is zero. Past this point the moment becomes negative, and in the case of the beam, free at the other end, increases gradually to a maximum and then decreases to zero. In the beam fixed at both ends, it increases to a maximum, then decreases to zero, then changes sign and becomes positive and increases to the other end. These points at which the moments are zero are *points of inflection*, because here the curvature changes from convex to concave, or the reverse.

They can be easily found from the equations for the moments by finding the value of x necessary to make the moments zero.

Thus, for a beam fixed at one end and supported at the other, uniform load, the inflection point is at a distance from the fixed end $x = \frac{1}{4}$. For both ends fixed, we make $\mathbf{M} = \frac{1}{12}\,p\,[l^2 - 6\,(l-x)\,x] = 0$, and find $x = \frac{1}{6}\,(3 \mp \sqrt{3})\,l = 0.21131\,l$ and $0.7887\,l$. The reader will also do well to discuss the *curves* of moments. He will find the moments represented by the ordinates to parabolas, and limited by straight lines similarly to Figs. 73 and 75, Pl. 18.

We shall give in the Supplement to Chap. XIV. much more general formulæ, from which, for one or both ends fixed or free, the moments and reactions at the supports may be found, *when any number of spans of varying length intervene*, for single load anywhere upon any span, or for load uniformly distributed over any span.

CHAPTER VIII.

APPLICATION OF THE GRAPHICAL METHOD TO CONTINUOUS GIRDERS—GENERAL PRINCIPLES.

80. Mohr's Principle.—Thus far, in addition to the general principles of the Graphical method, we have noticed more or less in detail its application to the composition and resolution of forces, and the corresponding determination of the strains in the various pieces of such framed structures as Bridge Girders, Roof Trusses, etc. We have also illustrated the graphical determination of the *centre of gravity* and *moment of inertia* of areas, as also of the bending moments and shearing forces for simple girders, including several important cases in practical mechanics. (See Art. 41.) Lastly, we have taken up the subject of Bridge girders more in detail, and developed in order the principles to be applied in the solution of any particular case. Although brief, it is hoped that this portion will be found sufficient to illustrate fully the method of procedure to be followed in practice.

As regards *simple* girders, the principles referred to are so easy of application that the reader will find no difficulty in diagraming the strains in any structure of the kind, as explained in the "practical applications" of Arts. 8 to 13; or he can find the maximum moment at any cross-section for given loading according to the last chapter. In the case of beams or girders *continuous* over three or more supports, however, we meet with difficulties which for some time were considered insuperable.

Thus *Culmann*, in the work which we have so often quoted, says:* "The determination of the reactions at the supports for a continuous beam, which depend upon the deflection, the law of which is given by the theory of the elastic line, is *impossible* by the graphical method, at least so far as at present developed. The theory rests upon the princip¹

---

* *Culmann's Graphische Statik*, p. 2

curvature of the deflected beam, for any cross-section, is inversely proportional to the moment of the exterior forces. Now the deflection at any point is so small, and the radius of curvature so great, that its construction is impracticable, and will so remain until Geometry furnishes us with simple relations between the corresponding radii of curvature of projected figures whose projection centre lies in the vertical to the horizontal axis of the beam. If such relations were known, we could by projection exaggerate the deflection of the beam until the radius of curvature became measurable. Since we are not yet able to do this, *we must have recourse to calculation.*" He then enters into a somewhat abstruse analytic discussion of the continuous girder, and deduces formulæ for the reactions at the supports. These being thus known, the graphical method is then applied.

Concerning this difficulty, Mohr * remarks that it has but little weight, and may be easily overcome if the same simplification of the graphical method is made which is considered allowable in the analytical investigation, viz., when we take instead of the exact value of the radius of curvative $\dfrac{\left(1 + \dfrac{dy^2}{dx}\right)^{\frac{3}{2}}}{\dfrac{d^2y}{dx^2}}$ as given by the calculus, the approximate value $\dfrac{1}{\dfrac{d^2y}{dx^2}}$.

Thus, let Pl. 14, Fig. 50 represent a perfectly flexible cord **A B D** loaded by arbitrary successive forces. The variation of these forces per unit of horizontal projection dx we represent by p. Take the origin of co-ordinates at the lowest point **B**. If the cord is supposed cut at **B** and **D**, we have at **B** a horizontal force **H**, and at **D** a strain S, which may be resolved into a horizontal force H_1 and a vertical force **V**. Since these forces are in equilibrium with the external forces, the conditions of equilibrium are

$$(1) \ldots \ldots \mathbf{H} = \mathbf{H}_1$$

and

$$(2) \ldots \ldots \mathbf{V} = \int_0^x p \, dx.$$

---

* *Zeitsch. des hannov. Arch.–u. Ing. Vereins.*—Band xiv., Heft 1.

Moreover,

$$(3) \quad \frac{dy}{dx} = \frac{V}{H_1} = \frac{\int_0^x p\,dx}{H}$$

Differentiating:

$$\frac{d^2y}{dx} = \frac{p\,dx}{} \quad \text{or}$$

$$(4) \quad H\frac{d^2y}{dx^2} = p.$$

Now, had we formed a force polygon by laying off the forces, then taken a pole at distance H and drawn lines from pole to ends of forces, the corresponding equilibrium polygon would, as we have seen, Art. 43, be tangent to the curve **A B D** at the points midway between the forces. The greater the number of forces taken, the shorter, therefore, the sides of the polygon; the nearer it will approach the curve **A B D**. This curve is therefore the equilibrium curve, found according to the graphical method. Its equation is given above by (4).

But the equation of the *elastic line* is, as is well known,

$$(5) \quad E\frac{d^2y}{dx^2} = \frac{M}{I},\ast$$

where E is the modulus of elasticity of the material, M the moment of the exterior forces, and I the moment of inertia of the cross-section.

Comparing now this equation with equation (4) above, we see that *the elastic line is an equilibrium curve whose horizontal force H is E, and whose vertical load per unit of length p is represented by the variable quantity $\frac{M}{I}$.*

This simple relation, first given by Mohr, renders possible the graphical representation of the elastic line, and not only solves graphically almost all problems connected with it, but in many cases simplifies considerably the analytical discussion also.

81. Elastic Curve.—If we choose the pole distance H at $\frac{1}{n}$th E instead of E, the ordinates of the elastic line will be n times too great. If the scale of the figure is, however, $\frac{1}{n}$th the

---

* *Stoney—Theory of Strains*, p. 146. *Wood—Resistance of Materials*, p. 98. Also Supplement to Chap. VII, Art. 11.

real size, then in the diagram the ordinates of the elastic line *will be given in true size.*

Equation (5) may also be written

$$E I \frac{d^2 y}{d x^2} = M ;$$

that is, *the elastic curve is an equilibrium curve, or catenary, whose horizontal force* H *is* E I *or* 1 *, and whose corresponding variable load per unit of length is* M *or* $\frac{M}{I}$ *respectively.*

If we divide, then, the *moment area* by verticals into a number of smaller areas, and consider these areas as *forces acting at their centres of gravity*, these forces determine, as we have seen (Art. 43), an equilibrium polygon which is *tangent to the elastic curve* at the verticals which separate the areas. Thus we can construct any number of tangents to the elastic curve; areas, which are positive or negative, must, of course, be laid off in the force polygon in opposite directions.

If we divide the moment area by lines which are *not* vertical [Pl. 14, Fig. 51], the directions of the *outer polygon sides* are the same as for vertical divisions, because the vertical height between the corresponding outer sides in the force polygon is in any case always equal to the total load.

The *two outer* polygon sides for *any* method of division *are, therefore, tangents to the elastic curve at the ends of the same.* Here also we can, of course, have negative areas.

82. Effect of End Moments.—A beam or girder continuous over three or more supports differs from a beam simply resting upon its supports, in that, in addition to the outer forces, we have acting at each intermediate support a *moment* or *couple.*

But, as we have seen, Art. 23, the effect of these moments or couples will be simply to *shift* the closing line of the equilibrium polygon through a certain distance. Thus [Pl. 14, Fig. 52 (*a*)], if the span l_1 were uniformly loaded and simply supported at the extremities **A** and **B**, the equilibrium curve, or curve of moments, would, as we know (Art. 44), be a parabola **A D B**. If, however, the beam is continuous, we have at **A** and **B** moments or couples acting, and the closing line **A B** is shifted to some position as **A′ B′**. If now we consider the moment area, we see that by the shifting of the closing line the former moment area, which we shall call the *positive* area, is diminished, while to the right and left we have negative areas **A A′ C** and **B B′ C**.

It is evident that these areas have also a corresponding action upon the elastic line. For a positive moment area this last is *concave* upwards, while for negative areas it is *convex* upwards. At the points of transition C and C' we have the *inflection points*. This follows easily if we only hold fast the manner in which the elastic line is constructed, viz., by dividing the moment area into laminæ and regarding the area of each as a force. The forces thus obtained must plainly act, some upwards and some downwards, and the corresponding equilibrium polygon or elastic line must be in part convex upwards and in part convex downwards, and hence at the points of transition we must have *points of inflection* where the moment is zero.

83. Division of the Moment Area.—We shall assume the cross-section of beam constant. Regarding the elastic line simply as an equilibrium polygon, we can apply the principle that the *order* in which the forces are taken is indifferent (Art. 6) when the resultant only is desired. Since in the consideration of a single span only the first and last sides are of importance, we can, so long as we consider a single span only, take then the laminæ or divisions of the moment area in any order we please. More than this, we can, as we have seen in Art. 81, divide the moment area into laminæ *not* vertical; for example, we may in any span distinguish *three* parts, one positive and two negative, and consider each as a force acting at the centre of gravity of the corresponding area. [This holds good only for constant cross-section. For variable cross-section the horizontal force **E I** is variable.] Still further, we can divide the moment area for a single span into a positive area, *which is precisely the same as for a non-continuous beam*, and into a negative area, which will be evidently a trapezoid.

This is of great importance. To understand it fully we refer to Pl. 14, Fig. 52. Here, in the second span, we see that the real moment area consists of a positive part, viz., the parabola **C D C'**, and two negative parts **A A' C** and **B B' C'**. Instead of these we may take the *entire* parabolic area **A D B** and the trapezoid **A A' B' B**, or, finally, instead of this trapezoid, we may take the two triangles **A A' B'** and **B B' A'**. The parabolic area is positive, the triangular areas are negative.

If we assume the load as uniformly distributed, the first area will be always parabolic, and we may, therefore, call it the *parabolic area*.

By this division of the moment area we have obtained a great advantage. While the three areas **C D C′**, **A A′ C** and **B B′ C′** are all three dependent upon the moments at the supports **A A′** and **B B′**, we have by this new division to do with three areas, of which the first is entirely *independent* of the moments at the supports, the second depends only upon that to the left, and the third only upon that to the right.

84. Properties of the Equilibrium Polygon.—Let us consider now the case of a beam over four supports, that is, of three spans—l_0, l_1 and l_2—the first and last being, as is usually the case, equal, and the two first loaded with both live and dead load, the last with dead load only. The parabolas for the vertical loads [Pl. 14, Fig. 52] may be constructed by means of a force polygon, or the ordinates at the centre calculated, and the parabolas then drawn. The moments at the supports are **A A′** and **B B′**. Although these are unknown, it is not necessary to assume them at first. They may be directly constructed.

Thus, if we conceive the moment areas in each span divided into positive parabolic areas and negative triangles, we have in the first and last span one, in the middle two triangles. If we consider these areas as forces acting at the corresponding centres of gravity, we shall obtain an equilibrium polygon *of the form* given in Fig. 52 (*b*). That is, this polygon must have *eight* sides, and its angles must be *somewhere* on the verticals through the centres of gravity of the parabolic and triangular areas. The parabolic areas act downwards, the triangular areas upwards. The problem is, *to make these last so great that this polygon shall pass through all the points of support.*

One of the properties of the polygon we have, therefore, just noticed, viz.: its angles must lie in the verticals through the middle points of the spans and through the points distant from **A** and **B** one-third of the spans on each side (*i.e.*, the centres of gravity of the triangles).

If we prolong the second and fourth sides of the polygon, they intersect in a point **M**, the point of application of the resultant of the two contiguous triangular area forces (Art. 44).

The areas of these two triangles are $\frac{1}{2}$ **A A′** l_0 and $\frac{1}{2}$ **A A′** l_1, that is, the areas are as the spans l_0 and l_1.

Then by the principle of Art. 18 the resultant divides the distance between the forces into two portions, which are to each

CHAP. VIII.] CONTINUOUS GIRDERS. 131

other as l_1 to l_0, or inversely as the forces. Since the entire distance is $\frac{1}{3} l_0 + \frac{1}{3} l_1$, the distance of the resultant or of the point of intersection **M** from **L** is $\frac{1}{3} l_1$; from **N** it is $\frac{1}{3} l_0$. The point **M**, therefore, must lie *somewhere* in the vertical at $\frac{1}{3} l_1$ from **L**, the point of application of the triangular area force for the span l_0. The verticals through the centres of the parabolic areas we call the parabolic or *middle verticals;* those through the centres of gravity of the triangular areas, the *third verticals;* those through a point as **M**, the point of application of the resultant of two contiguous triangular area forces, the *limited third verticals*. Upon these verticals two sides must always intersect.

85. Polygon for the Positive Moment Areas.—It will be found best to take as the reduction base for areas $\frac{1}{2} l_1$, *i.e.*, half the second span, and for pole distance $\frac{1}{3} l_1$. Reducing the areas of the parabolas to this basis, and considering the heights thus obtained as forces, we can form a force polygon with pole distance $\frac{1}{3} l_1$. It is not necessary to draw this polygon; our object is to find the corresponding equilibrium polygon. This last, since we consider the entire parabolic area as a force acting at its centre of gravity, consists of two lines which intersect in the vertical through the middle of the span. We prolong each of these lines and obtain two lines as shown in Pl. 14, Fig. 52 (*c*). The segments cut off by these lines from the verticals through the supports are the *moments of the parabolic areas with respect to the supports*. These moments we can easily find.

Thus, let the deflection of the parabola in the second span at the centre be f, then its area is $\frac{2}{3} f l_1$. This reduced to the basis $\frac{1}{2} l_1$ gives $\frac{4}{3} f$ as the force. The moment of this force with reference to the supports is $\frac{4}{3} f \times \frac{1}{2} l_1 = 2 f \times \frac{1}{3} l_1$. This moment is equal to the segment sought multiplied by the pole distance. This last is $\frac{1}{3} l_1$. The segment, therefore, is $2f$. We do not need, therefore, to draw the force polygon, but have simply to take off with the dividers the middle ordinate of the parabola $f = \frac{p l^2}{8}$, and lay it off *twice* on the verticals right and left through the supports, and join the four points thus obtained by lines crossing each other under the centre of the span. The equilibrium polygon for the positive parabolic area is then ready for the middle span. [We advise the reader to construct it for himself.]

For the two side spans the construction is different. Here the area of the parabola is $\frac{2}{3} f' l_0$, the reduced area $\frac{4}{3} f' \frac{l_0}{l_1}$, and the moment $\frac{4}{3} f' \frac{l_0}{l_1} \times \frac{l_0}{2} = 2 f' \frac{l_0^2}{3 l_1}$.

Dividing by $\frac{1}{3} l_1$, we have for the segment required

$$2 f' \left(\frac{l_0}{l_1}\right)^2.$$

Therefore, in the end spans, or generally in any span *not* equal to the standard span, or that which furnishes the constants $\frac{1}{2} l_1$ and $\frac{1}{3} l_1$, we must multiply the middle ordinate of the parabola by the square of the ratio of the "*standard*" span to the span in question, and *then* lay the product off twice upon the verticals through the supports. This multiplication is easily performed graphically. If from the middle of span l_0 we lay off l_1^2 horizontally and join the end with the end of f', then lay off l_0^2 in same direction and draw a parallel to the first line, the segment on f' will be $f' \frac{l_0^2}{l_1^2}$. For we shall have

$$l_1^2 : f' :: l_0^2 : x \quad \text{or} \quad x = f' \frac{l_0^2}{l_1^2}.$$

Since these cross-lines depend upon given quantities, they can be constructed for every span, and thus we have Fig. 52, c. They give us not only the moments over the supports, but also the moments for any point of the parabolic area forces. We shall hereafter make use of them.

86. Construction of the Fixed Points, and of the Equilibrium Polygon.—We have thus all the given and known quantities, have deduced the general properties of the equilibrium polygon, and will now endeavor by their aid to draw the polygon itself. We shall then be able to find the actual moments **A A'** and **B B'** at the supports. It is impossible at first to draw any single *side* of the polygon in true position, and we must, therefore, endeavor to find certain *points* of the same sufficient to determine it.

Lay off first the three spans, Pl. 14, Fig. 52 (*d*). Suppose the second side of the polygon prolonged till it intersects the vertical through the end support *a*, in a point **K**, Fig. 52 (*b*). This point is *known*. It is given by the moment of the parabolic area in the first span with respect to this end support. This moment we have already by the cross-lines in Fig. 52 (*c*). We have then simply to take it off in the dividers from (*c*) and lay it off from *d* to **K'** in Fig. 52 (*d*). We have now in Fig. 52 (*b*), *two* points of the polygon known, namely, the end support and **K**, *which last must be in the second side prolonged.*

The triangle **L M N** is now of special importance. Whatever may be the position of **K M** and **M N**, we have already seen that the intersection **M** must always lie somewhere in the *limited third vertical*. The first side **K M** must, however, *always pass through* **K**, a known point. The second *must pass through the support*, also a known point. The points **L** and **N** must, moreover, *always lie in the third verticals*, distant from **A**, $\frac{1}{3} l_1$ and $\frac{1}{3} l_0$ respectively.

If the line **K M** takes up various positions under these conditions, the line **M N** will *revolve about a fixed point which is given by the intersection of a line through* **K** *and the support* **A** *with* **M N**.

If, then [Fig. 52 (*d*)], we draw a line in *any* arbitrary direction through **K'**, and note the intersections **L'** and **M'** with the first *third vertical* and the *limited third*, then through **L'** and the support draw a line to intersection **N'** with second third vertical, and join **M' N'**, and finally through **K'** and the support draw a line intersecting this last in **I**, the point **I** thus determined is a *fixed point*, and remains the same *for any position of* **K' M'**. It is therefore a point on the fourth side **M N** of the polygon. For the triangle **L' M' N'** may have any position, yet so long as its angles lie in three parallel fixed lines, and two of the sides pass through two fixed points, the other side must also pass through a fixed point.* Out of all possible positions of the triangle, one of these positions must coincide with the polygon sides, and hence this fixed point is a *third* known point, since we have already **K′** and the end support.

Although, then, we are as yet unable to draw any of the sides of the polygon in true position and direction, still from the hitherto known properties we have deduced a new one. We know now a point through which the fourth side must pass. But this is not all. We proceed still further. The fourth and fifth sides must intersect upon the vertical through the centre of the second span. These sides, moreover, cut off upon any vertical the moment of the parabolic area with respect to any point in that vertical. We know this moment thus for the point **I** just found. It is found by taking the segment cut off, from the vertical through **I**, by the cross-lines for the parabolic areas found above in Fig. 52 (*c*).

Laying this segment off from **I**, we thus find **I′**, *a point in the fifth side prolonged*. From this point we proceed as before to find the next fixed point **I″**. We then lay off from **I″** the moment of the parabolic area for this point and find **I‴**, a point upon the eighth side. We can *now* draw the polygon itself.

Thus the eighth side passes, of course, through the last support and also **I‴**. It is therefore determined. Through the intersection of this line with the vertical through the middle of the span and the point **I″** the seventh side passes. The seventh side is therefore determined. Through the intersection of this with the *third vertical* and the support the sixth side passes and continues till it intersects the *third vertical* on the other side. Then from this point towards **I′** to intersection with vertical through centre of middle span. From this last point towards **I** to intersection with third vertical. From this last point again through support to intersection with third vertical on other side; then towards **K′** to intersection with vertical through centre of end span; and lastly, from this last point of intersection through the end support, and the polygon is complete as given in Fig. 52 (*b*).

In this manner, however, inaccuracies may occur. To avoid these we may start from the *right* end support and also find *four* fixed points as above. It is unnecessary to make the con-

---
* *This proposition the reader can easily prove geometrically or analytically.* See Art. 112.

struction. We see at once that we shall thus obtain in each end span two points, in the middle span *four points*, which last, being joined by lines crossing each other, give in the middle span two sides in proper position. It is also evident how the polygon may then be completed.

87. Construction of the Moments at the Supports.—Thus we are able to construct the equilibrium polygon, or rather the extreme tangents to the elastic line for each span. We have now to determine the two moments over the supports. This is very simple. The first moment to the left is cut off by the fourth side, the second by the fifth side of the polygon, *from the verticals through the supports*. We have therefore only to prolong these two sides, take off the segments in the dividers, and lay them off in Fig. 52 (*a*) in **A A'** and **B B'**. We have, then, in Pl. 14, Fig. 52 (*a*) the moments for the given case and loading at any point, as shown by the shaded area.

The proof is simple. The two lines **N M** and **N L** [Fig. 52, *b*] evidently cut off upon the vertical at the support the moment of the force acting at **N**. This force is the area of the triangle **A A' B'** equal to $\frac{1}{2}$ **A A'** l_1. This reduced to the basis $\frac{1}{2} l_1$ gives **A A'**. If we multiply this by the lever arm of the force, we have its moment. This moment is, however, equal to the segment **A A'** multiplied by the pole distance, and since this pole distance is itself $\frac{1}{3} l_1$, the segment itself to the assumed pole distance gives us the moment.

We see that it is not necessary to draw the line **N L** as it passes through the support. We have simply to prolong the side **M N** to intersection with the vertical through the support. It is to be observed that the moments at the supports are cut off *at* the supports only by those lines which pertain to the "standard" span, or that span from which we take our reduction basis and pole distance. For lines in the other spans the above does not hold good without modification. It is, however, always possible, at least for from two to five symmetrical spans, to observe the above conditions. In those cases where this is not possible, an easy graphical multiplication of the segments by the square of the ratio of the spans will give the moments. We see also the reason why, for four symmetrical spans, the *second* and not the first must be taken as the *standard* span.

If the construction of the moments over the supports is our sole purpose, as is in practice the case, the polygon need not be drawn. We have only to find our *fixed points*, and note the intersection of the sides with the verticals through the supports, without drawing the sides themselves. In the preceding Arts· we have purposely considered only the particular case of uniform loading, and have taken only three spans, in order to familiarize the reader with the nature of the problem and the method of its solution. In order to attain a clear understanding of the subject as thus far developed, he would do well to take some particular case, as, for instance, that of a girder of two or three or four spans of given length, the end spans being equal, and intermediate spans equal and say one-fourth longer than the ends, and work out by diagram the moments at the supports for a uniform load over the whole length of girder. For two spans the moment at the centre support should be $\frac{1}{8} p \, l^2$, l being the length of span, p the load per unit of length.

For three spans the moment at the two inner supports is $\frac{1 + n^3}{4\,(3 + 2\,n)} p\, l^2$, where $n\, l =$ the length of end spans. Thus, if $n = \frac{3}{4}$, we have $\frac{91}{1152} p\, l^2$. For four spans the moment at the second and fourth supports is $\frac{1 + 2\,n^3}{4\,(3 + 4\,n)} p\, l^2$, and at the middle support $\frac{1 + 2\,n - n^3}{4\,(3 + 4\,n)} p\, l^2$. By these formulæ the graphical results may be checked.

When the reader has thus become thoroughly familiar with the principles of the preceding Arts. and their practical application, he will be ready to resume at this point the more general development which follows.

88. The Second Equilibrium Polygon.—We see, therefore, that the *actual form* of the elastic line is not required to be known. Only the outer forces and their moments are sought, and to determine these it is sufficient to know the position of the *tangents* to the elastic line *at the supports*. Thus the first line of the equilibrium polygon [Fig. 52, *b*] being given in position, by the aid of the *middle*, *third*, and *limited third* verticals and the known point **K**, all the other sides may be drawn, and the moments at the supports found. We conceive, therefore, the *moment area* as the difference of the trapezoid $\mathbf{A}' \mathbf{A}'' \mathbf{B}'' \mathbf{B}'$ [Pl. 15, Fig. 53] and the parabolic area $\mathbf{A}'' \mathbf{C}'' \mathbf{B}''$, or equal to $\mathbf{A}'' \mathbf{C}'' \mathbf{B}''$ *minus* triang. $\mathbf{A}' \mathbf{A}'' \mathbf{B}'$ *minus* triang. $\mathbf{B}' \mathbf{B}'' \mathbf{A}''$. The area $\mathbf{A}'' \mathbf{C}'' \mathbf{B}''$ we call the *simple* or *parabolic moment area*.

If we indicate the moments at the supports $\mathbf{A}' \mathbf{A}''$ and $\mathbf{B}' \mathbf{B}''$ by $\mathbf{M}'$ and $\mathbf{M}''$, then, for a given span l,

$$\mathbf{A}' \mathbf{A}'' \mathbf{B}' = \tfrac{1}{2} \mathbf{M}' l \text{ and}$$
$$\mathbf{A}'' \mathbf{B}' \mathbf{B}'' = \tfrac{1}{2} \mathbf{M}'' l.$$

If we indicate further the height of a *rectangle* of base l and area $\mathbf{A}'' \mathbf{C}'' \mathbf{B}''$, that is, the *mean value of the moments of the corresponding simple girder* by $\mathfrak{M}$, we have

$$\text{area } \mathbf{A}'' \mathbf{C}'' \mathbf{B}'' = \mathfrak{M} l.$$

The verticals through the centres of gravity of the triangles divide the span into three equal parts. We call these the *third verticals*. The load $\mathfrak{M} l$ acts at the centre of gravity of the parabolic area $\mathbf{A}'' \mathbf{B}'' \mathbf{C}''$.

The four-sided equilibrium polygon $\mathbf{A\,U\,S\,V\,B}$ corresponding to these forces we call the *second equilibrium polygon*. The pole distance must be (Art. 81) $\dfrac{1}{n} \mathbf{E\,I}$. Instead of this, we take, which amounts to the same thing, the forces $\mathbf{G\,F} = \dfrac{1}{2} \mathbf{M}' \dfrac{l}{\lambda}$, $\mathbf{E\,H} = \dfrac{1}{2} \mathbf{M}'' \dfrac{l}{\lambda}$ and $\mathbf{F\,E} = \mathfrak{M} \dfrac{l}{\lambda}$, and the pole distance $b = \dfrac{\mathbf{E\,I}}{n\,\lambda}$, where λ is any assumed length. For λ we may take the arithmetical mean of all the spans, or, as we have seen, one of the actual spans. If the outer spans are both equal to l_0 and the other spans equal to l_1, we should naturally choose $\lambda = l_1$, since then the forces would be $\dfrac{1}{2} \mathbf{M}'$, $\dfrac{1}{2} \mathbf{M}''$ and $\mathfrak{M}$.

If the position of the tangents at the supports were known or found, the equilibrium polygon could be easily drawn as follows: Upon the two verticals distant each side of the centre $\mathbf{S}$ [Fig. 53] by the pole distance $b = \dfrac{\mathbf{E\,I}}{n\,\lambda}$ lay off the distance

$A = \mathfrak{M}\frac{l}{\lambda}$, and join the points thus obtained by two lines crossing each other. These *cross-lines* are the lines **O F**, **O E** of the force polygon. If now we make **U U'** and **V V'** equal to the ordinates of the cross-lines vertically under **U** and **V**, then the sides of the equilibrium polygon **U S** and **V S** prolonged, pass through **U'** and **V'**. This will at once appear from an inspection of Fig. 53.

In this form the equilibrium polygon was first represented by *Mohr*. (*Zeitschrift des Arch. und Ing. Ver. zu Hannover*, 1868.)

89. Determination of the Moments over the Supports.—If we draw in the force polygon, lines parallel to the four sides of the second equilibrium polygon, then the segments of the force line between the lines parallel to **A U**, **B V** [Pl. 15, Fig. 53] and those parallel to **S U**, **S V**, are respectively **F G** $= \frac{1}{2} M' \frac{l}{\lambda}$ and **E H** $= \frac{1}{2} M'' \frac{l}{\lambda}$. If we prolong **S U** and **S V** to intersections **M** and **N** with verticals through the supports, and represent **A M** and **B N** by y' and y'', we have from the similarity of the triangles **U A M** and **V B N** with **O G F** and **O H E**

$$y' : \frac{1}{2} M' \frac{l}{\lambda} :: \frac{1}{3} l : b \text{ and } y'' : \frac{1}{2} M'' \frac{l}{\lambda} :: \frac{1}{3} l : b,$$

hence

$$y' = \frac{M' l^2}{6 b \lambda} \quad y'' = \frac{M'' l^2}{6 b \lambda}.$$

The segments **A M** and **B N** are, therefore, proportional to the moments at the supports **M'** and **M''**.

These moments themselves can now be determined in various ways.

1st. It is in general best to choose the second pole distance $b = \frac{1}{6} \lambda$. We have then

$$M' = y' \left(\frac{\lambda}{l}\right)^2 \quad M'' = y'' \left(\frac{\lambda}{l}\right)^2 \text{ (Art. 70)}.$$

If, then, at a distance from **U** and **V** either way equal to $2 b \left(\frac{\lambda}{l}\right) = \frac{1}{3} \frac{\lambda}{l} \lambda$ we draw verticals, the segments $A_1 M_1$ and $B_1 N_1$ cut off from these verticals will evidently be equal to the moments required, viz., **M'** and **M''**.

CHAP. VIII.]   CONTINUOUS GIRDERS.   139

2d. If we take $\lambda = l$, we have at once $\mathbf{M}' = y'$ and $\mathbf{M}'' = y''$. In the inner spans, therefore, we have directly, as we have already seen, for the special case (Art. 72), when $\lambda = l$, the moments at the supports.

3d. For a span adjoining a span whose length is λ, we have the moment for the intervening support directly from this last span. If the inner spans have the same length l, and the two outer the same length l_1, we can accordingly, by making $\lambda = l$, obtain directly the moments at the supports.

90. Comparison with Girder fixed horizontally at both ends.—If the ends are fixed horizontally, the lines in the *force polygon* parallel to **A U** and **B V** coincide, *i.e.*, **H** and **G** fall together. Accordingly, if we designate the end moments now by $\mathfrak{M}'$ $\mathfrak{M}''$ we have (Fig. 53)

$$\frac{1}{2}(\mathbf{F\ G} + \mathbf{G\ E}) = \frac{1}{2}\mathbf{F\ E}\ \text{or}\ \frac{1}{2}(\mathfrak{M}' + \mathfrak{M}'') = \mathfrak{M}.$$

Therefore, in the *first* equilibrium polygon, *the moment areas on each side of the closing line are equal.*

Indicating the points for this case by the index 0 [Fig. 54, Pl. 15], we have the triangle $\mathbf{A\ U_0\ M_0}$ equal to $\mathbf{U_0\ V_0\ V'_0}$, and therefore $\mathbf{A\ M_0} = \mathbf{V_0\ V'_0} = \mathbf{V\ V'}$, as also, in like manner, $\mathbf{B\ N_0} = \mathbf{U_0\ U'_0} = \mathbf{U\ U'}$, or, taking $b = \frac{1}{6}\lambda$,

$$\mathbf{U\ U'} = \mathfrak{M}'\left(\frac{l}{\lambda}\right)^2\quad \mathbf{V\ V'} = \mathfrak{M}''\left(\frac{l}{\lambda}\right)^2.$$

Therefore, the ordinates between the cross-lines at the verticals passing through **U** *and* **V** *are, for girders fixed horizontally, proportional to the end moments* $\mathfrak{M}'$ *and* $\mathfrak{M}''$.

For $\lambda = l$, $b = \frac{1}{6}l$, and these ordinates give the moments directly.

If we draw through **U'** and **V'** a straight line intersecting the end verticals in **Q** and **R**, and prolong **N U'** and **N V'** to intersections **S** and **T**, then $\mathbf{Q\ M} = 2\ \mathbf{U\ U'}$, $\mathbf{Q\ S} = \mathbf{V\ V'}$, and hence $\mathbf{M\ S} = 2\ \mathbf{U\ U'} + \mathbf{V\ V'} = (2\ \mathfrak{M}' + \mathfrak{M}'')\left(\frac{l}{\lambda}\right)^2$; and in the same way $\mathbf{N\ T} = (2\ \mathfrak{M}'' + \mathfrak{M}')\left(\frac{l}{\lambda}\right)^2$.

Therefore, the segments cut off upon the end verticals by the cross-lines are proportional to $2\ \mathfrak{M}' + \mathfrak{M}''$ *and* $2\ \mathfrak{M}'' + \mathfrak{M}'$.

The quantities $\mathfrak{M}'$ and $\mathfrak{M}''$ being known, we can easily construct the cross-lines.

If we draw a line through **U** and **V** to intersections **O** and **P**, we have **O M** = **V V'**, **P N** = **U U'**. Therefore, **A O** and **B P** are equal to $(\mathfrak{M}'-\mathbf{M}')\left(\dfrac{l}{\lambda}\right)^2$ and $(\mathfrak{M}''-\mathbf{M}'')\left(\dfrac{l}{\lambda}\right)^2$, where **M'** and **M''** are the moments at the supports for a continuous girder, and $\mathfrak{M}'$ $\mathfrak{M}''$ those for a girder horizontally fixed at its ends.

CHAPTER IX.

CONTINUOUS GIRDER—LOADED AND UNLOADED SPANS.

91. Unloaded Span.—If the span is unloaded, we have to construct the second equilibrium polygon, only the two forces $\frac{1}{2}\mathbf{M}'\,l$ and $\frac{1}{2}\mathbf{M}''\,l$. If the position of the end tangents is known, the polygon is completely determined. If we prolong the middle side $\mathbf{U\,V}$ to intersections $\mathbf{M}$ and $\mathbf{N}$ with the end verticals [Pl. 15, Fig. 55], then, by the preceding Art., $\mathbf{A\,M} = \mathbf{M}'\left(\frac{l}{\lambda}\right)^2$, $\mathbf{B\,N} = \mathbf{M}''\left(\frac{l}{\lambda}\right)^2$; therefore, $\mathbf{A\,M}:\mathbf{B\,N}::\mathbf{M}':\mathbf{M}''$. If now we draw $\mathbf{A\,B}$ intersecting $\mathbf{U\,V}$ in $\mathbf{I}$, the moment at this point is zero. That is, *the intersection* $\mathbf{I}$ *of the line joining the supports with the middle side of the polygon is the point of inflection of the elastic line.*

92. Two successive Unloaded Spans.—Prolong the two middle sides $\mathbf{U\,V}$ and $\mathbf{U_1\,V_1}$ [Pl. 15, Fig. 56] of the equilibrium polygon for the two spans l_0 and l_1. The point of intersection $\mathbf{W}$ is a point in the resultant of the forces at $\mathbf{V}$ and $\mathbf{U_1}$. Since these forces are $\frac{1}{2}\mathbf{M_1}\,l_0$ and $\frac{1}{2}\mathbf{M_1}\,l_1$, we have $\mathbf{W_0\,V_0}:\mathbf{W_0\,U_0}::l_1:l_0$. But the horizontal projection of $\mathbf{V_0\,U_0}$ is $\frac{1}{3}(l_0 + l_1)$, therefore that of $\mathbf{V_0\,W_0}$ is $\frac{1}{3}l_1$ and of $\mathbf{U_0\,W_0}$, $\frac{1}{3}l_0$; while that of $\mathbf{B\,W_0}$ is $\frac{1}{3}(l_1 - l_0)$. The vertical through $\mathbf{W}$ we have called the *limited third vertical.* Its position is, as we see, easily found, and depends simply upon the length of the spans.

Let us now consider more closely the intersections $\mathbf{I}$ and $\mathbf{L}$ of the middle sides with the straight line joining the supports. We have

$$\mathbf{L\,U_0}:\mathbf{L\,W_0}::\mathbf{U_0\,U_1}:\mathbf{W_0\,W}.$$

But we have also
$$U_0 U_1 : V_0 V :: B U_0 : B V_0 :: l_1 : l_0;$$
hence
$$L U_0 : L W_0 :: I V_0 \times l_1 : I W_0 \times l_0.$$

The ratio of the parts into which $W_0 U_0$ is divided by the point L depends, therefore, for given spans only upon the ratio of $I V_0$ to $I W_0$, or upon the position of I alone.

If, therefore, we were to draw through the point I *another polygon*, the point L would be unchanged, or still more generally, *if the point I for different heights of the supports and different polygons moves in the same vertical, the point L will also move in a vertical.*

If the supports are in the same straight line, the points I and L are the points of inflection of the elastic line. We have therefore the principle, *that if for different polygons the inflection point I remains the same, the inflection point L remains also the same.*

The point I being given, we can easily construct the point L. We have only to draw through I at will any line intersecting the *third vertical* through V_0 and the *limited third* at, say, V and W. Through V and the support B draw a line to intersection U_1 with third vertical through U_0. Join now U_1 with W. The line $U_1 W$ cuts the line through the supports A and B in the point L. (See also Art. 86, Fig. 52, *d*.)

93. The "Fixed Points."—Suppose that, starting from the left support A [Pl. 15, Fig. 57], we have a number of unloaded spans. The end A then is an inflection point, since the moment there is zero. Starting from this point, therefore, we can construct, according to the preceding Art., the inflection point I_2 for the next span. Then starting from this we may construct the point I_3 for the third span, and so on. Since these points, *under the assumption that the supports all lie in the same straight line*, do not change their position, whatever may be the loading of the loaded spans, and whatever spans be loaded, we call them *fixed points*.

A second series of fixed points may be in similar manner constructed, when a number of spans from the right are unloaded, so that there are *two series of fixed points*. In the end spans the end supports are fixed points.

It follows directly from the construction that *the fixed points are always within the outer third of the span.*

CHAP. IX.] LOADED AND UNLOADED SPANS. 143

The construction of the fixed points is the *first* operation in the graphical treatment of the continuous girder.

The above construction was first given by *Mohr*.

94. Shearing Force, Reactions at the Supports, and Moments in the Unloaded Spans.—The moments in the unloaded spans are given, then, by the ordinates to a broken line whose angles lie in the support verticals, and which, for the case of supports on a level, passes through the corresponding fixed points.

It follows directly that the moments at the supports are *alternately positive and negative,* and increase from the end, so that any one is more than three times the preceding. (See Art. 111.)

Since now this polygon has alternately angles down and up, the *reactions at the supports must be alternately positive and negative.* From the corresponding force polygon it follows that they must *increase from the end.*

The shearing forces are, therefore, also alternately positive and negative, and increase from the end on.

95. Loaded Span.—Let now the span **A B** [Pl. 15, Fig. 58] be arbitrarily loaded. It can be proved here also, as in Art. 92, that the prolongation of the sides **U' V'** and **S U**, as also of **V'' U''** and **S V**, intersect in the *limited third* verticals.

When the supports are in a straight line, then, by the construction of Art. 92, the fixed points **I** and **K** are the intersections of **S V** and **S U** with **A B**. We can, therefore, at once assert, *that the sides* **S U** *and* **S V** *of the second equilibrium polygon pass through the fixed points* **I** *and* **K***, when the supports are on a level.*

For known position of the fixed points and for given load, it is, therefore, easy to draw the second polygon by drawing verticals **I I**$_1$ and **K K**$_1$ equal to the corresponding ordinates of the *cross-lines*. **S U** and **S V** pass, then, through **I K**$_1$ and **K I**$_1$ respectively. Then, by Art. 89, the moments at the supports may be determined.

Since **A I** $< \frac{1}{3} l$, **U** must lie to the right of **I**, and the angle **A U S** is concave downwards. Accordingly, the force at **U**, viz., $\frac{1}{2}$ **M'** l, acts upwards. The same holds good for **V**. *Hence,* the moments **M' M''** *for the loaded span are always positive.** If we draw the lines **M N** and **U V** cutting the middle vertical in **P** and **Q**, then, for $b = \frac{1}{3}l$, $\mathbf{P O} = \frac{1}{2}(\mathbf{M'} + \mathbf{M''})$ and $\mathbf{P Q} = \frac{1}{2}(\mathfrak{M}' + \mathfrak{M}'')$. (See Arts. 89, 90.) Since now the points **U** and **V** must lie under **A B**,

$$\mathbf{P O} < \mathbf{P Q'} \text{ or } \mathbf{M'} + \mathbf{M''} < \mathfrak{M}' + \mathfrak{M}''.$$

If the supports are not upon a level, it follows from this Art. and Art. 92, *that the intersections of* **S U** *and* **S V** *prolonged, with the prolongations of the lines* **A A'** **B B'**, *joining the supports of two adjacent spans, lie in the* VERTICALS THROUGH THE FIXED POINTS.

96. Two successive Loaded Spans.—Pl. 15, Fig. 59.

1. *Here also, as in Art. 92, we can prove that the prolongations of* **S V** *and* $\mathbf{S_1 U_1}$ *intersect in the limited third vertical.*

2. Draw through **B** a line which intersects **S V** and $\mathbf{S_1 U_1}$ in **I'** and $\mathbf{I'_1}$, and the verticals through **V, W** and $\mathbf{U_1}$ in $\mathbf{V_0, W_0}$ and $\mathbf{U_0}$.

Then
$$\mathbf{U_0 U_1 : V_0 V :: U_0 B : V_0 B :: } l_0 : l_1$$
$$\mathbf{V_0 V : W_0 W :: I' V_0 : I' W_0}.$$

Hence by composition
$$\mathbf{U_0 U_1 : W_1 W :: I' V_0} \times l_0 : \mathbf{I' W_0} \times l_1;$$
or since $\mathbf{U_0 U_1 : W_0 W :: U_0 I'_1 : W_0 I'_1}$
$$\mathbf{U_0 I'_1 : W_0 I'_1 :: I' V_0} \times l_0 : \mathbf{I' W_0} \times l_1.$$

If, then, the point **I'** moves in a vertical, the ratio **I'** $\mathbf{V_0}$ to **I'** $\mathbf{W_0}$ does not change, therefore the ratio of $\mathbf{U_0 I_1'}$ to $\mathbf{W_0 I_1'}$ also remains unchanged, and accordingly $\mathbf{I'_1}$ *must also move in a vertical*. If **I'** coincides with **I**, it follows from the construction of Art. 93 that the point $\mathbf{I'_1}$ becomes the fixed point $\mathbf{I_1}$. Hence, *the intersections* **I'** *and* $\mathbf{I'_1}$ *of verticals through the fixed points* **I** *and* $\mathbf{I_1}$ *with the sides* **S V** *and* $\mathbf{S_1 U_1}$, *or with the middle sides of the two polygons adjacent to the support, lie always in a straight line through that support, for any heights of supports.*

---

* A positive moment always indicates compression in lower flange.

This property of the second equilibrium polygon was first made known by *Culmann*.

97. Arbitrary Loading.—According to the above properties of the second equilibrium polygon, the general course of procedure for any given case of loading is then as follows [Fig. 60, Pl. 16]:

1. Construct all the *fixed points* $\mathbf{A\, I_2\, I_3}$ - - - $\mathbf{K_1 K_2}$, etc. (Art. 93), and draw verticals through them.

2. Construct the *cross-lines* for every span, Art. 88.

3. Make $\mathbf{A\, C}$ equal to $\mathbf{O_1 Q_1}$ as given by the *cross-lines*, and draw a line through $\mathbf{C}$ and $\mathbf{A_1}$ to intersection $\mathbf{D_2}$, with vertical through $\mathbf{I_2}$. Then make $\mathbf{D_2\, C_2}$ equal to $\mathbf{O_2 Q_2}$, and draw a line through $\mathbf{C_2}$ and $\mathbf{A_2}$ to $\mathbf{D_3}$, and so on. Precisely the same construction holds for the other way from the right end. Thus $\mathbf{A_4\, E_4}$ is equal to $\mathbf{R_4\, P_4}$, etc.

4. In this way we obtain for each of the middle sides of the second equilibrium polygon two points, $\mathbf{C}$ and $\mathbf{F_1}$, $\mathbf{C_2}$ and $\mathbf{F_2}$, etc.; $\mathbf{A}$ and $\mathbf{E_1}$, $\mathbf{D_2}$ and $\mathbf{E_2}$, and so on; so that now we can actually draw these middle sides.

The intersections of these lines with the support verticals give, according to Art. 88, the *moments at the supports*. For spans whose length is λ these moments are given directly; for other spans the construction of Art. 74 must be applied. The following simple construction may also be applied. Let $\mathbf{I\, K}$ be the intersections of the verticals through the fixed points, with the line $\mathbf{A\, B}$ joining the supports [Fig. 61, Pl. 16].

Make $\mathbf{I\, D'} = \mathbf{I\, D} \left(\dfrac{\lambda}{l}\right)^2$, $\mathbf{K F'} = \mathbf{K F} \left(\dfrac{\lambda}{l}\right)^2$, $\mathbf{C'\, D'} = \mathbf{O\, Q} \left(\dfrac{\lambda}{l}\right)^2$,

$\mathbf{E'\, F'} = \mathbf{R\, P} \left(\dfrac{\lambda}{l}\right)^2$, and draw $\mathbf{C'\, F'}$ and $\mathbf{E'\, D'}$. These lines cut the support verticals in $\mathbf{M'}$ and $\mathbf{N'}$, so that $\mathbf{A\, M'}$ and $\mathbf{B\, N'}$ are the moments.

By the construction errors accumulate from one span to the next, so that the diagram must be made with care. We have also several checks, viz.: 1. The intersection of the middle sides must lie in the vertical through the intersection of the *cross-lines*. 2. The prolongation of the middle sides must intersect in the *limited third* vertical. 3. The corresponding intersections of the middle sides with the third verticals must lie in a straight line through the support.

If any span is unloaded, the cross-lines coincide. The above method of construction holds good when the supports are not upon a level. If the difference of height of the supports is represented $\frac{1}{m}$ th of the real amount, the unit for the moment scale is $\frac{\lambda}{6\,m\,EI}$, as is easily seen by reference to Arts. 81 and 88.

The above method of construction of the moments at supports was first given by *Culmann*.

CHAPTER X.

CONTINUOUS GIRDER—SPECIAL CASES OF LOADING.

98. Total uniform Load.—If a span is loaded with a uniformly distributed load of p pounds per unit of length, the *simple moment area is a parabolic segment whose vertical axis passes through the centre of the span.* [Pl. 16, Fig. 62.]

The ordinate **D C″** is $\frac{1}{8} p \, l^2$, and hence the area

$$\mathfrak{M} \, l = \frac{2}{3} l \times \frac{1}{8} p \, l^2 = \frac{1}{12} p \, l^3 \text{ or } \mathfrak{M} = \frac{1}{12} p \, l^2.$$

It will be advantageous here to take $p \, \lambda^2$ as the unit of the moment scale; and therefore

$$\mathfrak{M} = \frac{1}{12} p \, \lambda^2 \left(\frac{l}{\lambda}\right)^2.$$

The vertical height of the cross-lines at the pole distance b from the middle is $\mathfrak{M} \frac{l}{\lambda}$. If, then, we take $b = \frac{1}{6} \lambda$, we have

$$\mathbf{O\,P} \text{ or } \mathbf{Q\,R} : \mathfrak{M} \frac{l}{\lambda} :: \frac{l}{2} : \frac{1}{6} \lambda;$$

and therefore

$$\mathbf{O\,P} = \mathbf{Q\,R} = 3 \, \mathfrak{M} \left(\frac{l}{\lambda}\right)^2, \text{ that is}$$

$$\mathbf{O\,P} = \mathbf{Q\,R} = \tfrac{1}{4} p \, \lambda^2 \left(\frac{l}{\lambda}\right)^4.$$

2. *Moments.*

If the moments **A′ A″** and **B′ B″** are known, we can find the end tangents of the *first* equilibrium polygon by drawing **A″ B″**, dropping a vertical through the middle **D**, and laying off $\mathbf{D\,E} = 2 \times \frac{1}{8} p \, l^2 = \frac{1}{4} p \, l^2 = \frac{1}{4} p \, \lambda^2 \left(\frac{l}{\lambda}\right)^2$. The lines **A″ E** and **B″ E** are, then, these end tangents. With the help of these we may easily construct the parabola.

3. *Shearing force.*

If we draw in the first force polygon, lines parallel to the end tangents and the closing line **A′ B′**, the distances on the force line are the reactions at the ends of the span. Instead of this we may lay off from **A** and **B**, **A G** and **B H** equal to the first pole distance a, and through **G** and **H** draw parallels to the end tangents, intersecting the verticals through **A** and **B** in **A′** and **B₁**. We thus obtain the reactions **A A₁**, **B B₁**. The ordinates to **A₁ B₁** then give the shearing force at any point.

If the line $p\lambda^2$ representing the moment units is equal to m, and that representing the force units $p\lambda = n$, then the first pole distance must be $a = \dfrac{p\lambda}{p\lambda^2}\lambda = \dfrac{n}{m}\lambda$.

Accordingly, it is now easy, from the general construction given in Art. 97, to construct the shearing force and moments for uniform or dead load of girder in any case. Let us pass on to an example illustrating more fully the above principles.

99. Example.—As an example of the application of the above principles, we take a girder of four spans, as given in Pl. 17, Fig. 63. The two interior spans are each 96 ft., the exterior spans 80 ft. each; that is, $l_1 : l :: 5 : 6$. Choose any scale of length convenient, as, for instance, 50 ft. to an inch, lay off the spans and construct first the fixed points. For this purpose we draw the third and limited third verticals. These last are easily found from the principle already deduced, that they must divide the distance between the third verticals into segments inversely as the corresponding spans [Art. 92]. Laying off, then, from the third vertical in the first span, $\tfrac{1}{5}l$ to the right, or from the third vertical in the second span $\tfrac{1}{5}l_1$ to the left, we have the first limited third vertical. The same at the other end gives the other. For the centre support, of course, the limited third, since the adjacent spans are equal, passes through the support itself. We can, therefore, now construct the *fixed points* according to Art. 93.

Let the load per unit of length p be $\tfrac{1}{2}$ ton per ft. Then taking λ [Art. 88] equal to l, we have $n = p\lambda = pl = 48$ tons and $m = p\lambda^2 = pl^2 = 4608$ ft. tons [Art. 98 (3)]. It remains to assume a scale of force. Let this be 20 tons per inch, then our moment scale is $20 \times 50 = 1000$ ft. tons per inch. The values of which we shall need to make use are, then, to scale

$$l_1 = 1.6 \text{ inches}, \quad \lambda = l = 1.92 \text{ inches},$$

$\left(\dfrac{l_1}{l}\right)^2 = 0.6944$ inches, $\left(\dfrac{l}{l_1}\right)^2 = 1.44$ inches, $\left(\dfrac{l_1}{l}\right)^4 = 0.4823$ inches

These values are repeated upon the Pl. for convenience of reference. Also, $pl = 48$ tons $= 2.4$ in. $= n$, $pl^2 = 4608$ ft. tons $= 4.608$ in. $= m$. For the first pole distance [Art. 98 (3)] we have $a = \dfrac{n}{m}\lambda = \dfrac{2.4}{4.608}l = \dfrac{25}{48}l = 1$ in. Second pole distance [Art. 98] $b = \dfrac{1}{6}\lambda = 0.32$ in.

According to Art. 98, we have now, for the cross lines,

$$\mathbf{OP} = \mathbf{QR} = \tfrac{1}{4}p\lambda^2\left(\dfrac{l}{\lambda}\right)^4 \text{ and } \mathbf{O'P'} = \mathbf{Q'P'} = \tfrac{1}{4}p\lambda^2\left(\dfrac{l_1}{\lambda}\right)^4.$$

Laying off these distances under the supports, we have thus the *cross-lines*.

We have next to construct the *second equilibrium polygon*. This, by the aid of the cross-lines and fixed points already constructed, we can easily do, as detailed in Art. 97 (3). Then the moments at the supports are given directly to moment scale in the interior spans, or we can find them from the end spans by laying off $\tfrac{1}{3}\dfrac{l}{l_1}l$ [Art. 89].

Finally, the moments thus found and laid off at the supports, we can construct the *moment curve* by making $\mathbf{D'E'} = \tfrac{1}{4}p\lambda^2\left(\dfrac{l_1}{\lambda}\right)^2$ and $\mathbf{DE} = \tfrac{1}{4}p\lambda^2\left(\dfrac{l}{\lambda}\right)^2$ [Art. 98 (2)], and thus drawing the end tangents and corresponding parabolas.

According to Art. 98 (3), we can then find the shearing forces by laying off $a = \dfrac{n}{m}\lambda$ and drawing parallels to the end tangents to intersection with verticals through supports, as shown in Fig.

We thus have both moments and shearing forces for uniform load. By careful attention to the above, the reader will have no difficulty in solving any case. We recommend him earnestly to perform the entire construction for himself, referring to the proper Arts. at every step. [*For convenience of size, we have not observed our scales strictly in the Figs. The reader should therefore not attempt to check results with the dividers.*]

100. Partial uniform Load.—1. When the girder is only partially loaded, as, for instance, a certain portion of the span βl from B, the *simple* moment area consists of a triangle A B C and a parabolic segment **C E B**. [Pl. 16, Fig. 64.] If in the first *force* polygon **B′ D′** is the total load upon the span, the line **O A′** parallel to the end tangent **A G** divides **B′ D′** in the same ratio as the end of the load divides the span, or denoting the length $B_0 C_0$ by βl.

$$B' A' : B' D' :: \beta l : l.$$

The intersection **G** of the end tangents lies in the vertical **G H**, which halves **B F**. Since the triangle **B C T** is similar to **O A′ B′**, we have

$$B T : B' A' :: \tfrac{1}{2}\beta : a;$$

or since $B' D' = p l,\ B' A' = p \beta = p l \dfrac{\beta}{l}.$

$$B T : p l \dfrac{\beta}{l} :: \tfrac{1}{2}\beta : a;$$

or $B T = p l \dfrac{\beta}{l} \cdot \dfrac{\beta}{2a}.$

It is therefore easy to construct **B T** as in Fig. 64, where $B_1 D_1 = p l,\ A_1 B_1 = B' A' = p l \dfrac{\beta}{l}$, and pole distance $= \tfrac{1}{2}\beta$; then $B_2 D_2 = B T$.

If **B T** is thus found, we can easily, when **A** and **B** are given, construct the end tangents, and then construct the first equilibrium curve itself.

2. Make $G K = \tfrac{1}{3} G I$, then the triangle **C K B** is equal to the parabolic area **C E B**. If, then, through **K** we draw a parallel to **C B**, intersecting **C G** in **L**, and through **L** the vertical **L M**, the triangle **A L B** is equal to the entire simple moment area. This last is therefore proportional to **L M**, or $\mathfrak{M} l = \tfrac{1}{2}\, l \times L M$; hence $\mathfrak{M} = \tfrac{1}{2} L M$. It is easily proved also that $F M = \tfrac{1}{3} F B$. Thus, as **G K** is $\tfrac{1}{3}$ of **G I**, **G L** is $\tfrac{1}{3}$ of **G C**; hence **M H** is $\tfrac{1}{3}$ of **F H**, and therefore **F M** is $\tfrac{2}{3}$ of **F H**, or $\tfrac{2}{3}$ of $\tfrac{1}{2} F B = \tfrac{1}{3} F B$. **L M** can therefore be easily drawn.

3. Let **N** be the middle of **A B**. Make **N O** equal to $\tfrac{1}{3} F N$. Then the centre of gravity of the triangle **A C B** is in the vertical through **O**, while the centre of gravity of the parabola is in **H G**. If through **L** we draw a parallel to **A B**, intersecting the vertical through **C** in **P**, the two areas are to each other as **F C** to **C P**. If in the vertical through **O** we make $O Q = \tfrac{1}{2} C P$, then, since $I H = \tfrac{1}{2} C F$, we have

$$I H : O Q :: F C : C P.$$

SPECIAL CASES OF LOADING.

The intersection **R** of the line **Q I** with **A B** lies, then, in the vertical through the centre of gravity of the simple moment area.

Thus the construction of the cross-lines is now easy.

4. It is most convenient in the application of the above to construct or calculate the distances of the cross-lines under the supports once for all, for load over various parts of the span. The necessary formulæ can be directly deduced. Thus the

Triangle $\quad \mathbf{B\,A\,T} = \mathbf{B\,T} \times \frac{1}{2} l = \frac{p \beta^2}{2\,a} \times \frac{1}{2} l.$

Triangle $\quad \mathbf{B\,C\,T} = \mathbf{B\,T} \times \frac{1}{2} \beta l = \frac{p \beta^2}{2\,a} \times \frac{1}{2} \beta l.$

Triangle $\quad \mathbf{B\,L\,T} = \mathbf{B\,T} \times \frac{1}{3} \beta l = \frac{p \beta^2}{2\,a} \times \frac{1}{3} \beta l.$

The entire area is equal to the triangle **A L B**.

But $\mathbf{A\,L\,B} = \mathbf{B\,A\,T} - \mathbf{B\,L\,T} = \frac{p \beta^2}{2\,a} \left(\frac{1}{2} l - \frac{1}{3} \beta l. \right)$

The triangle $\mathbf{A\,C\,B} = \mathbf{B\,A\,T} - \mathbf{B\,C\,T}$; hence

$$\mathbf{A\,C\,B} = \frac{p \beta^2}{2\,a} \times \frac{1}{2} (l - \beta\, l).$$

The parabolic area is equal to the entire area minus **A C B**, or parabolic area $= \frac{p \beta^2}{2\,a} \times \frac{1}{6} \beta l.$

F is distant from **B T** by a distance βl, **N** by a distance $= \frac{1}{2} l$, **N O** is $\frac{1}{3}$ of **N F**; hence **O** is distant from **B T**

$$\frac{1}{3}\left(\beta l - \frac{1}{2} l\right) + \frac{1}{2} l \text{ or } \frac{1}{3}(\beta l + l).$$

Therefore, the moment of the triangular area, with reference to the right support **B**, is

$$\frac{p \beta^2}{2\,a} \times \frac{1}{2}(l - \beta l) \times \frac{1}{3}(l + \beta l) = \frac{p \beta^2}{12\,a}(l^2 - \beta^2 l^2).$$

The moment of the parabolic area is

$$\frac{p \beta^2}{2\,a} \times \frac{1}{6} \beta l \times \frac{1}{2} \beta l = \frac{p \beta^4 l^2}{24\,a}.$$

The total moment, with reference to the right support, is, therefore,

$$\mathfrak{M}_r = \frac{p \beta^2 l^2}{24\,a} [2 - \beta^2].$$

In a similar manner we find for the left support

$$\mathfrak{M}_1 = \frac{p\,\beta^2\,l^2}{24\,a}[2-\beta]^2.$$

When $\beta = 1$ the span is completely covered, and we have, then, right and left

$$\mathfrak{M} = \frac{p\,l^2}{24\,a}.$$

If we compare this value with those for partial loading, we see that they differ only by certain coefficients, and that these coefficients depend only upon the length of the loaded portion. If, then, we have the distance between the cross-lines for *total* load, we have only to multiply by certain factors to obtain the distances for partial loading. For uniform or total load over the whole span, this distance is given by $\frac{1}{4}p\lambda^2\left(\frac{l}{\lambda}\right)^4$ (Art. 98).

If we divide this distance in certain proportions we have at once the distances for partial loading. These proportions are given by $(2-\beta^2)\beta^2$ for the right support, and $(2-\beta)^2\beta$ for the left, under the supposition that the load comes on from the right. The reverse is the case for load coming on from left. If we take $\beta = \frac{1}{4}, \frac{1}{2}, \frac{3}{4}$ of the span, we can calculate these proportions once for all. We thus have the following table:

	Support under load $(2-\beta^2)\beta^2$.	Support for unloaded end $(2-\beta)^2\beta$.
$\frac{1}{4}$ span loaded	$\frac{31}{256} = 0.1211$	$\frac{49}{256} = 0.1914.$
$\frac{1}{2}$ span loaded	$\frac{7}{16} = 0.4375$	$\frac{9}{16} = 0.5625.$
$\frac{3}{4}$ span loaded	$\frac{207}{256} = 0.8086$	$\frac{225}{256} = 0.8789.$

The division of the distance between the cross-lines for uniform load over the whole span into these proportions is easily accomplished graphically. Thus, from the end of the line to be divided, draw a line in any direction, and lay off upon it the six numbers above, to any convenient scale. Join the end of the last division with the end of the line to be divided, and then draw parallels through the other points.

It is important to observe some definite system of numeration, otherwise, especially in the first attempt at construction, confusion is apt to arise.

We can thus find the cross-lines for any position of the load, and for each position can, if we wish, draw the equilibrium polygon and determine the moments according to the general method of Art. 97.

101. Concentrated Load.—The simple moment area is in this case a triangle [Pl. 16, Fig. 65] whose area is $\frac{1}{2} l h$, h being the height **C D**. Therefore

$$\mathfrak{M} = \frac{1}{2} h.$$

If from the centre of the span **E** we lay off $\mathbf{E\,F} = \frac{1}{3}\mathbf{E\,D}$, **D** being the point of application, a vertical through **F** passes through the centre of gravity.

As the height **C D** is proportional to $\mathfrak{M}$, we may take **C D** as second force polygon. Since $h = 2\,\mathfrak{M}$, the distance of the pole **N** must be $2 \times \frac{1}{6} l = \frac{1}{3} l$, when $\lambda = l$ (Art. 89). Draw **N P** parallel to **A B**, then is $\mathbf{N\,P} = \frac{1}{3}\mathbf{A\,B}$.

Parallel to **N C** and **N D** we may draw the *cross-lines*. A simple construction may be given for them when they are made to pass through **A** and **B**. Let **A M** and **B L** be the cross-lines. Then the triangle **S B M** is similar to **N C D**, and

$$\mathbf{B\,M : C\,D :: B\,F : N\,P}.$$ But

$\mathbf{B\,F} = \mathbf{B\,E} + \tfrac{1}{3}\mathbf{E\,D} = \tfrac{1}{2}\mathbf{A\,B} + \tfrac{1}{3}(\tfrac{1}{2}\mathbf{A\,B} - \mathbf{A\,D}) = \tfrac{1}{6}(2\,\mathbf{A\,B} - \mathbf{A\,D})$
and $\mathbf{N\,P} = \tfrac{1}{3}\mathbf{A\,B};$ therefore,

$$\mathbf{B\,M : C\,D :: (2\,A\,B - A\,D) : A\,B}.$$

Make $\mathbf{D\,G} = \mathbf{A\,B}$, then $\mathbf{B\,G} = 2\,\mathbf{A\,B} - \mathbf{A\,D}$.

The point **M** is accordingly found by drawing a straight line through **G** and **C**, **D G** being equal to **A B**. In the same way make $\mathbf{D\,H} = \mathbf{A\,B}$, and draw a line through **H** and **C**. We thus obtain **L**.

The prolongations of **M C** *and* **L C***, therefore, intersect* **A B** *prolonged in the points* **G** *and* **H***, distant from* **D** *by* **A B**.

If, then, we have to investigate a concentrated load in various positions, we draw the first equilibrium polygon **C X** and **C Y** [Pl. 16, Fig. 66], and lay off in the same the closing lines (for the simple moment area) for the different positions of the load. The distances cut off on the vertical through C by these lines give, then, the various values of h or $2\mathfrak{M}$. If we draw from these intersections lines to the pole **N**, at the distance $\frac{1}{3} l$ from **D C**, the cross-lines are parallel to these lines. It will be best to keep for each pair the common line **P Q** parallel to **N C**. When the point of application of the load divides the span into two equal parts, the point of intersection of the cross-lines divides the middle third of **P Q** into equal parts.

The centre of gravity of the simple moment area cannot pass beyond the middle third of the span. Since any load can be considered as made up of a number of concentrated loads, it follows generally that *for any method of loading the centre of gravity of the simple moment area lies between the third verticals.*

CHAPTER XI.

METHODS OF LOADING CAUSING MAXIMUM STRAINS.

102. Maximum Shearing Force—Uniformly distributed Moving Load.—Suppose, first, the span in question loaded with a concentrated weight. The simple moment area is $A' C' B'$. [Pl. 18, Fig. 67.]

In the force polygon let $O A_1$, $O B_1$ and $O C_1$ be respectively parallel to $C'A'$, $C'B'$ and $A'B'$. Then $C_1 A_1$ and $C_1 B_1$ are the reactions at A and B. Since, according to Art. 80, the moments AA' and BB' are always positive, and the middle sides $A'S$, $B'S$ pass through the fixed points I and K, it follows from the construction of the preceding Art. that the intersections O and P of the sides $A'C'$ and $B'C'$ of the first equilibrium polygon with the closing line AB must always lie within AI and BK. That is, *the points of inflection O and K are always between the fixed points and the ends.* Therefore A', B' and the point C' must lie on opposite sides of the closing line AB, and consequently C_1 in the force polygon must lie *between* A_1 and B_1.

Accordingly the shear $A_1 C_1$ at A is positive, and the shear $C_1 B_1$ at B is negative.

Let the distance of the load from the left support be l_1, from the right support l_2. The load itself is P, and the moment AA' at the left support M', BB' at the right M''. Required, the shearing force S at a point distant x from the left support. The partial reaction at the left is R'. Then

$$R'l = M' - M'' + P l_2,$$

$$\text{or } R' = \frac{M' - M''}{l} + \frac{P l_2}{l}.$$

M' and M'' are always positive. If, therefore, $M' > M''$, R' is positive; if $M' < M''$, then $M'' - M' < M'' + M'$, or, since by Arts. 75 and 80, $M' + M'' < \mathfrak{M}' + \mathfrak{M}''$, $M'' - M' < \mathfrak{M}' + \mathfrak{M}''$. Now it can be easily proved analytically that for a girder horizontally fixed,* $\mathfrak{M}' = \frac{P l_1 l_2^2}{l^2}$ and $\mathfrak{M}'' = \frac{P l_1^2 l_2}{l^2}$, hence $\mathfrak{M}' + \mathfrak{M}''$

---

* Supplement to Chap. VII., Art. 18.

$= \frac{P\, l_1\, l_2}{l}$ since $l_1 + l_2 = l$. Therefore $\mathbf{M}'' - \mathbf{M}' < \frac{P\, l_1\, l_2}{l}$.

Since, however, $l_1 < l$, we have also $\mathbf{M}'' - \mathbf{M}' < P\, l_2$. Hence, if $\mathbf{M}'$ is $< \mathbf{M}''$, we have also $\mathbf{R}'$ positive. $\mathbf{R}'$ is therefore *always* positive whatever may be the position of the load. In the same way it may be shown that $\mathbf{R}''$ is *always negative*.

If now the load is to the right of the point distant x from the left support, then for this point the shearing force $\mathbf{S}' = \mathbf{R}'$, and is therefore *positive*. If the load is to the left of this point, the shearing force $\mathbf{S} = \mathbf{R}' - \mathbf{P} = \mathbf{R}''$, and is therefore negative. $\mathbf{S}$ for any point is therefore *positive* or *negative*, according as the load lies *right* or *left* of this point. Hence for a uniform load we deduce directly—

The shearing force at any point is a positive or negative maximum when the load extends from this point to the right or left support respectively.

The same principle holds good for the simple girder.

2. Thus far we have considered the load in the span itself. Suppose now the load is in some other span, and the span in question is unloaded, then

$$\mathbf{R}' = \frac{\mathbf{M}' - \mathbf{M}''}{l} \quad \mathbf{R}'' = \frac{\mathbf{M}'' - \mathbf{M}'''}{l}.$$

As we pass away from the loaded span the moments at the supports are alternately positive and negative, and each is greater than the one following (Art. 94). Since the moments $\mathbf{M}'$ and $\mathbf{M}''$ are alternately positive and negative, $\mathbf{R}'$ will have the same sign as $\mathbf{M}'$, and $\mathbf{R}''$ as $\mathbf{M}'''$, and generally $\mathbf{R}_m$ as $\mathbf{M}_m$.

Adopting, then, the notation shown in Pl. 18, Fig. 68, we have for the span l_{m-1}

$$\mathbf{R}_m = \frac{\mathbf{M}_{m-1} - \mathbf{M}_m}{l_{m-1}},$$

where $\mathbf{R}_m$ has the same sign as $\mathbf{M}_m$.

In the same way

$$\mathbf{R}_{m+1} = \frac{\mathbf{M}_m - \mathbf{M}_{m+1}}{l_m};$$

and therefore

$$\frac{\mathbf{R}_{m+1}}{\mathbf{R}_m} = \frac{1 - \dfrac{\mathbf{M}_{m+1}}{\mathbf{M}_m}}{\dfrac{\mathbf{M}_{m-1}}{\mathbf{M}_m} - 1} \times \frac{l_{m-1}}{l_m}.$$

But $\frac{M_{m+1}}{M_m}$ is negative and greater than 2 (Art. 94). Therefore in the preceding expression the numerator is positive and > 3. Further, $\frac{M_{m-1}}{M_m}$ is negative and less than $\frac{1}{2}$, hence the denominator of the above expression is negative and $< \frac{3}{2}$. Therefore $\frac{R_{m+1}}{R_m}$ is negative and $> 2\frac{l_{m-1}}{l_m}$, that is, *the shearing forces at the supports are alternately positive and negative, and increase* (when $2\,l_{m-1}$ is not less than l_m) *towards the loaded span.* We have then

$$-\frac{R_{m+1}}{R_m} > 2\frac{l_{m-1}}{l_m}.$$

For any span, then, the shear at the left support **R'** will be positive when the left adjacent span is loaded, the right adjacent span unloaded, and all the other spans each way alternately loaded. The shear **R'** will be negative when the remaining spans are loaded. Hence:

The shearing force is a maximum (positive) at any point when the load extends from this point to the right support, and the other spans are alternately loaded, the adjacent span to the right being unloaded, that to the left, loaded. The negative maximum, on the contrary, occurs when the load extends from the point to the left support, when the right adjacent span is loaded and the left unloaded; the other spans alternately loaded.

Pl. 18, Fig. 69, gives these two cases.

In practice such a loading can never occur. If we suppose the rolling load divided into two portions only, the above rule reads as follows:

The shearing force at any point will be a positive maximum when the load reaches from the right support to this point, and when the left adjacent span is covered. The negative maximum occurs when the load reaches from left support to the point, and the right adjacent span is covered.

103. Maximum Moments.—1*st. Loaded Span.* Let a weight act at the point D, Pl. 18, Fig. 67. Then **A U S V B** is the second, **A' C' B'** the first equilibrium polygon, and **A A', B B'**

are the moments at the supports. We have already seen that the inflection points **O** and **P** (for which the moment is zero) lie outside of the fixed points. We can therefore assert that WITHIN *the fixed points the moments are negative wherever the weight may be placed.* From the Fig. we see at once that the inflection points **O** and **P** move to the right or left as the weight moves to the right or left. Accordingly when for the weight at **D** the moment at **O** is zero, the moment at this point will be positive when the load moves to the right of **D**, negative when it moves to the left of **D**.

Hence for the maximum moment we have at once the following principle:

For any point **O** *outside the fixed points the moment will be a positive or negative maximum when the load reaches from the point* **D**, *where a load must be placed to cause the moment at* **O** *to be zero, to the right or left support respectively. For the negative maximum, therefore, the load reaches from* **A** *to* **D**; *for the positive, from* **D** *to* **B**.

If the point **O** is given, it is indeed possible to determine by construction the point **D** to which the load must reach. It is, however, simpler to *assume* **D** *and then construct* **O**.

If we choose for the different positions of **D** an arbitrary length for **C′ D′** (Fig. 67), so that the point **C′** falls in a parallel **Q R** to **A′ B′** (Fig. 70), and, moreover, take **D** at equal intervals, then the points **L** and **M** will be at equal distances (Fig. 67), and hence the points **I** and **K** (Fig. 70), in which the verticals through the fixed points are intersected by the lines **A′ M** and **B′ L** (Fig. 67), will be at equal distances. We have, then, the following simple construction [Fig. 70, Pl. 18]:

Between the verticals through the supports draw two parallel lines **A B** and **Q R** at any convenient distance apart, and divide **Q R** into a number of equal parts; four or five are sufficient. Draw lines from **A** to **R** and the middle **S** intersecting the vertical through the fixed point **I** in I_1 and J_2. In the same way find K_1 and K_2. Divide $I_1 I_2$ and $K_1 K_2$ into the same number of equal parts as **Q R** has been divided into, and join these points in reverse order by lines. The intersection of these lines with the lines drawn from **A** and **B** to the points upon **Q R** give the points **O**, for which the moment is a maximum when the load is limited by the corresponding point upon **Q R**.

This construction was first given by Mohr.

104. Determination of the Maximum Shearing Forces.
—According to the general method of construction given in Art. 97, we can now determine by reference to Arts. 98 and 99, which treat of total and partial distributed loading, the shearing forces corresponding to the methods of loading which cause maximum strains.

As a review of the preceding principles, we take the same example as before, as given in Pl. 19, Fig. 71. Here again the reader should construct the Figs. for himself. The scales are as before, Art. 99.

Fig. a shows the method of loading for positive maximum shear in first span; and the second Fig. below, the same for the second span. [Art. 102.]

We first find, precisely as in Fig. 63, the shearing forces in the third and fourth spans for the total loads over those spans, and lay off the shear thus obtained in the first and second spans, as indicated by the broken lines in Fig. b in those spans. Thus having first found the fixed points, which we may here take directly from Fig. 63, we construct as in that Fig. also the crosslines for total load in third and fourth spans. Thus laying off 54 equal to $O_1 P_1$, and drawing a line through 4 and support to intersection with vertical through fixed point in second span [Fig. 60], we determine D', and then from the cross-lines find at once D''. In like manner, supposing for the moment the load on the other two spans, we have $e\, a$, $a\, D$, $D\, a$ and $a\, F'$, and then at once F''. $F''\, D'$ cuts off then the moment at the right support, and joining 1, 2 and 5, we find, according to Art. 98 (2), precisely as in Fig. 63, the end tangents; and then from these, with the first pole distance a, find the shear. This is given by the broken lines in third and fourth spans. Lay off these lines in first and second spans, remembering, since the shears at the supports alternate, that the positive shear at left of fourth span must be laid off as negative (down) at right end of first, etc. [See Fig. 63.]

Now for the positive shears in first and second spans for the different methods of loading, we have only to determine the direction of the tangents through end of load [Fig. 64, Art. 100]. The lengths of the segments cut off upon the verticals through the supports by these tangents are given for first and second spans by Figs. f and e, for each position of load. [Art. 100 (1).]

We have first, then, to construct the *cross-lines* [Art. 100 (4)] as shown in Fig. *d*.

Take first the second span. For this span, as shown in Fig. *a*, the first span is fully loaded, as also the last. Make, therefore, $e\,a = O_1\,P_1$, draw *a* **D**, and we thus find the point **D**, common to all the middle sides of the second equilibrium polygon for different positions of load. So also make on right $54 = O_1\,P_1$ and draw $4\,D'$, and then, since third span is empty, $D'\,F$, and we thus have **F**. Now from **D** and **F** lay off the cross-line distances, as $D\,a$, $D\,b$, $D\,c$, equal to $e\,a$, $e\,b$, $e\,c$, etc. Draw lines through these points and **D** and **F** respectively, and note their points of intersection *a b c d* with the verticals through the supports. [NOTE. *Be careful to preserve an orderly notation.*] These points give the *moments* for each position of load in second span. Take the length *a a* from Fig. *e* and lay it off from *a* on the right support vertical, and join the end with *a* on left. This is the tangent for full load in second span. A parallel to it at distance *a* from left gives the shear in Fig. *b*. Then lay off *b b* taken from Fig. *e* on right, and join with *b* on left. This is tangent for load over three-fourths second span from right. A parallel to it at distance *a* from foot of perpendicular one-fourth of span from left cuts off shear for this position of load. So for tangents *c c*, *d d*. We thus obtain the curve for positive shears in second span. The negative shears are obtained by subtracting these from the shear already found for full load. We thus have the lower curve, and the shear diagram for second span is complete.

For first span, only the third is loaded. We lay off, then, 78 equal to the distance between cross-lines corresponding, draw $8\,F_1$, and thus find F_1. Lay off now at left end *e d*, *e c*, *e b*, draw lines from these points through second support, and note intersections with vertical through **D**. Through each of these intersections draw a line through F_1, and produce to intersections *a b c d* with vertical through second support. It is from these *last* points that the distances *a a*, *b b*, etc., taken from Fig. *f* must be laid off respectively in order to find the tangents *e a*, *e b*, *e c*, *e d*, *e e*.

Parallels to these tangents above in Fig. *b* give, as before, the positive shear for each position of load. The negative shear is, as before, found by subtracting the positive from total load shear. Thus shear diagram for first span is complete.

Of course, the same circumstances can hold good for third and fourth spans as for first and second, except that positive shears on one side of centre support are the corresponding negative shears upon the other, and *vice versa*.

Following carefully the above with the aid of the Fig., the reader cannot fail to grasp the method. An independent construction for a similar case will make both principles and details familiar. Once thoroughly understood, the method is rapid, accurate, simple, and of general application.

105. Determination of the Maximum Moments.—In like manner, it is easy, according to the general construction given in Art. 97, and referring to Arts. 98 and 99, to determine the maximum moments. In Fig. 72, Pl. 20, we have the same example as before, concerning which we have but little additional to remark. Fig. 64, Art. 100, shows that the end tangents give the moments within the unloaded portion of the girder. These tangents are constructed precisely as before in the several spans, except it will be noticed that in the first span we have made use of the construction given in Art. 97, Fig. 61. Thus the point $\mathbf{F'}$ is determined so that $\mathbf{K F'} = \mathbf{K F} \left(\frac{l}{l_1}\right)^2$, and thus the moments are measured directly at the end vertical. Also upon the left support vertical we have laid off the distances between the cross-lines in the second span multiplied by $\left(\frac{l_1}{l}\right)^2$.

The only thing new in the Pl. is Fig. *c*, which, as we have seen in Art. 103, Fig. 70, gives the points at which the positive moment is a maximum for each position of load. The positive moments can then be taken directly off upon the verticals through these points, and are limited by the horizontal through the supports and the tangents, as above.

Thus, at second support the vertical distance to *a*, for first span, gives the moment at the support. Lay it off in Fig. *a* from the support line. For a load over ¾ span, we see at once the point for which the positive moment is a maximum from Fig. *c*. Follow up the vertical through this point. The distance on this vertical in Fig. *b*, between the support line and tangent $b\,b_1$, gives the moment to be laid off in Fig. *a* upon this vertical. So, for load over ½ span, we have next vertical and tangent $c\,c_1$, and so on. We thus obtain the curve for

11

positive moments at the right end of first span, $a\,b\,c\,d\,e$. From e draw a line to left support. At the other supports, in like manner, we determine the positive moments, and join the points $e\,e$ in second and third spans, and e and right support in fourth.

We have already seen (Art. 103) that within the fixed points the moments are negative wherever a load may be placed. The maximum, therefore, occurs for full load. We have, therefore, found for 3d and 4th spans the parabola for full load, precisely as in Fig. 63. These parabolas are given in broken lines in the Fig. a. By subtraction of the positive moments outside the fixed points from the positive moments at the same points for total load given by these parabolas, we obtain directly the lower curves as far as the points e in each span.

The second parabola (partly full, partly broken) is all that is needed to complete our Fig. To obtain this we have simply, in 2d and 3d spans, to make the vertical through centre of line $e\,e$ equal to its length already laid off for total load, viz.: $\tfrac{1}{4} p \lambda^2 \left(\dfrac{l}{\lambda}\right)^2$, Art. 98 (2), produce $e\,e$ to intersections with support verticals, and join these intersections with the extremities of the first verticals above. We can then construct the parabola, which completes our diagram, and gives us Fig. a.

106. Practical simplifications.—In practice, the constructions given in Figs. 63, 71 and 72, admit of many simplifications, which, in order to avoid confusion at first, have been disregarded. The whole solution, given for the sake of clearness in three separate Plates, can be performed upon a single sheet, since the Fig. for the second equilibrium polygon in Figs. 63, 71 and 72 may be combined in one. Indeed, the lines necessary for the construction of the maximum shearing forces can be applied directly to the determination of the maximum moments. It is therefore unnecessary to divide the construction into separate sheets.

2. The cross-lines in the end spans can be omitted, since all that is required are the distances to be laid off upon the end verticals, and these when found can be laid off at once.

3. We can apply the second equilibrium polygon directly in order to find the moments for the dead and moving load. Thus the transferring of ordinates from one Fig. to another is avoided.

4. It is evidently unnecessary to actually draw all the various lines. We need only to mark the different points of intersection.

5. The construction for dead and live loads can be performed at once, thus avoiding the necessity of a subsequent addition.

107. Approximate Practical Constructions.—If the successive steps of the preceding development are carefully followed, the method will be found simple and easy of application. Indeed, the complete and accurate solution of the difficult problem of the continuous girder by a method purely graphical, is the most important extension of the system since the date of Culmann's treatise, and well illustrates the power and practical value of the Graphical Method.

Humber gives the following constructions, "which may be relied upon for safety without extravagance." * As rapid means of obtaining approximate results, they may not be without value to the practical engineer, and we therefore append them here. It must be remembered that the constructions hold good ONLY *for end spans three-fourths the length of the others.*

I. Beam of Uniform Strength, continuous over one Pier, forming two equal Spans, subject to a fixed Load uniformly distributed, and also to a moving Load. Maximum Moments.* Pl. 18, Fig. 73.

The greatest moment at the pier (positive) will be when both spans are fully loaded.

The greatest negative moment will obtain in the loaded span when the other span bears only the fixed load.

(A moment is *positive* when the upper fibres or flanges are *extended, negative* when the upper flange is *compressed*.)

Construction.—Let **A B C** be the beam. On **A B** draw the parabola whose centre ordinate **D E** is $(p + m) \dfrac{l^2}{8}$, and on **B C** the parabola whose centre ordinate **G F** is $\dfrac{p\, l^2}{8}$.

At the pier **B** erect the perpendicular $\mathbf{B\,H} = \dfrac{(p + m)\, l^2}{6}$, and make $\mathbf{B\,L} = \dfrac{(2p + m)\, l^2}{12}$. Join **A H**, **A L**, and **L C**.

---

* "Strains in Girders, calculated by Formulas and Diagrams." Humber. New York: D. Van Nostrand, publisher.

Then the vertical distances between the parabolic arc **A E B** and the lines **A H** and **A L**, the *greatest* being taken, will give the maximum moments—*positive* in the first case and *negative* in the last. The points of inflection approach as near the pier as **K** and recede as far as **M**.

If $\frac{pl^2}{2}$ is *less* than $\frac{(2p+m)l^2}{12}$, the beam must be *latched down* at the abutments. The load comes on from the left; p and m are the loads per unit of length of the permanent or fixed and the moving or live loads.

Shearing Forces (Pl. 18, Fig. 74).—The maximum shearing force at *either abutment* will obtain when its span only sustains the moving load. The maximum shear at the *centre pier* will obtain when both spans are fully loaded.

Construction.—Lay off $AC = \frac{l}{12}(4p+5m)$ and $AD = \frac{l}{3}(p+m)$. At **B** lay off $BF = $ twice **A D**. Take a point **M** distant $\frac{1}{8}l$ from **A**, and join **D** and **F** to **M**. Draw **C N** parallel to **D M**. Sketch in a curve similar to that dotted in the figure, giving an additional depth to the ordinates at the point of minimum shear of $m\frac{l}{8}$. Then the vertical ordinates between **A B** and **C O P F** may be considered to give the maximum shearing force for either span.

II. Beam as above—continuous over three or more Piers. $l_1 = $ end spans. $l = $ the other spans.

Moments.

The maximum moment (positive) will obtain when only the two adjacent spans, and every alternate span from them, are simultaneously loaded with the total load—the remaining spans sustaining only the fixed load.

The maximum moment at the centre of any span will obtain when it and the alternate spans from it are fully loaded—the remaining spans sustaining only the fixed load.

Construction (Pl. 18, Fig. 75).—Let **A B C** be part of the beam. On **B C** draw the parabola whose centre ordinate $EF = \frac{(p+m)l^2}{8}$, and on **A B** the parabola whose centre ordinate $CD = \frac{(p+m)l_1^2}{8}$. At **B** and **C** make $BH = CL = $

$\frac{l^2}{3}\left(\frac{2p}{7}+\frac{m}{2}\right)$. Join **A H** and **L**. Make **B G** $=$ **C S** $=\frac{3\,p\,l^2}{32}$.

Join **A G** and **S**. The maximum vertical ordinates between the two parabolas and the lines **A H**, **A G**, **H L**, and **G S**, as shown in the figure, give the maximum moments.

The points **I, K** and **O, P** or **M, N** show the limits of deviation of the *points of inflection*.

If $\frac{p\,l_1^2}{2}$ is *less* than $\frac{3\,(p+m)\,l^2}{32}$, the beam will require to be *held down* at the abutments.

If the beam be continuous for *three spans only*, l in the expression for **B H** $=\frac{l^2}{3}\left(\frac{2p}{7}+\frac{m}{2}\right)$ must have a value given to it $=\frac{l+l_1}{2}$.

Shearing Forces.

The maximum shear at any pier (**B** or **C**) will obtain simultaneously with the maximum moment over that pier.

Construction.—Pl. 18, Fig. 76.

Let **A B C** be part of the beam. First, for any inner span as l—. At **B** and **C** erect **B G** $=$ **C H** $=\left(\frac{p}{2}+\frac{2\,m\,l}{3}\right)$. Make **B D** and **C E** each $=\frac{l}{2}\,(p+m)$. Join **D** and **E** to midspan **F**, and draw **G K** and **H K** parallel to **D F** and **F E** respectively.

Second, for either end span as l_1—. At **B** erect a perpendicular $=\frac{2\,l_1}{3}\,(p+m)$, which, if $l_1=\frac{3}{4}\,l$, will coincide with **B D**. At **A** make **A L** $=\frac{1}{2}$ **B D**. Join **D** and **L** to **M** distant $\frac{1}{3}\,l_1$ from **A**. Make **A O** $=\frac{l_1}{2}\,(p+m)-\frac{3\,p\,l^2}{32\,l_1}$, and draw **O N** parallel to **L M**. Sketch in curves as shown by the dotted curves in the figure, giving additional depth to the ordinates there of $\frac{m\,l_1}{8}$ and $\frac{m\,l}{8}$ respectively. Then the vertical distances between **O** a b **D** and **A B** give the maximum shearing forces for either end span, and those between **G** c d **H** and **B C**, the shearing forces for the remaining spans.

If the beam be continuous for *three spans only*, **B G** and **C H** must be made equal to $\frac{(13\ p + 16\ m)\ l}{32} + \frac{L^2}{3\ l}\left(\frac{2\ p}{7} + \frac{m}{2}\right)$, where $L = \frac{l + l_1}{2}$. Further, the value given to **B D** for the inner span

$$= \frac{2\ l_1}{3}(p+m) - l(p+m) = (p+m)\frac{2\ l_1 - 3\ l}{3}.$$

108. Method by Resolution of Forces—Draw Spans.—The most usual cases of continuous girders which occur in practice are *draw* or *pivot spans*, which when shut must be considered as continuous girders of two spans. The graphical method becomes for such cases short and easy of application. In the case of *framed structures* of this character, it may, however, be more satisfactory to first find the maximum shearing forces (Art. 104), and then follow the reactions thus obtained through the structure from end to end by the method of Arts. 8–13. As a check upon the accuracy of the work, we may apply the "method of sections" referred to in Art. 14. In either case we must, of course, start from an end support where only two pieces intersect and the moment is zero.

Still again, we may find the reactions by calculation, and then apply the method of Arts. 8–13. In the case of two spans only, the formulæ for the reactions are sufficiently simple, and the ready and accurate determination of the strains offers, therefore, no difficulty.

We shall give here, therefore, the analytical formulæ requisite for our purpose, referring the reader to treatises upon the subject for their demonstration.*

Formulæ for Reactions.—*Continuous Girder of two unequal Spans, l and n l.* 1st. Concentrated weight **P**, in first span l, distant a from left end support. Reaction at left end support:

$$\mathbf{A} = \frac{\mathbf{P}}{2\ (1+n)}\left[(2+2\ n) - (3+2\ n)\ \frac{a}{l} + \frac{a^3}{l^3}\right].$$

Reaction at middle support:

$$\mathbf{B} = \frac{\mathbf{P}}{2\ n}\left[(2\ n+1)\ \frac{a}{l} - \frac{a^3}{l^3}\right].$$

---

* Bresse—*La Flexion et la Resistance*, and *Cours de Mécanique Appliquée*. Weyrauch—*Theorie der Träger*. Collignon—*Théorie Élémentaire des Poutres Droites*, etc. Also Supplement to Chap. XIII.

Reaction at right support:

$$C = \frac{P}{2n(1+n)}\left[-\frac{a}{l}+\frac{a^3}{l^3}\right].$$

2*d*. Uniformly distributed load extending to a distance β from left support. Load per unit of length $= p$.

$$A = \frac{pa}{2(1+n)}\left[(2+2n)-(3+2n)\frac{a}{2l}+\frac{a^3}{4l^3}\right].$$

$$B = \frac{pa}{2n}\left[(2n+1)\frac{a}{2l}-\frac{a^3}{4l^3}\right].\ *$$

$$C = \frac{pa}{2n(1+n)}\left[-\frac{a}{2l}+\frac{a^3}{4l^3}\right].$$

For two *equal* spans we have only to make $n = 1$ in the above equations. For a uniform load over whole span $\beta = l$.

From the above formulæ we can find the reactions for any case, and then proceed as indicated above.

109. By means of the graphical method, as we have now seen, we are enabled to solve completely the problem of the continuous girder, and that too without the aid of analytical formulæ, tables, or tedious computation. The method can also be applied to continuous girders of *variable cross-section*, or of uniform strength. We shall not, however, proceed further with the development of the method in this direction. The preceding will, we think, be found to contain all that is practically serviceable. For the application of the method to girders of variable cross-section, we refer the reader to *Winkler*—"*Der Brückenbau*," Wien, 1873—where will be found a thorough presentation of the subject, both analytically and graphically, to which we are greatly indebted in the preparation of the preceding pages. Plates 17, 19 and 20, are, with but few alterations, reproduced from that work.

---

* These formulæ are demonstrated in *Van Nostrand's Eng. Mag.*, July 1875.

CHAPTER XII.

CONTINUOUS GIRDER (CONTINUED)—COMBINATION OF GRAPHICAL AND ANALYTICAL METHODS.

110. In the present chapter we shall develop a method for the solution of continuous girders *not* purely graphical, but based upon the method of resolution of forces illustrated in Arts. 8–13, together with well-known analytical results, which method for accuracy, simplicity, and ease of application will, we think, be found superior to any hitherto proposed. The method is, of course, applicable only to *framed* structures, but for such cases is the most satisfactory of any with which we are acquainted.

111. The Inflection Points being known, the Shearing Forces and Moments at the Supports can, by a simple construction, be easily determined.—1*st. Loaded Span*—Fig. 77, Pl. 21.—Thus in the span **B C** $= l$, let the distance of the weight **P** from the left support be a, and let i and i' be the distances of the inflection points from **B** and **C** respectively. Then if through any point **P** of the weight we draw lines, as **P D**, **P E**, through i and i', intersecting the verticals at **B** and **C** in the points **D** and **E**, *the vertical ordinates between these lines and* **B C** *will be proportional to the moments*. For, as we see from the *force polygon*, the equilibrium polygon must consist of two lines as **D P**, **P E**, parallel to **O 0** and **O 1**, and because of the moments at the ends, the closing line **D E** is *shifted* to **B C** (Art. 23). Since the moments at the points of inflection are *zero*, the ordinates to **P D** and **P E** to *pole distance* **H** *will give the moments*. Now the points of inflection being known, and **P D** and **P E** drawn, we can easily find the pole distance **H** and the shearing forces **L 0** and **1 L** by laying off **P** vertically, and drawing from its extremities lines parallel to **P D** and **P E** intersecting in **O**. A perpendicular through **O** upon **0 1** gives **H** and the reactions **L 0** and **1 L**. In other words, we have simply *to decompose* **P** *along* **P D** *and* **P E**.

The construction, then, is simply as follows: Take any point on the direction of **P**, and draw **P D**, **P E** through the points of inflection. Lay off **P** to the scale of force as **A P**, and draw **A O** parallel to **P E** or **P D**. We have thus the *pole distance* **H**, and the shearing forces **P H** and **H A** at **B** and **C**.

B D or **C E** to the scale of distance, multiplied by **H** to the scale of force, give the moments at **B** and **C**. That is, **D** and **E** may be regarded as the points of application for **H**. The forces along **P D** and **P E** considered as acting at these points are held in equilibrium by the reactions **P H** and **H A** = **L** 0 and 1 **L** and **H**. Since **H**, acting as indicated in the figure with the lever arm **B D** or **C E**, causes *tension* in the upper fibres, the moments at **B** and **C** are *positive*.

2d. Unloaded Span—Fig. 78, Pl. 21.

As we have already seen in Art. 93, the inflection points in the unloaded spans *are independent of the load*, and are found by the simple construction there given for the "*fixed points.*" Since each fixed point lies within the outer third of the span, we have in Fig. 78 the broken line $a\,b\,c$, referred to in Art. 94, where the moments are alternately positive and negative, and increase from the end, so that any one is more than twice the preceding. Lines drawn parallel to these lines in the force polygon, cut off from the force line the reactions at the supports. Thus, in Fig. 78, $c\,b$ in the force polygon gives the *reaction* at **D**; $a\,b$ the reaction at **C**, and if **B** were an *end* support—that is, if $b\,a$ went through **B**—a **H** would be the reaction at **B**. For the resultant shear at **D**, we should then have $a\,\mathbf{H} - a\,b + c\,b = \mathbf{H}\,c$. So for any number of spans; the inflection points in the loaded span being known, we can easily find the *fixed* or inflection points in the other spans, which are independent of the load, and depend only on the length of these spans. Then draw the broken line $a\,b\,c\,\mathbf{P}\,d$. Then find the *pole distance* **H** by laying off $c\,\mathbf{P} = \mathbf{P}$ to scale, and drawing $c\,\mathbf{O}$ parallel to $c\,\mathbf{P}$, and through the point **O** thus determined drawing **H O**. Then find the reactions at the other supports, or the shear at any support, by lines in the force polygon parallel to $a\,b$, $b\,c$, etc. Thus the *shear* at **B** is the distance **H** a cut off by **H** and **O** a parallel to $a\,b$. Since the shear at **D** is plus and alternates from **D**, we have at **B** the shear + **H** a. The shear at **C** is − **H** b; at **D**, + **H** c, etc. **O** c being parallel to **P** c; **O** a, to $b\,a$; **O** b, to $c\,b$, etc.

112. Inflection Verticals.—Draw a line from **P** through the support **D**, and through its intersection with cb draw a vertical 3 (Fig. 78). This vertical we call the inflection vertical.

The equation of the line cb is
$$y = \frac{m_1}{nl - i} x + m_1,$$
where $m_1 = DC$, $nl = CD$, $i = Ci'$. The origin being at **D**.
For the line C,
$$y = -\frac{m_1}{i_1} x + m_1,$$
where $i_1 = Di_1$.

If in this last equation we make $x = a$, we have for the ordinate at **P**,
$$\frac{m_1(i_1 - a)}{i_1},$$
and hence for the line **PD**,
$$y = \frac{m_1(i_1 - a)}{i_1 a} x.$$

For the intersection of **PD** with bc then
$$\frac{m_1}{nl - i} x + m_1 = \frac{m_1(i_1 - a)}{i_1 a} x.$$

Hence
$$x = \frac{i_1 a (nl - i)}{(i_1 - a)(nl - i) - i_1 a} \quad \ldots \quad (1)$$

We see at once that the value of x is *independent* of m_1 or DC, hence the intersection of **PD** and cb *lies always in the same vertical, whatever be the position of* **PC**. In other words, if the three sides of a triangle pass always through three fixed points (i', **D**, i_1), and two of the angles (**P** and c) be always in the same verticals, the *third angle must also always lie in the same vertical.*

For the distance of the inflection vertical on the other side of the loaded span (beyond **E**), we have similarly
$$x = \frac{i_2 i_3 (l - a)}{(l - a)(i_3 + i_2) - i_3 i_2} \quad \ldots \quad (2)$$
where l is the loaded span and i_3 the distance of the inflection point to the right of **E**.

Equations (1) and (2) give the distances of the *inflection verticals* from the supports **D** and **E**.

113. Beam fixed horizontally at both ends—Supports on level.—Consider the span **D E** (Fig. 78) as *fixed* at the supports so that the tangent to the deflection curves at **D** and **E** is always horizontal. Conceive the span prolonged right and left beyond the supports a distance equal to the span l. It is required to find the position of the *inflection verticals*.

From equation (1) of the preceding Art. we have, since $n = 1$, $i = 0$,

$$x = \frac{i_1 \, a \, l}{(i_1 - a) \, l - i_1 \, a},$$

and from equation (2), since $i_3 = l$,

$$x = \frac{i_2 \, l \, (l - a)}{(l - a)(l + i_2) - l \, i_2}.$$

Now for a beam fixed at the ends the distances of the points of inflection are

$$i_1 = \frac{a \, l}{l + 2a} \text{ and } -i_2 = \frac{l \, (l - a)}{3 \, l - 2 \, a}.$$

Substituting these values in the equations above, we have $x = -\frac{l}{3}$ and $x = +\frac{l}{3}$. That is, *the position of the inflection verticals is in this case independent of the load, and always equal to $\frac{l}{3}$ from the supports.*[*]

This remarkable property of the beam fixed at both ends enables us to find the inflection points by a construction similar to that for the *fixed points* in the unloaded spans, as given in Art. 86.

Thus we have simply to draw from **C** distant l from **A** (Pl. 21, Fig. 79) a line in *any* convenient direction, as **C** b intersecting the inflection vertical **I**, which is distant from **A**, $\frac{1}{3} \, l$, at a.

Through a and the fixed end support **A** draw a line to intersection with the weight **P**. Then draw **P** b. *The intersection i_1 of this last line with **A B** is the inflection point.* A similar construction gives i_2.

We can now find the reactions and moments. Thus **H O** to

---

[*] This important result, which renders possible a complete graphical solution of this case, has, so far as we are aware, never before been published.

the scale of force, multiplied by **A** *b* to the scale of distance, gives the moment at **A**, while **H G** is the reaction at **A** (Fig. 79).

114. Beam fixed at both ends—Example.—Since when the points of inflection are once determined, we may draw **P** *b* or **P** *c* at *any* inclination (Fig. 79), provided we afterwards find the corresponding *pole distance* **H O**; if **A** *b* or **B** *c* be made *equal to the height of the truss*, **H O** *will be the strain in the upper or lower flange at the wall* (the flange in question being always that for which there is no diagonal at its union with the wall). Thus in Pl. 21, Fig. 80, we lay off **D E** = *l*, draw the vertical **I** at $\frac{1}{3}$ *l* from **D**, and for the given position of the load **P** find the inflection point i_2 by the preceding Art. A similar construction on the other side gives i_1. Now laying off **P M** equal by scale to the weight **P**, and decomposing it along **P D** and **P C**, we find **O H** the *pole distance* which to the scale of force will give directly the strain in the lower flange **B** *m* at the wall, *provided* **P D** *is made to pass through the intersection of the upper flange with the wall.* If the triangulation were reversed, **O H** would be the strain in the *upper flange* at the wall. In any case it is the strain in that flange *at whose junction with the wall there is no diagonal.*

The reaction at **D** is also **H M**, at **C** it is **P H**. Lay off then **B B'** in Fig. 80 (*a*) equal to **P M**, and make **B A** = **H M** and **A B'** = **P H**. Now draw *m* **B** parallel to **O H** and *m* **A** parallel to **O M**, and produce both lines to intersection at *m*. Then *m* **B** to scale of force is evidently the strain in the *lower end flange at the wall*. We assume the following notation.*

Let **A** represent all the space above the girder, **B** all the space below, and *a b c d*, etc., the spaces within the girder included by the flanges and diagonals. Then, for instance, **A** *b* is the first upper flange at the left, **B** *a* the first below; *a b* the first diagonal at the left, and so on.

Now draw in Fig. 80 (*a*), *m l* and **A** *l* parallel to the corresponding lines in the frame, and we have at once the strains in these pieces to scale. Following round the triangle according to our rule (Arts. 8–13) from *m* to **A**, **A** to *l* and *l* to *m*, we

---

* See an excellent little treatise on "*Economics of Construction in Relation to Framed Structures*," by R. H. Bow, to whom this method of notation is due.

find **A** l *tension* and $l\,m$ *tension*. [The strains in the upper flanges must *always* be tension, since the moments at the supports for loaded span are always positive.] Moreover, the Fig. thus far shows that **A** l, $l\,m$ and $m\,$**B** are in equilibrium with the shear **B A** = **H M**, as evidently should be the case; hence the strain in **B** m is *compressive*.

We have thus the strains in the three pieces at the right, and can proceed from these to find all the others. Thus the strains in **B** k and $k\,l$ are in *equilibrium* with $m\,l$ and **B** m. Lines parallel to **B** k and $k\,l$, therefore, which close the polygon commenced by **B** m and $m\,l$, give us the strains in **B** k and $k\,l$. Observe that the line $l\,k$ *crosses* **A B**, thereby making **B** k opposite in *direction*, consequently in *strain* from **B** m. This may also be seen by following round the triangle $m\,l\,k$ **B**, remembering that, as $m\,l$ is *always* found to be in tension, it must act *away* from the new apex, that is, from m to l. We thus find $k\,l$ in compression, and k **B** acting *away* from this apex, or in *tension*, therefore of opposite strain from the preceding flange **B** m, which, as we have seen, is in *compression*. The reason is obvious. The inflection point i_2 falls in the flange **B** k. If the beam were *solid*, the strain at i_2 would be zero; to the right of i_2 we should have compression, to the left, tension. In the *framed* structure the strains can only change *at the vertices*. The *crossing* of **A B** by $l\,k$ indicates such change, and **B** k gives its amount by scale.

Now taking the upper apex, we have here **A** l and $l\,k$ in equilibrium with $k\,h$ and **A** h. As we already know, $k\,l$ is in compression. We must, therefore, now take it acting from k to l, and following round the triangle we find **A** h *compression*, and $h\,k$ *tension*. From h on, the traverses between **A** h and **B** k produced towards the right [Fig. 80 (*a*)] will give the diagonals, while the upper and lower flanges will be given by the distances to them from **A** and **B** respectively, *until we arrive at the weight* **P**. Observe the influence of the weight. We have $k\,h$ and **B** k in equilibrium with $h\,g$ and $g\,$**B**, *and also the weight* **P** = **B**′ **B**. We must take, therefore, **A B**′ = **P H**, and then draw $h\,g$ and **B**′ g. Distances to the *right* of **B**′ along **B**′ g are compressive lower flanges, to the left, tensile; while to the right of **A** we have compressive upper, and to the left of **A** tensile upper flanges. The two diagonals at the weight $k\,h$ and $h\,g$ are in tension. From $h\,g$ on, the diagonals are alternately tension and compression. Moreover, the diagonal $e\,d$ passes through **A**, that is **A** d is zero. *The weight* **P** *causes no strain in* **A** d, and for this one position of **P**, **A** d might be omitted from the structure. The reason is again obvious. The point of inflection i_1 *coincides* with the apex under **A** d. Since at i_1 the moment of rupture is zero, if the flange **A** d were *cut* there would be no tendency to motion. We have at i_1 the shearing force only, **A B**$'$ giving the strains in the diagonal $e\,d$ and $d\,c$. The upper flange **A** b', we see again, is in *tension*, which is also shown by its lying to the *left* of **A**.

Thus we have the strains in every piece by a very simple construction for any position of **P**, without any calculation whatever. The method in this case is purely graphical. We have only to find the points of inflection and then proceed as above.

Fig. 80 (*b*) gives the strains for the same girder and position of weight **P**, merely *supported at the ends*. For this case **P D** in Fig. 80 not only passes through **D**, but **P O** also passes through the upper left-hand corner at **C**. Hence **A B** will be *less* than **H M**, and **A B**$'$ greater than **P H**. Moreover, the end lower flanges **B** a and **B** m no longer act, and must be removed. Starting now with the reaction **B**$'$ **A** [Fig. 80 (*b*)], we go along to the weight, from which point at $h\,k$ we go *back* towards the force line, and the reactions are such that the last diagonal must pass exactly through **B**, just as in Fig. (*a*) $e\,d$ passed through **A**, because the points of inflection or zero moments are now at the *ends* **C** and **D**. A careful comparison and study of the two cases and their points of difference will be advantageous to the reader.

115. Counterbracing.—The objection may arise that the above method applies only to a system of bracing such as represented in the Fig., where the diagonals take *both* compressive and tensile strains. In case, as in the Howe or Pratt Truss, for instance, we had vertical pieces as also *two* diagonals in each panel, then the strain in any diagonal as $m\,l$ and flange as **B** m, even if found, are apparently in equilibrium with *three* pieces, viz., $k\,l$, **B** k and a vertical strut or tie at the intersection of these pieces. Hence, having only two known strains and *three* to be determined, the method would seem to fail, as any number of polygons may be constructed with sides parallel to the forces, and hence the problem is indeterminate (Art. 9).

Now in any framed structure of the above kind, the counter ties are inserted to prevent the deforming action of the *rolling load only*. For the *dead load* but *one* system of triangulation is required, and the strains in every piece due to this dead load can therefore easily be determined.

We have then only to determine the strains in the *same pieces* due to the rolling load also. If now in any diagonal the strain due to this rolling load *exceeds* the constant strain due to the dead load, and is of opposite character, and if the diagonal *is to be so constructed as to take but one kind of strain*, then a counter diagonal must be inserted in that panel, and proportioned to this excess of strain only. For instance, if a diagonal takes only the compressive strain (a condition which is easily secured in practice) due to the dead load, and the live load *would* cause in that diagonal a tensile strain, then the *excess* of this tensile strain over the constant compressive strain due to the dead load must be resisted by a *counter* diagonal, which also takes *compressive strain only*. The method is precisely the same as by calculation (see Stoney and other authors on the subject), and we only notice the point here, as in all our examples we have taken a single system of triangulation only—a system which, we may here remark in passing, has many advantages, and is worthy of more general attention * than it has hitherto obtained.

[See also on this point Art. 10 of Appendix.]

116. Beam fixed horizontally at one end, supported at the other—Supports on Level.—In this case, equation (1), Art. 112, becomes for left end fixed, since $n = 1$, $i_2 = 0$, $i = 0$,

$$x = \frac{i_1\, a\, l}{(i_1-a)\, l - i_1\, a}.$$

But for this case the distance of the point of inflection from the fixed end is

$$i_1 = \frac{(2\, l-a)\, l\, a}{(2\, l-a)\, a + 2\, l^2}.\dagger$$

---

* See "*A Treatise on Bracing.*" By *R. H. Bow*. *D. Van Nostrand, publisher*.

† The values of the distance of the inflection points which we assume above as known, may easily be deduced by the theory of elasticity. See Supplement to Chap. VII., Arts. 16 and 19. See *Wood, Strength of Materials; Bresse, Mécanique Appliquée;* or other treatises on the subject.

Inserting this value of i_1 in the value for x above, we have
$$x = -\frac{l\,(2\,l-a)}{5\,l-a}$$
for the distance of the *inflection vertical* to the left of the left-hand support, which is supposed fixed.

Now this is the equation of an hyperbola, as shown in Pl. 21, Fig. 81, whose vertex is at f, the distance $\mathbf{A}f$ being $2\,l$, whose assymptotes are respectively parallel and perpendicular to the span, the perpendicular distance of $\mathbf{E}$ above the span $\mathbf{AB}$ being $\frac{5}{2}\,l$, and which intersects $\mathbf{AB}$ at $\frac{2}{5}\,l$ from $\mathbf{A}$. The ordinate $d\,e$, $\mathbf{A}\,d$ being equal to l, is $\frac{1}{3}\,l$. The diameter passes through $\mathbf{E}$ and f, and $\mathbf{E}\,f$ is, therefore, the semi-transverse axis. The hyperbola can, therefore, be easily constructed. We need only to construct that portion between $\mathbf{AB}$ and the point e.

The construction for the point of inflection i_1 is, therefore, simply as follows:

Lay off $\mathbf{A}\,h$ vertically upwards and equal to the distance of the weight $\mathbf{P}$ from $\mathbf{A}$, and draw the horizontal $h\,b$ to intersection b with the curve. Now make $\mathbf{A}\,a = l$ and draw $a\,b$ to intersection c. Draw $b\,\mathbf{A}$ to intersection $\mathbf{P}$ with weight, and then $\mathbf{P}\,c$ intersects $\mathbf{AB}$ at the point of inflection i_1. Decomposing $\mathbf{P}$ along $\mathbf{PB}$ and $\mathbf{P}\,c$, as in Art. 114, we have at once the *reactions* at $\mathbf{A}$ and $\mathbf{B}$. Here also we see that, by a construction purely graphic and abundantly exact, we can find the *inflection point* and the reactions.

The method detailed in Art. 114 can then be applied to determine the various strains in the different pieces. It is unnecessary to give an example, as the process is precisely similar. We have simply in this case to start with the reaction at the free end $\mathbf{B}$ and follow it through. Observe only that, as this reaction must be *less* than for a girder with free ends for the same position of $\mathbf{P}$, the point h will lie *nearer* the force line $\mathbf{B}'\mathbf{AB}$ (Fig. 80, b), hence $l\,m$ will not pass exactly through $\mathbf{B}$, but will lie to the *right* of it, giving thus a *reversal* of strain in the flanges, as by reason of the inflection point should be the case.

Instead of constructing the hyperbola, we may calculate its ordinates from the equation for x above, for different values of a.

Thus, for

$a=0 \quad a=\tfrac{1}{4}l \quad a=\tfrac{1}{2}l \quad a=\tfrac{3}{4}l \quad a=l$
$x=-0.4\,l \quad x=-0.388\,l \quad x=-0.375\,l \quad x=-0.357\,l \quad x=-0.333\,l$

This will be sufficient to construct the curve in any given case. The *inflection vertical* moves, therefore, between the narrow limits of $x = \tfrac{3}{8}l$ and $x = \tfrac{1}{3}l$, or within $\tfrac{1}{15}$th of the span, as the load passes from **A** to **B**.

Inasmuch as all that is needed for the determination of the strains in the various pieces are the *reactions at the supports*, and (for girder fixed at *both* ends) the *moments* at the supports also, and as the formulæ for the two cases above are very simple, we may determine these quantities at once by interpolation of the given distance of the weight **P** in the formulæ, and then apply the graphical method for the strains, as illustrated in Art. 114.

Thus, for a *horizontal beam fixed at both ends*, we have for the moment at the left support **A**,

$$M_A = \frac{Pa}{l^2}(l-a)^2.$$

At the right support **B**,

$$M_B = \frac{Pa^2}{l^2}(l-a).$$

For the *reaction* at the left,

$$R_A = \frac{P}{l^3}(l^3 - 3a^2l + 2a^3).$$

For the *reaction* at the right,

$$R_B = \frac{P}{l^3}(3a^2l - 2a^3).$$

In the case of a horizontal beam fixed at left end and merely resting upon the right support, we have

$$M_A = \frac{P}{2l^2}(3a^2l - 2al^2 - a^3), \qquad M_B = 0,$$

$$R_A = \frac{P}{2l^3}(2l^3 - 3a^2l + a^3), \qquad R_B = \frac{Pa^2}{2l^3}(3l-a),$$

a being always the distance of the weight **P** from the left. These formulæ are simple, and easily applied to any case.

· We may also observe that in Figs. 79 and 81 the ordinates to the lines **P** b, **P** c, and **P** c, **P B**, from **A B**, are *proportional to the moments* (Art. 110). These ordinates to the scale of distance, multipied by the pole distance to scale of force, give the moments at any point. Our construction, therefore, gives the moments also at every point, and we may thus check the results obtained by Art. 114 by the results obtained by the method of moments.

117. Approximate Construction.—It will be readily seen that the portion of the hyperbola in Fig. 81, Pl. 21, needed for our construction, is *nearly straight*. In most cases it will be practically exact enough to lay off $\frac{2}{3} l$ to the left of **A**, and $\frac{1}{4} l$ also to the left of **A** at a vertical distance equal to l, and join the two points thus obtained by a straight line. This line can be taken instead of the curve, and the construction is then the same as above. The error due to thus considering the curve as a straight line is greatest for a weight in the middle of the span, where it does not exceed $\frac{1}{100}$th of the span for the position of the *inflection vertical*, and diminishes from the centre both ways.

118. Girder continuous over three Level Supports— Draw Spans.—This case is perhaps of the most frequent practical occurrence, and an accurate and simple method of solution is therefore very desirable.

In the first place, the formulæ for the reactions are very simple and easy of application. Thus, for left end support **A**, the load being in the *second span*, or to the right of the middle support **B**,

$$R_A = \frac{P}{4 l^3} (3 a^2 l - 2 a l^2 - a^3);$$

for the reaction at middle support,

$$R_B = \frac{P}{2 l^3} (2 l^3 + a^3 - 3 a^2 l);$$

for reaction at right end **C**,

$$R_C = \frac{P}{4 l^3} (2 a l^2 + 3 a^2 l - a^3);$$

where a is always the distance of the weight **P** from the middle support.* We are therefore already in a position to solve completely the case under consideration. We have only to

---

* As already remarked, the development of the formulæ assumed in this chapter must be sought for in special treatises on the subject. We assume them as known, and then apply them graphically as above.

See also Supplement to Chap. XIII.

find the reactions and follow them through by the method of Art. 114.

From the above reactions we can, however, easily determine the distance of the *inflection point*. This will, of course, be found only in the loaded span, at a distance from the middle support.

$$x = \frac{l\,a\,(2\,l - a)}{4\,l^2 + 2\,a\,l - a^2} = i.$$

We can find the values of x corresponding to different values of a, and thus plot the curve for the inflection points. Thus, for

$a = 0 \quad a = \tfrac{1}{4}l \quad a = \tfrac{1}{2}l \quad a = \tfrac{3}{4}l \quad a = l$
$x = 0 \quad x = \tfrac{7}{71}l \quad x = \tfrac{9}{19}l \quad x = \tfrac{15}{18}l \quad x = \tfrac{1}{4}l.$

This curve being drawn for any particular case, we can easily find the position of the inflection point for any given value of a, and hence the *reactions*, and then find the strains in the various pieces.

Thus, in Pl. 21, Fig. 82, the curve **B** e d being drawn, we can at once find the inflection point i for any position a of the weight **P**. We have simply to make **B** b = a and draw $b\,e$. $b\,e$ is the distance of the point of inflection from **B**. We can now, as explained above, draw *any* line as **P** i, and then **P C** and **A** h. The ordinates to the broken line **A** h **P C** from **A C**, to the scale of distance, multiplied by the *pole* distance **H** to scale of force, will give the moments at any point. Moreover, **H E** is the *shear* at **B**. **E** a is the *reaction* at **B**, **H** a the reaction at **A**, and **H P** the reaction at **C**. The reactions at **B** and **C** are, of course, positive or *upwards*, that at **A** negative or downwards. Hence **E** a − **H** a + **H P** = **P**, as should be.

The value of x for the inflection vertical is by Art. 112

$$x = \frac{i\,a\,l}{(i - a)\,l - i\,a},$$

or, substituting the value of i above,

$$x = -\frac{a\,l\,(2\,l - a)}{2\,l^2 + 5\,a\,l - 2\,a^2}.$$

Since, therefore, in this case the value of x is no simpler than that for i given above, it will be preferable to plot the first curve directly as represented in Fig. 82.

119. Approximate Construction.—In practice it will be

found abundantly accurate to assume the curve for i between the required limits, as a *parabola* whose equation is $i = x = \frac{a(2l-a)}{5l}$. The greatest error for $a = \frac{l}{4}$ will then be about $\frac{1}{100}$ of the span, and decreases both ways to $a = o$ and $a = \frac{l}{2}$. From $a = \frac{l}{2}$ to $a = l$ the parabola coincides closely with the true curve. The difference for $a = \frac{3}{4}l$ is only $\frac{1}{500}l$, and we have, therefore, a very simple practical construction for both reactions and moments. We have only (Fig. 82) to erect a vertical at the centre support **B** and make it equal to l, and then construct a parabola passing through **B** whose ordinate $cd = \frac{1}{5}l$. The horizontal ordinates to this parabola for any vertical value of a, give the distance out from **B** of the inflection points. For the load in first span **A B**, of course this parabola lies on the other side of **B** c, and $b\ e$ is laid off to the left. The remainder of the construction is as in Art. 118 for the reactions and moments. When great accuracy is required, we can find the reactions from the equations of Art. 118. In any case, the reactions being given, we can follow them through the structure by the method of Art. 114, and thus determine the strains in every piece due to every position of each apex load. A tabulation of these strains will then give by inspection the maximum strain in any piece due to the live load. *All* the weights taken as acting simultaneously will then give the strains due to uniform total live load. The strains due to dead load will be multiples or sub-multiples of these. Thus if total live load causes, say, 100 tons compression in a certain piece, and if the dead load is $\frac{3}{2}$ of the live load, then we shall have $\frac{3}{2}$ of $100 = 150$ tons compression in the same piece *due to the dead load* alone. If now the live load causes a maximum *tension* in the same piece of 200 tons, then the piece must be made to resist both tensile strain of $200 - 150 = 50$ tons and compressive strain of $150 + 100 = 250$ tons. If a diagonal, the *counter tie* is strained 50 tons, while the maximum strain on the diagonal is 250 tons compression. It is only necessary, therefore, to find the strains due to each weight of the live load. From the tabulation we can, then, by means of the *ratio* of the dead to live load, find the strains due to dead load alone, and then by a comparison of the two find the maximum compressive and tensile strains. If the maximum strains due to live load are of opposite kind, but *less* than the *constant* strains due to dead load, we shall need no counterbracing. The resultant strains will then always be of the character given by the dead load. If greater, we must counterbrace accordingly. The process is the same as by the methods of calculation, and the reader may refer to Stoney—Theory of Strains—for illustrations.

120. The "Tipper," or Pivot Draw, with secondary central Span.—We have said that a pivot draw may be considered as a beam continuous over three supports. In practical construction this statement needs some modifications which deserve special notice. Thus practically that portion of the beam over the central support forms a *short secondary* span **D D** [Fig. 83, Pl. 22] the reactions at the supports **D** and **D** being *always equal* and of the same character. If a weight acts, say, on the first span **A B**, and the beam itself is considered without weight, the end **C** must be held down, that is, the reaction there is *negative*. Now as the weight **P** deflects the span **A B** (Fig. 83), it causes one secondary support **D** to sink, and the other to rise an equal amount. In practice **D** and **D** may be the extremities of the turn-table, and the reactions are then evidently different from those given by the formulæ of Art. 118.

If in this case we take a as the distance of the weight **P** from the *left* support **A**, the reaction for load in **A B** will be given by the following formulæ:

Where the ratio $\frac{a}{l} = k$, a being the distance of the weight **P** from the *left support* **A** (for load in the span **A D**), $l =$ span, **A D** = **D C**, and $n\, l =$ span **D D**, and where the constant $(4 + 8\, n + 3\, n^2)$ is put for convenience $=$ **H**, then

$$R_A = \frac{P}{2\,H}\left[\, 2\,H - (10 + 15\, n + 3\, n^2)\, k + (2 + n)\, k^3 \,\right]*$$

$$R_D = R_D = \frac{P}{2\,H}\left[\, (6 + 9\, n + 3\, n^2)\, k - (2 + n)\, k^3 \,\right]$$

---

* See Supplement to Chap. XIII., Art. 6.

$$R_c = \frac{P}{2 H}\left[(2+n) k^3 - (2+3n+3n^2) k \right]$$

These reactions, it will be observed, when added together $R_A + 2 R_D + R_C$ are equal to P, as should be the case.

By the application of these formulæ, which are for any particular case by no means intricate, we can find the reactions at A and C as also at D or D; and then starting, say, from A, can follow the reaction there through the frame by the method of Art. 114. A negative reaction indicates that the support tends to *rise*, and unless more than counterbalanced by the positive reaction due to uniform load, the end where this negative reaction occurs must be latched down.

121. Supports in Pivot Span are not on a level—Reactions for live load, however, are the same as for level supports.—The three supports of a pivot span should *not* be on a level. It is evident that if this were the case, the first time the draw is opened the two cantilevers *deflect* and it would be difficult to shut it again. The centre support should therefore be raised until the reactions at the end supports are zero, that is, until they *just bear*. The centre support is then raised by an amount *equal to the deflection of the beam when open, due to the dead load*. Even when shut, then, there are no reactions at the end supports except when the moving load comes on. Now this being the condition of things, it may seem strange to assert that *these* reactions are *precisely the same* as for three *level supports*, and yet such is the fact. If the beam originally straight were *held* down at the lower ends by negative reactions, then the reactions would have to be investigated for supports out of level, and a load would diminish these negative reactions, or might even cause them to become positive. But such is not the state of things. The end reactions are *in the beginning zero*, and any load gives, therefore, *at once* positive reaction at its end support. This positive reaction is just *what it would be for the same beam over three level supports*.

An analytical discussion of the case would be out of place here, but assuming the expression to which such a discussion would lead us, we may show that this is so.

Thus, for a beam over three supports A, B, and C, *not* on a level, c_1 being the distance of A below B, and c_2 the distance of C below B, the modulus of elasticity being E and the moment of inertia **I**, we have for the moment at the centre support **B** due to any number of weights in both spans,

$$4 M_B l = - \left[\frac{c_1 + c_2}{l}\right] 6 E I - \frac{1}{l} \Sigma P a (l-a)(l+a)$$
$$- \frac{1}{l} \Sigma P a (l-a)(2l-a),^*$$

a being always measured from the left support.

Now in this expression the last two terms are precisely the same as for supports on a level; the influence of the different levels is contained in the first term on the right only. Now by the supposition, c_1 and c_2 must be taken equal to *the deflection due to the dead load*, and the value of this term will therefore be entirely independent of the live load, which enters only in the last two terms.

A particular case may perhaps render this plainer. If a girder of two equal spans over three level supports is uniformly loaded, the reaction at an end support is, as is well known, ⅜ths of the load on one span.

Now let us take the girder over three supports *not* on a level, and from our formula above find the reaction at one end due to uniform load when c_1 and c_2 have the proper values given to them. First the dead load $p\,l$ over each span causes a deflection at each end of the two cantilevers $= \frac{1}{8}\frac{p\,l^3}{E I}$.† This, then, is the value for c_1 and c_2 in the formula. Now let us take an additional moving load of $m\,l$ over the whole beam, and with this value of c_1, c_2 find the reaction. We have from our formula

$$4 M_B l = - \tfrac{3}{2} p l^3 - \tfrac{1}{2}(p+m) l^3,$$
or $$M_B = - \tfrac{1}{2} p l^2 - \tfrac{1}{8} m l^2.$$

Now we have by moments,

$$R_A \times l - (p+m)\frac{l^2}{2} = + M_B;$$

hence, inserting the value of M_B above,

$$R_A = \tfrac{3}{8} m l.$$

---

* *Theorie der Träger:* Weyrauch. Also Supplement to Chap. XIII., Art. 3.
† Supplement to Chap. VII., Art. 13.

That is, the reaction at **A** is due to the moving load alone, as evidently should be the case, and is, moreover, *just what it should be for a girder with level supports;* viz., ⅜ $m\,l$. (See also Appendix, Art. 18, Ex. 5.)

The raising of the centre support, then, will not affect our construction for the reactions as given in Figs. 81 and 82, provided there are only three supports.

We have deemed it well thus to call special attention to the considerations of the last two articles, both on account of their practical importance and because they are not brought out clearly, nor indeed, so far as we are aware, even alluded to in any treatise upon the subject.*

122. Beam continuous over four Level Supports.—We thus see that a draw or pivot span is more properly considered as a beam of *three* spans instead of two, of which the centre span is very small compared to the end spans; it may be only two or three panels long. Moreover, we must often in practice consider the beam as a "tipper," and therefore apply the formulæ for reactions of Art. 120. If, however, by reason of the method of construction, as often happens, for instance, by the under portion of the beam coming in contact with the frame below, this *tipping* of **D D** (Fig. 83) is confined between certain limits, beyond which the supports must be considered *fixed*, it will be necessary to find the reactions as for a beam over four *fixed* supports, and determine the corresponding strains in this case also.

Comparing, then, the strains obtained each way, we take only the maximum strains from each.

The formulæ for the reactions at the *fixed* supports **A B C D** are as follows (Pl. 22, Fig. 84):

1st. Load **P** in left end span **A B** at a distance a from left support **A**, the *end* spans being $n\,l$ and the *centre* span **B C**$=l$.

We put $k = \dfrac{a}{n\,l}$, and **H** $= 3 + 8\,n + 4\,n^2$. Then

$$\mathbf{R}_A = \frac{\mathbf{P}}{\mathbf{H}} \left[\mathbf{H} - (\mathbf{H} + 2\,n + 2\,n^2)\,k + (2\,n + 2\,n^2)\,k^3 \right]$$

---

* Clemens Herschel, in his treatise upon "Continuous, Revolving Drawbridges" (Little, Brown & Co., Boston, 1875), notices this fact for the first time.

$$R_B = \frac{P}{H}\left[(3+10n+9n^2+2n^3)k - (2n+5n^2+2n^3)k^3\right]$$

$$R_C = \frac{P}{H}\left[-(n+3n^2+2n^3)k + (n+3n^2+2n^3)k^3\right]$$

$$R_D = \frac{P}{H}\left[nk - nk^3\right]$$

These reactions add up, as they should, equal to P.

In practical cases of pivot spans, we have only to consider the the outer spans; as a load in the middle span $BC = l$ rests directly upon the turn-table. The above formulæ are then all we need. For a load in the right end span the same formulæ hold good, only remembering to put now R_D in place of R_A, R_C in place of R_B, R_B in place of R_C, and R_A in place of R_D.

If, however, neglecting the particular case of pivot spans, we suppose the middle span $BC = l$ loaded, we have—a being now the distance of P from B, and k being now $\frac{a}{l}$ instead of $\frac{a}{nl}$, as above, H remaining the same.

2d. Load in BC.

$$R_A = \frac{P}{nH}\left[-(3+4n)k + 6(n+1)k^2 - (3+2n)k^3\right]$$

$$R_B = \frac{P}{nH}\left[nH + (3+4n-6n^2-4n^3)k - (6+15n+6n^2)k^2 + H k^3\right]$$

$$R_C = \frac{P}{nH}\left[(2n+6n^2+4n^3)k + (3+9n+6n^2)k^2 - H k^3\right]$$

$$R_D = \frac{P}{nH}\left[-2n k - 3 k^2 + (3+2n)k^3\right]$$

These reactions should also add up to P, as is the case. The number n may be taken at pleasure, so that the end spans may be as much larger or less than the centre spans as is desired.

H, P and the quantities in the parentheses, it will be observed, are for any given case, constants which may be determined and inserted once for all.

We have, then, only to insert the values of k for different positions of the load P. Thus the equations for any particular case are very simple and easy of application.

123. Construction.—We may, if desired, apply our method of construction to the determination of the reactions. Thus from the above reactions we may easily determine general expressions for the inflection points. For the case of a load in $CD = nl$ (Pl. 22, Fig. 85), we have, when i is the distance of the inflection point from C,
$$- R_D \times (nl - i) + P(a - i) = 0;$$
whence
$$i = \frac{Pa - R_D nl}{P - R_D}.$$

For the inflection point distant i from **B** in the unloaded span,
$$R_A(nl + i) - R_B i = 0;$$
hence
$$i = \frac{R_A nl}{R_B - R_A}.$$

For the second case of load in $BC = l$, we have for the inflection point between **B** and **P**
$$- R_A (nl + i) + R_B i = 0, \text{ or}$$
$$i = \frac{R_A nl}{R_B - R_A}.$$

For the point between **P** and **C**
$$- R_D (nl + i) + R_C i = 0, \text{ or}$$
$$i = \frac{R_D nl}{R_C - R_D}.$$

The insertion of the proper values of the reactions for each case, as given above, will easily give general expressions for the inflection points, which the reader may, if desired, deduce for himself.

Our construction is, then, as follows [Pl. 22, Fig. 84]:

1st Case. Load in C D.

Having found i_1, draw a line at *any* inclination, as $c_1 d$ through i_1, intersecting **P** at d, and the vertical through C at c_1. Then lay off B i and draw d D, $c_1 b$ and b A.

Make $dc = $ **P** by scale, and c D drawn parallel to $c_1 d$ then gives the pole distance **H**. The ordinates, then, to the broken line A b c_1 d D taken to scale of distance, multiplied by **H** to scale of force, give the moments at every point. Moreover, **H** d is the reaction at **D**. Draw D b parallel to $c_1 b$, then $c b$ is the reaction at **C**. In like manner $a b$ is the reaction at **B**, and **H** a the reaction at **A**. The moment at **C**, and reactions at C and

D, are *positive*. Reaction and moment at B *negative;* reaction at A positive; as a little consideration of what the *curve of the deflected beam* must be, will show. The *shear* at C is, therefore, $+H\,a - a\,b + b\,c = +H\,c$. The *shear* at B is $-H\,b$ or $+H\,a - a\,b$, and so on. The *shear* being always given by the segment *between* H and lines parallel to A b, $b\,c_1$, $c_1\,d$ and d D.

2d Case.—Pl. 22, Fig. 85.—Load in B C.

Having found the distance B i_1 from the equation for this distance of the point of inflection above, we lay off B $c_1 =$ B i_1 and thus draw c_1 E at an angle of 45°. Finding then the value of C i_2 from its equation above, we can draw E c_2 and then c_2 D and c_1 A. The construction is then the same as before. Thus H is the pole distance, H c the negative reaction at D, H b the negative reaction at A, and $c_1\,b$, c E the positive reactions at B and C. The *shear* at B is H c_1, etc. Thus the outer forces are completely known for a weight at any point. It will, however, in general, in practice, be found more satisfactory to use the formulæ for the reactions which we have given than to find these reactions by the above construction.

We shall now illustrate the preceding principles by an example taken from actual practice.

124. Draw Span—Example.—In Pl. 22, Fig. 86, we have given to a scale of 20 ft. to an inch the elevation of one of the trusses of the pivot draw over the Quinnipiac River at Fair Haven, Conn.*

Length of span A B $= 89.88$ ft. B C $= 21.666$ ft., divided into seven panels of 12.84 ft. and two of 10.833 ft. respectively.

Height at B and C, 16 ft.; at A and D, 12.1 ft. Diagonal bracing as shown in Fig. Line load 9 tons per panel.

In this case $n = \dfrac{21.666}{89.88}$ or $n = 0.24106$, hence the equations of Art. 120 become

$$A = P\left(1 - 1.1298\,\frac{a}{l} + 0.1836\,\frac{a^3}{l^3}\right)$$

$$B = C = P\left(0.6836\,\frac{a}{l} - 0.1836\,\frac{a^3}{l^3}\right)$$

---

* Designed and erected by Clemens Herschel, C.E., and probably the only structure of the kind in this country for which the strains have been accurately and thoroughly determined. For the above data I am indebted to M. Merriman, assistant engineer in charge.

$$D = -P\left(0.2374\frac{a}{l} - 0.1836\frac{a^3}{l^3}\right).$$

Now $\frac{a}{l}$ is $\frac{1}{7}, \frac{2}{7}, \frac{3}{7}, \frac{4}{7}$ ths, etc., according to the position of the weight at 1st, 2d, 3d apex from end. So also $\frac{a^3}{l^3}$ is $\frac{1}{343}, \frac{8}{343}, \frac{27}{343}$, etc. The above equations for the reactions, then, may be written

$A = P (1 - 0.1614\, b + 0.000535\, b^3),$

$B = C = P (0.09766\, b - 0.000535\, b^3),$

$D = -P (0.03391\, b - 0.000535\, b^3),$

where b has the values 1, 2, 3, 4, etc., for P_1, P_2, P_3, P_4.

Thus, if we wish the reactions due to a weight P_4 of 9 tons at the fourth apex, as shown in Fig., we have only to make $P = 9$ and $b = 4$, and we find at once $A = 3.498$ tons, $B = C = 3.207$ tons, $D = -0.912$ tons. The sum of all these reactions exactly equals P, as should be.

The middle supports are supposed raised by an amount equal to the end deflections of the open draw, therefore the strains due to *dead load* are easily found, as in the "braced semi-arch," Art. 9.

The reactions due to live load, according to Art. 121, will not be affected by this raising of the supports.

To find the strains due to P_4, we draw the force line $E_2 F$ [Fig. 86 (a)] by laying off $P_4 = 9$ tons down from F to F_1, then $F_1 E_2$ *downwards* equal to the *negative* reaction at D, viz., -0.912 tons. Then from E_2 lay off *upwards* $E_2 E_1 =$ to *positive* reaction at $C = +3.207$ tons. Then $E_1 E =$ reaction at $B = +3.207$ tons, and finally E F equal to reaction at $A = +3.498$ tons, which should bring us back exactly to point of beginning F, since the reactions and the weight P must be in equilibrium. [*Note.*—When we wish to begin at the left end of the frame, it is best, as in this case, to lay off the reactions in order, commencing at the *right.*] We have taken the scale of force 4 tons per inch.

The weight P_4 acts upon the triangulation drawn full in the figure. Using now the notation of Art. 114, and representing all the space *above* the truss by E, all *below* by F, we have at A the reaction E F [Fig. 86 (a)] in equilibrium with E 1 and F 1, and drawing parallels to these lines from E and F, we find the strain in F 1 = 3.54 tons *tension*, and E 1 = 5.1 *compression*.

So we go through the truss and find the strains in every piece. Heavy lines in the strain diagram denote compression. We see at once that for this position of the weight, *all* the upper flanges in span **A B** are compressed, the last lower flange **F** 7 is also compressed, and all the other lower flanges are in tension. At the point of application of the weight P_4, the two diagonals 3 4 and 4 5 are in tension, and either side they alternate in strain as far as **C** or diagonal 8 9. Diagonals 8 9 and 9 10 are both tension, and then the strains alternate to support **D**. All the upper flanges of the right half are tension and increase towards the middle. All the lower are compression and likewise increase towards the middle.

If we go through the whole truss from **A** to **D**, the last diagonal 1 5, 1 6 should evidently pass exactly through E_2, thus *checking* the accuracy of the construction. The diagonal 6 7 *crosses* the force line, thus causing the strain in the lower flange to change from tension in F_1 5 to compression in F_1 7. The point of inflection, therefore, falls to the *right* of diagonal 5 6.

The reaction at **B** diminishes greatly the strain which would otherwise take effect in 7 8 and **E** 8; while the reaction at **C** *reverses* the strain which would otherwise take effect in 9 10 and diminishes **E** 10. We recommend the reader to follow through carefully the strain diagram, Fig. 86 (*a*).

A series of figures similar to Fig. 86 (*a*) (in the present case seven separate figures) will give completely the strains due to the rolling load. A table may then be drawn up containing the strains due to *dead load*, and the maximum strains due to live load in every piece, and the total maximum tension and compression in every piece may then be found. [*Compare Art.* 12, *Fig.* 7.]

For the supports *fixed*, instead of **B** and **C** tipping, the process is precisely similar, except that we have to make use of the formulæ of Art. 122. The reaction at **A** will then be somewhat *less* than in the present case; the inflection point is therefore found *further* from the right support **B**; it may be even to the *left* of diagonal 5 6, in which case (see Fig. 86, *a*) we should have *tension* in upper flange **E** 6. The reaction at **B** would then be still positive, but greater than **E E**$_1$, while **C** would be *negative* and no longer equal to **B**, and **D** would be *positive*. We should thus have 7 8 *tension* and **E** 8 tension; **F** 7, as before, compression, 8 9 compression, and 9 10 compression, and **E** 10 compression; while **F** 9 would be tension. From 9 10 to the right the diagonals would alternate in strain, the compressed upper flanges, as also the tensile lower flanges, would diminish towards **D**, and the last diagonal should pass exactly through new position of E_2, thus closing the strain diagram and checking the work. The reader will do well to construct the diagram.

The strains should be found for *both* cases, and the maximum strains taken from each, which, compared with the permanent strains due to the dead load, will give the total maximum strains.

We have taken for convenience of size too small a scale for the frame to ensure good results. With a large and accurately constructed *frame diagram*, dealing as we do with only single weights, and consequently small strains, the above *force* scale of 4 tons per inch would give very accurate results.

If the strains due to uniform load (no end reactions) are found by addition of the strains for each apex load diagramed separately, the same scale may be employed; but if all the loads are taken as acting together (Fig. 5, *b*), a smaller scale for strains will have to be adopted, as the *force line* will otherwise be too long. [See Art. 16 of Appendix for the method of calculation.]

125. Method of passing direct from one Span to next. —By inspection of Fig. 86 we see that we might find the strains in the intermediate span **B C** *without* first going through the whole of **A B** or **C D**. Thus, if we knew the *moment* at **B**, this moment, divided by depth of truss at **B**, would give the strain in flange **F** 7 for the system of triangulation indicated by the full lines. If then we knew also the *shear* at **B** = **P** − **A** − **B** = $E_1 F_1$ (Fig. 86, *a*), we could at once lay off F_1 7 and $E_1 F_1$ (Fig. 86, *a*), and then proceed to find **E** 8 and 7 8, just as before. In the same way the moment at **C**, divided by height of truss at **C**, would give us strain in **F** 9, and with *shear* at **C** = **P** − **A** − **B** − **C** = $E_2 F_1$ = **D**, we could find **E** 10 and 9 10, as before. As we know already, a load anywhere upon a *beam* causes *positive* moments at a fixed end—*i.e.*, makes upper flange over support *tension* and lower flange *compression*. But as we see from the last case, owing to the triangulation, the last upper flange may *also* be compression (see **E** 6 in Fig. 86) if the inflection point lies between diagonal 5 6 and the support. The known moment gives, then, the character of the strain *only for that flange which does not meet a diagonal at the support*. The moment at **B**, therefore, being positive, gives us compression here in lower flange, because, for the system of triangulation corresponding to the weight, that flange does not meet a diagonal at **B**. For a weight upon the other system of triangulation (dotted in Fig.), the same moment would give us the *tension* in **E** 7. The construction assumes equilibrium between **F** 7, 7 8, and **E** 8, and the shear at **B**; that is, between the pieces cut by an ideal section to the right of **B** through the truss and the shear at that section. That this is so is shown by the strain diagram, since there we see that the strains in these pieces form a closed polygon with the shear at $B = E_1 F_1$. This must evidently be so if these are the only pieces cut by such a section, since then the horizontal components of the strains in these pieces must balance, and the resultant vertical component must be equal and opposite to the shear.

It is important to know *which side* of $E_1 F_1$ to lay off F_1 7, since, if we had laid it off in this case to the right, we would have obtained a very different value for E_1 8. For this purpose we have only to suppose the strain in the flange (either upper or lower, as the case may be) to be applied at the point of junction or apex of the other two pieces, and then lay it off *in the direction with reference to that apex corresponding to the known character of its strain*. The direction of the shear is always known from the reactions.

Thus in our Fig. the shear between **B** and **C** acts *down* from E_1 to F_1, because P_4, which also acts down, is greater than the sum of the upward reactions at **A** and **B**. The strain in **F** 7 is also known to be compressive, and therefore, in following round the strain polygon commencing from E_1 to F_1, it must act *towards* apex at 7. We must, therefore, lay it off to the left of $E_1 F_1$. In similar manner, for the other triangulation, the strain in flange **E** 7 is, in span **B C**, in equilibrium with 7 8 (dotted diagonal) and **F** 8 and shear $E_1 F_1$, and is, moreover, known to be tension. Consider it acting then at **B**; and then, since it is tension, we go round the polygon from E_1 to F_1, and then to the *right* of $E_1 F_1$, or *away* from **B**, the point at which it is supposed to act.

Now for the case of the "*tipper:*" the reaction at **D**, and therefore the moment at **C**, is also positive. The lower flange **F** 9 is therefore compression, or for the dotted system of triangulation **E** 9 is tension. The shear to the left of **C**, $E_1 F_1$ acts *down*, since $-P + A + B = -E_1 F_1$. Consider **F** 9 acting at apex 9, and then, since it is compression, it must act *towards* 9 (from right to left), and passing down then from E_1 to F_1, we must lay off **F** 9 to the *left* of $E_1 F_1$. For similar reasons, for the other system, **E** 9 must be laid off to the right.

For *fixed* supports **B** and **C**, the moments alternate from **B**, and the moment at **C** is therefore negative—that is, gives compression above and tension below. Flange **F** 9, for the system of triangulation of **P**, would then be *tension* instead of compression, as above; **P** will, however, still be greater than **A** + **B**, and hence the shear is to be laid off *down*, and **F** 9 must be laid off to the *right*.

If, then, it were required to find the strains in the span **B C**, preceded and followed, it may be, by many others, it is sufficient to know the moment and shear at one support. We can then commence and continue the strain diagram, without being obliged to go off to a distant free end and trace all the strains through till we arrive at the span in question.

126. Method of procedure for any number of Spans.— The general method of procedure which we advise, is then as follows. Let us take any number of spans, say *seven* [Pl. 22, Fig. 87], and let it be required to find the maximum strains in the span **D E**. It is not, as we have just seen, necessary to commence at the extreme end **A** or **H**, and follow the reaction there through, from span to span, till we arrive at **D**. As we have seen from the preceding Art., we may start directly from **D**, provided we know the moment and shear there. Now, since a load in any span causes *positive* moments and reactions at the two ends of that span, and since either way from these ends the moments and shear at the other supports *alternate* in character [Art. 102], any and all loads in **A B** cause positive moments and reactions at **D**. So also for loads in **C D** and in **F G**. Loads in **B C**, **E F** and **G H**, on the other hand, cause *negative* moments and reactions at **D**. [See Fig. 87.]

To find the maximum positive moment and shear at **D** due to the other spans, we must then suppose the method of loading shown in Fig. 87 (*a*). For the maximum negative moment and shear at **D**, we have the system of loading shown in Fig. 87 (*b*).

Now these two moments and shears being once known, we can find by diagram and tabulate the respective strains in every piece of the span **D E**. Thus dividing the moment at **D** for either case by the height of truss, we have at once the strain in either upper or lower flange at **D** depending upon the system of triangulation as explained in Art. 125. With this strain and the shear at **D** properly laid off to scale, we can commence the strain diagram precisely as though we had traced all the loads through from the extreme end **A** or **H** to **D** or **E**.

We must next find and tabulate the strains in **D E** due to each apex load in the span itself, and for this we must know to begin with the moments and shears for each separate load.

[*Note.*—Distinguish carefully between *shear* and *reaction* at a support. The shear at **D**, or at a point just to right of **D**, is the algebraic sum of all the reactions and weights between that point and **A**. See also Fig. 84 (Art. 123), where the reaction at **B** is $-b\,a$, but the *shear* at **B** is $-b\,a + \mathbf{H}\,a = -\mathbf{H}\,b$. So also the reaction at **C** is $+b\,c$, but the *shear* at **C** is $+b\,c - b\,a + \mathbf{H}\,a = \mathbf{H\,C}$, etc.]

Conceiving now that we have found and tabulated the strains due to the first and second systems of loading as shown in Fig. 87, and also the strains for each load **P** in **D E**, the sum of these strains will give the strains due to live load over the whole length of girder, and taking the proper proportion of these, we shall have the strains due to the dead load. Combining then these strains with those first found, we can easily find the total maximum strains which can ever occur in **D E**.

Such is the method of procedure we advise for many spans, in order to find the maximum strains in any one span. The method is not, however, *strictly* correct, and does not give the theoretical maximum strains in the span required to be solved. The reason is obvious. The method gives correctly the maximum moment and shear *at the left support* of the span in question, but does not give the true maximum at *other* points of that span.

Thus, as will be seen by reference to the table of the following Art., the maximum tension in **A** e occurs really for *all* the spans, except the one in question, loaded, while according to the above method we should have taken only the 1st, 3d, and

6th spans loaded, and should have considered the other "exterior" spans as causing compression in the upper flange. This, although true for the end flanges, and indeed all upper flanges as far as the inflection points, is not strictly true for the flanges between the inflection points. The error is not great, more especially as even the strains thus found can never be realized in practice. It is exceedingly improbable that moving trains will ever in practice occupy just such positions as those supposed. We recommend, then, the above method as giving safe and reliable results, while it makes the table much smaller and economizes much labor.

In order to find the *true* maximum strains, we must find the strains in every piece of the span in question due to load over *each* exterior span. We thus have a column in our table for each of these spans, or, in the present case, *six* columns, instead of only two, as by the above method. We leave the reader to adopt this method or not, as he chooses, and shall content ourselves with illustrating by an example the method of finding the *true* maximum strains. This method, though more laborious than the above, is by no means more difficult. We have only to consider the effect of each exterior span by itself, instead of the combined effect of several.

127. Example.—Let us take, as an illustration of the preceding, the girder shown in Fig. 87, of seven equal spans, and seek the maximum strains which can ever occur in the middle span **DE**. Let Fig. 88, Pl. 23, represent the span **DE**—length 80 feet, divided into 4 panels; and let the live load per panel be 40 tons,* the uniform load being *half* as much, or 20 tons per panel. Height of truss, 10 ft.

Now the quantities, which for the present we must suppose known or already found, we give below. How these quantities are found will be the subject of the next chapter.

For 1st span loaded........moment = + 61.55 ft. tons at **D**.
 shear at **D** = + 0.97 tons.
For 2d span loaded........moment at **D** = − 184.65 ft. tons.
 shear at **D** = − 4.74 tons.
For 3d span loaded........moment at **D** = + 677.05 ft. tons.
 shear at **D** = + 10.67 tons.

---

* A very great load: half the resulting strains would give more nearly the strains in a single truss.

CHAP. XII.]   METHODS COMBINED.   195

For 5th span loaded........moment at $D = -181.38$.
                           shear at $D = -10.67$ tons.
For 6th span loaded........moment at $D = +49.46$ ft. tons.
                           shear at $D = +2.91$ tons.
For 7th span loaded........moment at $D = -16.48$ ft. tons.
                           shear at $D = -0.97$ tons.

Also for the loads in the span DE itself:

For the 1st load P_1........moment at $D = +158.92$ ft. tons.
                             shear at $D = +36.17$ tons.
For the 2d load P_2........moment at $D = +271.96$ ft. tons.
                            shear at $D = +25.88$ tons.
For the 3d load P_3........moment at $D = +202.36$ ft. tons.
                            shear at $D = +14.16$ tons.
For the 4th load P_4........moment at $D = +62.88$ ft. tons.
                             shear at $D = +3.82$ tons.

In Fig. 88 we have found by diagram the strains due to P_3. [For notation, see Art. 114.] We lay off to scale the shear 14.16 upwards, since it is positive, and then, since the moment 203.36 at D is positive, and hence the strain in Aa must be tension, we lay off $Aa = \dfrac{203.36}{10} = 20.3$ tons to the right of BA (Art. 125). With BA and Aa thus given, we can rapidly and accurately find all the other strains. Thus from our diagram we have, representing tension by minus and compression by plus:

$Aa = -20.4$ $Ac = +8.0$ $Ae = +36.4$ $Ag = +24.4$
$Ak = -27.2$
$Bb = +6.0$ $Bd = -22$ $Bf = -50.8$ $Bh = +1.2$ tons;

and for the diagonals:

$ab = +19.6$ $bc = -19.6$ $cd = +19.6$ $de = -19.6$, etc.

Heavy lines in the diagram represent compression. In a manner precisely similar, we can find the strains due to each of the other "*interior*" weights, as also to the "*exterior*" loads upon the other spans. Suppose all these strains thus found. Then the method of tabulation is as follows:

196   GRAPHIC AND ANALYTIC   [CHAP. XII.

Places	Live Load in 4th Span.				Exterior Loading.						Dead Load = ½ Live.	Total Maximum Strains.	
	P_1	P_2	P_3	P_4	L_1	L_2	L_3	L_5	L_6	L_7		Tens. −	Comp. +
Aa	−15.6	−27.2	−20.4	−8.4	−8.15	+16.4	−67.71	+18.1	−19.7	+1.6	−62.5	225.6	......
Ac	+16.4	+24.4	+8.0	+1.2	−4.21	+8.9	−46.2	−8.3	−11.6	−0.3	+11.8	38.6	85.9
Ad	+8.8	+38.4	+38.4	+8.8	−2.27	−0.65	−24.8	−24.8	−0.65	−2.27	+17.6	37.9	107.9
Ae	+1.2	+8.0	+24.4	+16.4	−0.3	−11.6	−8.3	−46.2	+8.9	−4.21	+11.8	38.6	85.9
Af	−6.4	−20.4	−27.2	−15.6	+1.6	−19.7	+18.1	−67.71	+18.4	−6.15	−62.5	225.6	......
Bc	−20.4	+1.2	+6.0	+2.8	+5.19	−13.6	+56.9	−7.4	+18.4	−0.6	+22.4	19.6	109.3
Bd	−12.8	−50.8	−22.0	−5.2	+3.24	−4.12	+35.5	+14.0	+5.43	+1.3	−17.7	112.6	77.1
Bf	−5.2	−22.0	−50.8	−12.5	+1.3	+5.43	+14.0	+35.5	−4.12	+3.24	−17.7	112.6	77.1
Ba	+2.6	+6.0	+1.2	−20.4	−0.6	+1.49	−7.4	+56.9	−13.6	+5.19	+22.4	19.6	109.3
ab	+51.2	+36.4	+19.6	+5.2	+1.37	+6.75	+15.2	−15.2	+6.75	−1.37	+56.2	......	191.8
bc	−5.2	−36.4	−19.6	−5.2	−1.37	+6.75	−15.2	+15.2	+6.75	+1.37	−28.0	112.4	......
cd	−5.2	+36.4	+19.6	+5.2	+1.37	+6.75	+15.2	+15.2	+6.75	−1.37	+28.0	......	112.4
de	+5.2	+19.6	−19.6	−5.2	−1.37	+6.75	−15.2	+15.2	+6.75	+1.37	0.0	48.0	48.0
ef	−5.2	−19.6	+19.6	+5.2	+1.37	+6.75	+15.2	−15.2	+6.75	−1.37	0.0	48.0	48.0
fg	+5.2	+19.6	+36.4	+5.2	−1.37	+6.75	−15.2	+15.2	+6.75	+1.37	+28.0	......	112.4
gh	−5.2	−19.0	−36.4	+5.2	+1.37	+6.75	+15.2	−15.2	+6.75	−1.37	−28.0	112.4	......
hk	+5.2	+19.6	+36.4	+51.2	−1.37	+6.75	−15.2	+15.2	+6.75	+1.37	+56.2	......	191.8

We first find and tabulate the strain in each piece of the span in question due to each apex load in that span, thus obtaining the four columns for "*interior*" loads.

We then find and tabulate the strains in each and every piece for load over each of the other spans in succession, as given by columns for **L₁**, **L₂**, **L₃**, etc.

Now for dead load, if this is $\frac{1}{2}$ of the live or any other fraction of the live load, we simply have to add algebraically all the other columns horizontally, and divide the resultant sum by $\frac{1}{2}$ or the proper fraction, whatever it is. We thus obtain the column for dead load. The table now gives us at once the maximum strains in every piece, as well as the position of loads which cause these maximum strains. We can also tell at once whether any piece needs to be counterbraced or is subject to strains of two kinds. Thus the dead load of course acts always, and in **A**a, for instance, causes a tension of 62.5 tons. Now all the interior loads P_1, P_2, P_3, etc., we see from our table, also cause tension in **A**a, as do also the loads of the 1st, 3d, and 6th spans. The maximum tensions, since all these loads *may* act simultaneously, is therefore the sum or 225.6 tons tension. On the other hand, the only loads which can cause compression in the piece **A**a are those in the 2d, 5th, and 7th spans.

If these three spans are all fully loaded simultaneously, the united compression in **A**a is *less* than the tension in the same piece, due to the dead load which always acts in that piece. Hence their united action will only diminish the dead load strain, but cannot overcome it. This piece, then, need not be counterbraced. The greatest strain it will ever have to bear, under any possible loading, is tensile *only*, and equal to 225.6 tons.

Again, for **A**c we have a dead load compression of 11.8 tons, which may be increased by other loads, viz., all interior loads and load in span 2 to 85.9 tons compression. Loads in all other spans, except the second, cause tension in **A**c. Adding up all these tensions, and subtracting the dead load compression, we have 38.6 tons tension remaining. The piece **A**c then is subject to 38.6 tons tension and 85.9 tons compression, and must be made to resist both. We also know at once from the table what weights, and where placed, give these two strains. So for each and every piece the two columns for total maximum strains are at once made out from the preceding. The preceding are easily found, provided we only know the moment and shear at the left support for each apex load in the span **D E** itself, and for each exterior span fully loaded with live load. The formulæ for these two cases, which are all sufficient for solution, will be given in the next chapter.

We see, therefore, that there will be as many columns for exterior loading as there are spans, less one; in addition to which there are as many columns for interior loads as there are apex loads in the span itself. For long spans, and many of them, this gives a large table. The solution becomes more tedious, but not more difficult.

In such cases the method referred to in Art. 126 is to be preferred. It gives safe and reliable results, if not strictly maximum results. It reduces all the "exterior loading" columns to only two, and thus materially shortens the labor of solution.

As in the example above, we have taken a middle span, observe that the strains of P_1 and P_4, P_3 and P_2, are the same, as should be, only in reverse order. Thus, strain in **A** a due to P_4 is the same as in **A** k due to P_1, and so on.

128. Method of Moments.—We can very easily check our results by the method of moments of Art. 14.

Thus for load over the 1st span we have the moment at $D = + 61.55$ ft. tons and the shear $= + 0.97$. We have then for the upper flanges of the 4th span for full live load over the 1st span (see Fig. 88):

$$\mathbf{A}\,a \times 10 = -61.55 \qquad \text{or } \mathbf{A}\,a = -6.15$$
$$\mathbf{A}\,c \times 10 = -61.55 + 0.97 \times 20 \qquad \text{or } \mathbf{A}\,c = -4.21$$
$$\mathbf{A}\,e \times 10 = -61.55 + 0.97 \times 40 \qquad \text{or } \mathbf{A}\,e = -2.27$$
$$\mathbf{A}\,g \times 10 = -61.55 + 0.97 \times 60 \qquad \text{or } \mathbf{A}\,g = -0.3$$
$$\mathbf{A}\,k \times 10 = -61.55 + 0.97 \times 80 \qquad \text{or } \mathbf{A}\,k = +1.6$$

For the lower flanges, in like manner:

$\mathbf{B}b \times 10 = +61.55 - 0.97 \times 10$ or $\mathbf{B}b = +5.18$
$\mathbf{B}d \times 10 = +61.55 - 0.97 \times 30$ or $\mathbf{B}d = +3.24$
$\mathbf{B}f \times 10 = +61.55 - 0.97 \times 50$ or $\mathbf{B}f = +1.3$
$\mathbf{B}h \times 10 = +61.55 - 0.97 \times 70$ or $\mathbf{B}h = -0.6$

For the diagonals, since the angle made by these with the vertical is $45° = \theta$ and hence sec. $\theta = 1.414$, we have:

$$ab = 0.97 \times 1.414 = +1.37 \quad bc = -1.37, \text{ etc.}$$

We can thus fill out the column for L_1.

In similar manner also we fill out the column for L_2, L_3, etc.

So also for each of the apex loads in the span itself. Thus for P_1, we have moment at $D = +158.92$ and shear at $D = +36.17$. We have then (Fig. 88).

$\mathbf{A}a \times 10 = -158.92$, or $\mathbf{A}a = -15.89$ tons.
$\mathbf{A}c \times 10 = -158.92 + 36.17 \times 20 - 40 \times 10$, or
$\quad \mathbf{A}c = +16.4$
$\mathbf{A}e \times 10 = -158.92 + 36.17 \times 40 - 40 \times 30$, or
$\quad \mathbf{A}e = +8.8$, etc.

So also for under flanges:

$\mathbf{B}b \times 10 = +158.92 - 36.17 \times 10$, or $\mathbf{B}b = -20.2$.
$\mathbf{B}d \times 10 = +158.92 - 36.17 \times 30 + 40 \times 20$, or
$\quad \mathbf{B}d = -12.6$, etc.

Also for the diagonals we have:

$$ab = 36.17 \times 1.414 = +51.1,\ bc = (40 - 36.17)1.414 = +5.2,$$
$$cd = -5.2, \text{ etc.}$$

We can thus fill out the column for P_1, and in similar manner the columns for P_2, P_3, etc.

We have only to remember that a plus moment at the support gives compression in lower flange and tension in upper, and that a plus shear acts upwards, a minus shear downwards. Also, that for the diagonals, the shear at any apex, multiplied by the secant of the angle with the vertical, gives at once the strain. In the present case the angle is 45°; therefore the secant is 1.414.

The shear at any point is equal to the shear just to the right of the left support, minus all the weights, if any between the support and point. Thus, as above, for P_1 we have for diagonal bc, the shear $36.17 - 40$, or a downward shear of 3.83, since the weight of 40 acts down. This downward shear causes compression in bc, since it acts at the top of bc. For ab we have 36.17 acting up at the *foot* of ab, and therefore also causing compression. The diagonals which meet at the weight are always either both tension or both compression, according as the weight acts at the top or bottom. Right and left from the weight the diagonals alternate in sign.

The calculation, then, of the strains in any span of a continuous girder, as also the diagraming of these strains, is simple and easy, and offers no more real difficulty than the case of a simple truss, *provided we know or can find the moments and shearing forces at the supports for the various cases of loading.* The method of finding these necessary quantities will form the subject of the next chapter.

The reader will do well to compare the strains in the above table with those for the same simple girder similarly loaded. Considerable saving of material, so far as indicated by strain, will be found. On the other hand, some of the flanges are subjected to both tensile and compressive strains.

It is to be well noted that a small sinking of one of the supports of a continuous girder may, if the sinking is equal to the deflection, make one span of what before was two. A very slight sinking, then, may cause very great changes in the strains, and hence, in structures of this kind, it is imperative that the foundations must be secure from settling. If this condition is complied with, then, where long spans are desirable, the continuous girder may be preferable to a succession of single spans.

If the greatest negative reaction at any support, as **D**, is greater than the constant positive reaction at that point due to the dead load, the girder will require to be latched or held down at that support.

CHAPTER XIII.

ANALYTICAL FORMULÆ FOR THE SOLUTION OF CONTINUOUS GIRDERS.

129. Introduction.—As we have seen in the preceding chapter, the complete and accurate determination of the strains in the continuous girder, both for uniform and moving loads, is easy, *provided we can find the moments and shearing forces at the supports for the various states of loading, and for each apex load.* Now this we are able to do with mathematical accuracy, and without much labor. The formulæ necessary for the purpose, when put into proper shape for use, are neither difficult of application nor more complicated than many which the practical engineer is often called upon to manipulate. Since the publication of *Clapeyron's* paper * in 1857, in which, for the first time, his well-known method was developed, and his celebrated "theorem of three moments" made known, the subject has engaged the attention of many mathematicians. In 1862 *Winkler* † first developed a general theory, and gave general rules for the determination of the methods of loading causing greatest strains, together with tables for the maximum values of the moments, shearing forces, etc., for various numbers of spans of varying length. In the same year *Bresse* ‡ followed with a similar work. In 1867 *Winkler* § gave a general analytical theory, and, finally, in 1873 *Weyrauch* ∥ has treated the subject with a degree of completeness and thoroughness which leaves but little to be desired. He discusses the subject in its most general form, for any number of spans of varying length,

---

* *Clapeyron*—Calcul d'une poutre élastique reposant librement sur des appuis inégalement espòces.—Compte rendus, 1857.
† Beiträge zur Theorie der continuirlichen Brückenträger—*Civil Ingenieur*, 1862.
‡ *Bresse*—Cours méchanique appliquée. Paris, 1862.
§ *Winkler*—Die Lehre von der Elasticität und Festigkeit. Prag. 1867.
∥ *Weyrauch*—Allgemeine Theorie und Berechnung der continuirlichen und einfachen Träger. Leipzig, 1873.

and for all kinds of regularly and irregularly distributed and concentrated loads—both for constant and varying cross-section of girder. His formulæ are mathematically exact, and for given loading are free from integrals.

The above is but a very imperfect sketch, and we have named but a few of the many writers who have been occupied with the subject. *Clapeyron's* Theorem above alluded to, as originally given by him, applied only to uniform load over whole length of girder, or over an entire span. But as early as *Bresse's* Treatise, it had been extended to include concentrated and local loads as well, and *Winkler* has also given a very complete and practical discussion of the subject.

Notwithstanding the labors of these and many other mathematicians, there seems to be a wide-spread idea, even among those who are supposed to have considerable familiarity with mathematical literature, that the results deduced are *unpractical*. It is not uncommon to meet with even recent publications* in which it is stated that the authorities pass over such problems with "judicious silence;" that the mathematical investigations are intricate, and the formulæ deduced troublesome in application; that even a "partial solution of the problem by mathematical calculation is attended with considerable difficulty, and that a complete solution for the bending moment and shearing force at every section, under moving partial and irregular loads, taxes the powers of the best mathematicians, and is well-nigh impossible, so far as any practical application of them by the engineer is concerned." How far such ideas are justified may be seen from the following pages. That the authors and works above referred to can only be read by good mathematicians is not to be denied. It may also be admitted that the subject is an intricate one, and when treated mathematically in its most general form the results are naturally in an unpractical shape. But that these results are, therefore, worthless, or that the formulæ, when applied to any particular case, are "too intricate for practical use," by no means follows.

The desirability of formulæ for the application of our graphical method as developed in the preceding chapter; the erroneous ideas prevalent on the subject which we have just noticed;

---

* *Graphical Method for the Analysis of Bridge Trusses:* Greene. D. Van Nostrand, publisher, New York, 1875.

204   CONTINUOUS GIRDER   [CHAP. XIII.

and the deplorable fact that the "authorities" do but too often treat the subject with "judicious silence," and that, therefore, there exists in our engineering literature no collection of practical and useful formulæ for this important class of bridges, though such formulæ are, and have been for years, free to all for the asking,—all these facts may serve as apology for the introduction of the present chapter, in a work which professedly treats only of Graphical methods. The apologies of those who professedly treat the subject analytically, and have yet omitted such formulæ, are not so numerous.

We propose to give the analytical results necessary for complete solution of a girder of uniform cross-section over any number of level supports, with all spans of different lengths: for uniform load over whole length from end to end of girder, for uniform load over any single span, and for concentrated load in any single span at any point of that span. These three cases, as we have seen in the preceding chapter, are sufficient for the *complete solution* of framed Bridge Trusses.*

Many of these results are here given for the first time, at least in their present shape, in any published treatise, though, as remarked, some of them in more or less practical form have long been common property for all who may have desired to make use of them.

The formulæ only will be given, in such shape and with such illustrations of their application that, we trust, they will be found free from complexity, and of considerable practical importance. In the Supplement to this chapter a demonstration of the formulæ is presented.

130. Notation.—The notation which we shall adopt is as follows [see Fig. 89, Pl. 23]:

---

* The formulæ for concentrated loads are alone all that is really necessary. Their addition gives, as we have seen in our tabulation, Art. 127, the strains for uniform load also. In fact, for strict accuracy, *only* single isolated loads should be considered, as the results given by the formulæ for uniform load are not perfectly accurate. This may be seen from the well-known fact that, for a girder fixed at one end, and supported at the other, the reaction at the free end for a load in the middle is $\frac{5}{16}$ of the load, while if the same load were uniformly distributed, the reaction is $\frac{3}{8}$ths, or $\frac{6}{10}$ of the load. The difference, however, for any practical case, where there are a number of panels, is very slight.

Whole number of spans is indicated by s;

Hence, whole number of supports is $s+1$—numbered from left to right.

Number of any support in general, always from left is m.

The supports *adjacent to a loaded span* left and right are indicated by r and $r+1$.

When extreme end spans vary in length from the intermediate, they are always denoted by nl, where n is a given fraction or ratio for any particular case. Thus, if intermediate spans are all 70 feet and end spans 50 feet, $l = 70$, $nl = 50$, and $n = \frac{5}{7}$ths.

When spans next to ends also vary, they are similarly denoted by pl.

All other spans are of equal length and denoted by l.

The length of the span in which the load is supposed to be is in general l_r, where the value of r for any particular case indicates the number of the loaded span from left.

A concentrated load is indicated by **P**.

Its distance from nearest left-hand support, by a.

The ratio of a to length of loaded span l_r, is $k = \dfrac{a}{l_r}$.

Moment at any support in general is **M**$_m$, where m may be 1, 2, 3, r, $r+1$, s, etc., indicating in every case the moment at corresponding support from left.

In same way reaction at any support is **R**$_m$, shear **S**$_m$.

At supports adjacent to loaded span, then, we have **M**$_r$, **M**$_{r+1}$, **R**$_r$, **R**$_{r+1}$, **S**$_r$, **S**$_{r+1}$, for the moments, reactions, and shears at those supports.

A dead uniform load is u per unit of length.

A uniform live load, w per unit of length.

wl_m, then, indicates a uniform live load over any span.

These comprise all the symbols we shall have occasion to use. By reference to Fig. 89, the reader can familiarize himself with their signification, and will then find no difficulty in understanding and using the following formulæ. Certain symbols which we shall use for expressions of frequent occurrence, will be best explained as we have occasion to introduce them.

131. "Theorem of Three Moments."—This remarkable Theorem, due to *Clapeyron*, expresses a relation between the moments at any three consecutive supports, both for uniform load over whole length of girder from end to end, and for uniform load over the whole of any single span. It may be written as follows:

$$M_m\, l_m + 2\, M_{m+1}\, (l_m + l_{m+1}) + M_{m+2}\, l_{m+1} = \frac{w+u}{4}\, [l_m^3 + l_{m+1}^3].$$

If we suppose only one of the two adjacent spans as l_m to contain the full live load w, while all the spans are of course covered with the dead load u, the above equation becomes

$$M_m\, l_m + 2\, M_{m+1}\, [l_m + l_{m+1}] + M_{m+2}\, l_{m+1} = \frac{w+u}{4}\, l_m^3 + \frac{u}{4}\, l_{m+1}^3.$$

If both spans bear the same uniform load u alone,

$$M_m\, l_m + 2\, M_{m+1}\, [l_m + l_{m+1}] + M_{m+2}\, l_{m+1} = \frac{u}{4}\, [l_m^3 + l_{m+1}^3].$$

If the spans are *equal*, the above two equations become respectively

$$M_m\, l + 4\, M_{m+1}\, l + M_{m+2}\, l = \left[\frac{w + 2u}{4}\right] l^3$$

and

$$M_m\, l + 4\, M_{m+1}\, l + M_{m+2}\, l = \frac{u\, l^3}{2}.$$

Now in every continuous beam, whose extreme ends are not fixed, two moments are always known, viz., those at the extreme supports, which are always *zero*. Hence, by the application of this theorem, we can form in any given case as many equations as there are unknown moments, and then, by solving these equations, can determine the moments themselves.

132. Example—Total uniform Load—all Spans equal.— Thus let it be required to find the moments at the supports for a beam of seven equal spans, uniformly loaded over its whole length. The moments at the end supports M_1 and M_8 are zero. We have then, by the application of the last equation above, the following equations:

For the first three supports 1, 2 and 3, $m = 1$, and

$$4\, M_2\, l + M_3\, l = \frac{u\, l^3}{2},\; \text{or}\; 4\, M_2 + M_3 = \frac{u\, l^2}{2}.$$

For supports 2, 3 and 4, $m = 2$, and

$$M_2 + 4\, M_3 + M_4 = \frac{u\, l^2}{2}.$$

For supports 3, 4 and 5, $m = 3$, and

$$M_3 + 4 M_4 + M_5 = \frac{u\,l^2}{2},$$

or since in this case the moments equally distant each way from the middle are equal, this last equation becomes

$$M_3 + 4 M_4 + M_4 = \frac{u\,l^2}{2}.$$

We have therefore three equations between three unknown moments, M_2, M_3 and M_4, and by elimination and substitution can easily find

$$M_1 = M_8 = 0 \quad M_2 = M_7 = \frac{15}{142} u\,l^2 \quad M_3 = M_6 = \frac{11}{142} u\,l^2$$

$$M_4 = M_5 = \frac{12}{142} u\,l^2.$$

If, as in our example of Art. 127 in the preceding chapter, we take $u = 1$ ton per ft., $l = 80$ ft., then $u\,l^2 = 6400$, and the moment at the fourth support becomes 540.8. If the height of truss is ten feet, this gives [Fig. 88] 54.1 tons strain in the upper flange **A** a. By reference to our tabulation, Art. 127, we see that this agrees closely with strain in **A 1** due to uniform load, found in a manner entirely different, viz., by summation of the strains due to first case of loading, and the several loads in the span itself, and serves therefore as a check upon our results.

133. Triangle of Moments.—For the benefit of the practical engineer, who may object to the algebraic work involved in elimination of the unknown moments from the equations above, when the number of spans is great, we offer the following tabulation, from which he may easily and directly determine the moments at the supports for any desired number of spans *without formulæ or calculation.*

Thus, if we were in the above manner to find the moments for a number of spans, and tabulate our results as given in the annexed table, an inspection of the table will show us that we can produce it to any extent desired without further calculation.

MOMENTS AT SUPPORTS—TOTAL UNIFORM LOAD—ALL SPANS EQUAL.

Coefficients of $u\,l^2$ given in triangle.

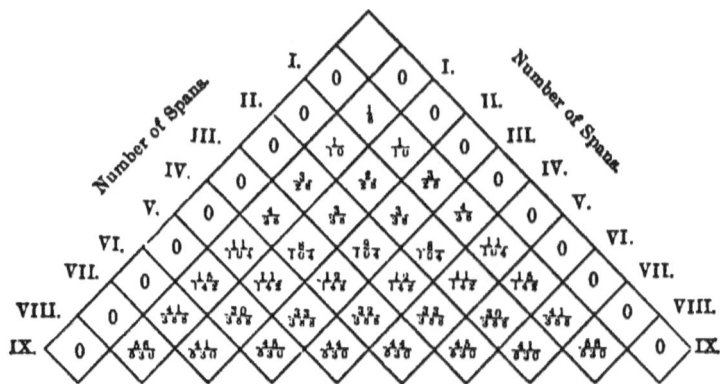

The Roman numerals along the sides of the triangle indicate the number of spans, and the horizontal line to which they belong give the moments. Thus, for our example of seven spans just worked out, we have the extreme moments M_1 and $M_8 = 0$,

$$M_2 \text{ and } M_7 = \frac{15}{142}u\,l^2,\ M_3 \text{ and } M_6 = \frac{11}{142}u\,l^2,\text{ etc.}$$

Now, a simple inspection of this table will show us that for any *even* number of spans, as VIII., for example, the numbers in the horizontal line are obtained by multiplying the fraction above in any *diagonal* column, both numerator and denominator, by 2, and adding the numerator and denominator of the fraction preceding that.

Thus, $\dfrac{15 \times 2 + 11}{142 \times 2 + 104} = \dfrac{41}{388}$; $\dfrac{11 \times 2 + 8}{142 \times 2 + 104} = \dfrac{30}{388} = \dfrac{15 \times 2}{142 \times 2}$

in the other diagonal column; $\dfrac{12 \times 2 + 9}{142 \times 2 + 104} = \dfrac{33}{388}$ or

$= \dfrac{11 \times 2 + 11}{142 \times 2 + 104}$ in the other diagonal column; and so on.

For any *odd* number of spans, as IX., we have simply to add numerator to numerator and denominator to denominator, the two preceding fractions in the same diagonal column.

Thus, $\dfrac{41+15}{388+142} = \dfrac{56}{530}$, $\dfrac{33+12}{388+142}$, or $\dfrac{30+15}{388+142} = \dfrac{45}{530}$,
and so on.

We can, therefore, independently of the theorem and analytical method by which the above results were deduced, produce the table to any required number of spans.*

134. Total uniform Load—all Spans equal—Reactions.—The moments being known, the reactions at the supports can be very easily found.

Thus, the reaction at the first or last support is

$$R_1 = \frac{ul}{2} - \frac{M_2}{l}, \quad R_{s+1} = \frac{ul}{2} - \frac{M_s}{l};$$

at any other support

$$R_m = \frac{6\,M_m}{l} + \frac{1}{2} u\,l.$$

Thus, in our example in Art. 127, we find

$$R_1 = \frac{56}{142} ul, \quad R_2 = \frac{161}{142} ul, \quad R_3 = \frac{137}{142} ul, \quad R_4 = \frac{143}{142} ul.$$

Hence, the *shear* at the fourth support is

$$\frac{56}{142} ul + \frac{161}{142} ul + \frac{137}{142} ul + \frac{143}{142} ul - 3\,ul = \frac{71}{142} ul,$$

or when $ul = 80$ tons, $\dfrac{71}{142} ul = 40$ tons.

Multiplying this shear by 1.414 (the secant of the angle with vertical), we find for the strain in diagonal $a\,b$ (Fig. 88) due to uniform load $+$ 56.5 tons, the same nearly as already found in our tabulation.

135. Triangle for Reactions.—The reactions for a number of spans being found, and tabulated, as above, in the case of the moments, we shall have a triangular table precisely similar to the one above, in which the same rule holds good for odd and even numbers of spans.

---

* The above relations between the moments can be shown analytically to be a result of the properties of the well-known "Clapeyronian numbers." For the table above, as also the others which follow, we are indebted to the kindness of Mr. Mansfield Merriman, Instructor in Civil Engineering in the Sheffield Sci. School of Yale College. They are given, so far as we are aware, in no treatise upon the subject yet published.

REACTIONS AT SUPPORTS—TOTAL UNIFORM LOAD—ALL SPANS EQUAL.

Coefficients of ul given in triangle.

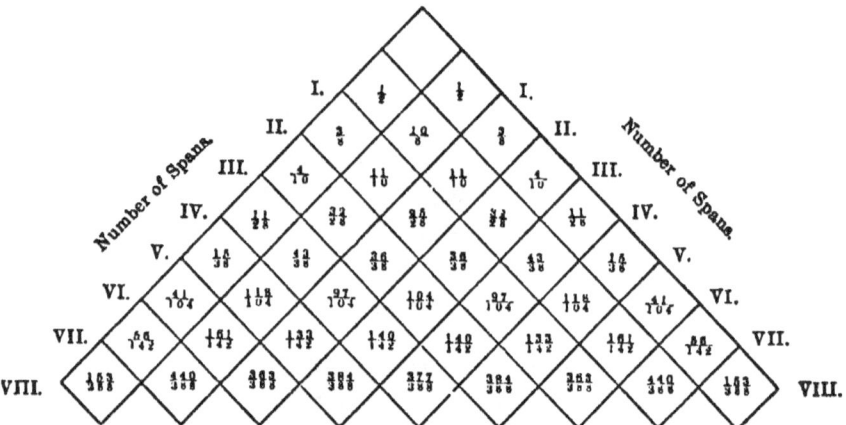

We are thus able to find both moments and reactions at the supports for *any* number of spans, so far as uniform loading is considered, and may then either diagram the strains in the various pieces or calculate them as explained in Arts. 127 and 128. No formulæ are required. Any one who understands the method of moments as applied to simple girders can, by the aid of the two tables above, find accurately the strains in every piece of a girder, continuous over as many equal spans as is desired, and *uniformly loaded over its entire length*, all supports being on the same straight line.

As we have seen, Art. 127, this is one of the cases which must be considered in order to find the maximum strains in any span,* and the results above given for its solution will, we trust, be found by the practical engineer to be neither "complex" nor "difficult of application."

136. Clapeyronian Numbers.—In the analytical discussion of continuous girders, certain numbers having many remarkable properties play a very important *rôle*.

We have seen that the theorem of three moments furnishes us with as many equations between the moments as there are moments to be determined. For a small number of supports,

---

* See note to Art. 129.

these equations can be solved by the ordinary rules of algebra; but for a great number, or in the general analytic discussion of *any* number, we must have recourse to a special artifice. Thus we multiply our equations, beginning with the last, by numbers indicated by $c_1, c_2, c_3, \ldots c_{s-1}$, and then choose these numbers such that, by the addition of all the equations, all the $\mathbf{M}$'s, with the exception of $\mathbf{M}_1$, disappear. We thus easily determine $\mathbf{M}_1$ without the tedious process of substituting from one equation to the other, through the entire list.

The following relations must then evidently hold between these numbers, as is evident from the theorem of three moments of Art. 131:

$$2 c_1 (l_{s-1} + l_s) + c_2 l_{s-1} = 0.$$
$$c_1 l_{s-1} + 2 c_2 (l_{s-2} + l_{s-1}) + c_3 l_{s-2} = 0.$$
$$\cdots\cdots\cdots\cdots\cdots\cdots\cdots\cdots$$
$$c_{s-3} l_3 + 2 c_{s-2} (l_2 + l_3) + c_{s-1} l_2 = 0.$$

If the first number is chosen at will, say ± 1, the other numbers can be found from these equations.

Now in the present case of all spans equal, we have between any three of these numbers the relation:

$$c_{m-1} + 4 c_m + c_{m+1} = 0.$$

If we take the first, $c_1 = 0$, and the next, $c_2 = 1$, we have for the others the following values:

$c_1 = 0$	$c_4 = +15$	$c_7 = -780$	$c_{10} = +40545$
$c_2 = +1$	$c_5 = -56$	$c_8 = +2911$	$c_{11} = -151316$
$c_3 = -4$	$c_6 = +209$	$c_9 = -10864$	$c_{12} = +564719$

These are the so-called *Clapeyronian numbers*. They alternate, as we see, in sign, and each is numerically 4 *times the preceding minus the one preceding that*. We shall always indicate these numbers by the letter c, the index denoting the particular one. Thus, c_7 is the seventh number, counting 0 and 1 as the two first.

No table of these numbers is needed. The index being given, any one can write down the series for himself, till he arrives at the desired number.

137. Uniform Live Load over any single Span—Moments at Supports of Loaded Span.—These numbers being premised, we can now give the following formulæ for the moments at the supports r and $r+1$ of the uniformly loaded span:

For the left support,

$$M_r = -\frac{1}{4} w l^2 \left[\frac{c_r\, c_{s-r+2} + c_r\, c_{r-r+}}{c_{s+1}} \right].$$

For the right support,

$$M_{r+1} = -\frac{1}{4} w l^2 \left[\frac{c_r\, c_{s-r+1} + c_{r+1}\, c_{s-r+1}}{c_{s+1}} \right].$$

These formulæ, it will be seen, are very simple and easy of application.

Thus, for seven spans, load over the fourth from left, we have $s = 7$, $r = 4$, and hence

$$M_4 = M_5 = -\frac{1}{4} w l^2 \left[\frac{c_4\, c_5 + c_4\, c_4}{c_8} \right].$$

Both moments are equal, as should be the case for a middle span. Inserting now the proper values for the Clapeyronian numbers from the preceding Art., we have

$$M_4 = M_5 = -\frac{1}{4} w l^2 \left[\frac{15 \times -56 + 15^2}{2911} \right] = \frac{615}{11644} w l^2.$$

So for any desired number of spans, the values of r and s being known, the corresponding Clapeyronian numbers can be easily found, and, inserted in our formulæ, give us at once the moments at the supports.

Turning again to our example, Art. 127, and making $w = 2$ tons, and $l = 80$ ft., we have $w l^2 = 12800$, and therefore $M_4 = 676$, and dividing by depth of truss $= 10$ ft., we find the strain in Aa (Fig. 88) 67.6 tons, nearly what we may find by the summation of the strains due to the loads P_{1-4} in our tabulation.

138. Triangle of Moments—Uniform Live Load over any single Span.—If from the above formulæ we find the moments at supports for a number of spans, and tabulate as before, we shall have a triangle of moments similar to those already given, which may be produced to include any desired number of spans. We have only to observe that the numerator or denominator of any fraction in the table follows the law of the Clapeyronian numbers—that is, is four times the preceding in the same diagonal column minus the one preceding that.

MOMENTS AT SUPPORTS OF LOADED SPAN.—UNIFORM LIVE LOAD OVER ANY SINGLE SPAN.

Coefficients of $w\,l^2$ given in triangle.

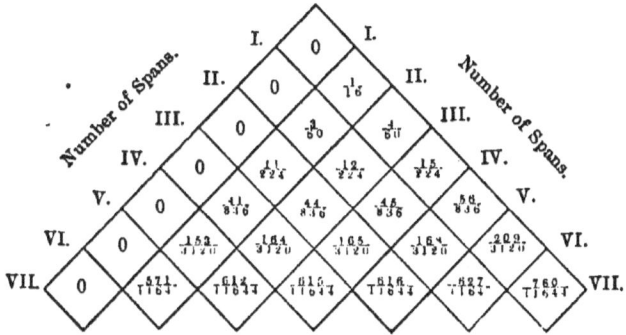

Thus for seven spans, $\dfrac{153 \times 4 - 41}{3120 \times 4 - 836} = \dfrac{571}{11644}$, and so on.

The triangle above gives the moments for uniform load over any span, both right and left. For *left* supports we have simply to count the span from right to left. Thus for seven spans for load in the sixth span from left, we have moment at left-hand support $= \dfrac{627}{11644}\,w\,l^2$, counting the spans from left to right in triangle. For the moment at *right* support of same loaded span, we count *six the other way* from right to left, and find $\dfrac{571}{11644}\,w\,l^2$.

139. Moments at Supports of Unloaded Spans.—The triangle and formulæ above give the moments at the supports of the loaded span only, both positive—that is, always tending to cause tension in upper flange and compression in lower.

If m represents the number of any support counting from the left, the moments at *any* support generally may be found by the following formulæ:

When $m < r+1$, $\mathbf{M}_m = \dfrac{1}{4}\,w\,l^2 \left[\dfrac{c_m\,c_{s-r+2} + c_m\,c_{s-r+1}}{c_{s-1} + 4\,c_s} \right]$.

When $m > r$, $\mathbf{M}_m = \dfrac{1}{4}\,w\,l^2 \left[\dfrac{c_r\,c_{s-m+2} + c_{r+1}\,c_{s-m+2}}{c_{s-1} + 4\,c_s} \right]$.

If we make in these formulæ $m = r$ in the first, and $m =$

$r+1$ in the second, we obtain the formulæ of Art. 137. For any other support left of r, or right of $r+1$, we have only to give the proper values to m, s and r for any given case, and find the corresponding Clapeyronian numbers.

140. Practical Rule, and Table.—The moments at the supports of the loaded span having been found by the formulæ of Art. 137, or the triangle of moments of Art. 138, instead of using the above formulæ, we may find the moments at the other supports as follows:

For all supports left of the loaded span: Commencing at the left end support, place over each support the Clapeyronian numbers

$$1 \quad 4 \quad 15 \quad 56 \quad 209 \quad 780, \text{ etc.}$$

Take the last number thus obtained, *before* reaching the left support of the loaded span, as a common denominator. Then the moment at the left end is of course zero. At the second support 1, at the third 4, at the fourth 15, at the fifth 56, etc., all divided by this common denominator, will express the fractional part of the moment at the left support of the loaded span, which the moment at the support in question is. For the moments at the supports right of the loaded span, proceed similarly, only count from the *right end*.

Thus, for a girder of ten spans, sixth span from left loaded: The moments M_6 and M_7 due to load being found, suppose we wish the moments left of M_6. Commencing at left end, number the supports 1, 4, 15, 56, 209. (Let the reader draw a figure representing the case.) The number 209 is the last before reaching the sixth support. We take this, therefore, for a common denominator. Then we have $M_1 = 0 \quad M_2 = \dfrac{1}{209} M_6$

$$M_3 = \frac{4}{209} M_6 \quad M_4 = \frac{15}{209} M_6 \text{ and } M_5 = \frac{56}{209} M_6.$$

So for supports to the right of support 7, we have

$$M_{11} = 0 \quad M_{10} = \frac{1}{56} M_7 \quad M_9 = \frac{4}{56} M_7 \quad M_8 = \frac{15}{56} M_7.$$

Remembering that M_6 and M_7 are both positive, and that the moments *alternate* either way from these supports, we find easily the proper signs for the moments right and left.

We can now, therefore, find the moments at **D** and **E** due to the first and second cases of loading of Fig. 87 (Art. 126).

CHAP. XIII.]   ANALYTICAL FORMULÆ.   215

Let us take the first case. For load on **A B**, we have from our triangle or formulæ the moment at $\mathbf{B} = \dfrac{780}{11644}\, w\, l^2$. At **D**, then, we have $\dfrac{56}{780} \times \dfrac{780}{11644}\, w\, l^2 = \dfrac{56}{11644}\, w\, l^2$.

For load on **C D**, we have at once from triangle the moment at $\mathbf{D} = \dfrac{616}{11644}\, w\, l^2$.

Finally, for load on **F G**, we have for moment at **F**, from triangle $= \dfrac{627}{11644}\, w\, l^2$, and therefore at **D**, $\dfrac{15}{209} \times \dfrac{627}{11644}\, w\, l^2 = \dfrac{45}{11644}\, w\, l^2$.

All these moments at **D** are positive; we have therefore, for the first case of loading, the total moment at $\mathbf{D} = + \dfrac{717}{11644}\, w\, l^2$.

If we make $l = 80$ ft. and $w = 2$ tons, we find the moment at $\mathbf{D} = 788$, and dividing by 10 we obtain 78.8 tons as the strain in **A** a, Fig. 88, corresponding with our tabulation, Art. 127.

Table for all the Moments.

All the moments may be found from a simple table similar to the following, which will be found perhaps preferable to the triangle of Art. 138.

TABLE FOR MOMENTS.—UNIFORM LOAD IN SINGLE SPAN.

Support counted from Left.   Denominator Δ.

1	2	3	4	5	6	Loaded Span from Right.	Spans.	¼ Δ.
0	1	4	15	56	209	I.	1	1
0	3	12	45	168	627	II.	2	4
0	11	44	165	616		III.	3	15
0	41	104	615			IV.	4	56
0	153	612				V.	5	209
0	571					VI.	6	780
							7	2911

This table, it will be observed, can be produced to include any number of supports desired. The law of the Clapeyronian numbers runs both horizontally and vertically. The smaller table gives the denominator, the larger the numerator of the coefficient of $w\,l^2$ for any case. Thus, for seven spans we have *four* times $2911 = 11644$ for the common denominator. For load on second span from right, moment at sixth support from left, we have then directly $\dfrac{627}{11644}\,w\,l^2$; for fourth support from left, $\dfrac{45}{11644}\,w\,l^2$, the same as above.

For load in fifth span from right, the table gives us at once $0, -\dfrac{153}{11644}\,w\,l^2$ and $+\dfrac{612}{11644}\,w\,l^2$, for supports 1, 2 and 3. For the other supports, since if now we were to continue counting from left *we should have to pass a loaded support*, we must count the loaded span from *left*, and count the supports in reverse order. For fifth support from *right*, then, the number required is at intersection of III. (instead of V.) and 5, or $\dfrac{616}{11644}\,w\,l^2$, as found above. Thus the tables above cover all cases, giving supports at loaded span itself, as also right and left of this span. We have only to remember to count supports from left, and loaded span from right, for all supports *left* of load, and inversely for all supports right of load. [*The reader should always, when using Table, make a sketch of the given number of spans, indicate the loaded span, and number the supports.*]

141. Reactions at Supports—Live Load over single Span.—For the *reactions* at ends of loaded span, we have

$$R_r = 6\,\frac{M_r}{l} + \frac{1}{4}\,w\,l, \quad R_{r+1} = 6\,\frac{M_{r+1}}{l} + \frac{1}{4}\,w\,l.$$

For reactions at extreme ends, when end spans are loaded, or when $r = 1$ or $r = s$, $R_1 = -\dfrac{M_2}{l} + \dfrac{1}{2}\,w\,l, \quad R_s = -\dfrac{M_s}{l} + \dfrac{1}{2}\,w\,l.$

When any other spans are loaded, or

when $r > 1$ and $< s$, $R_1 = -\dfrac{M_2}{l} \quad R_{s+1} = -\dfrac{M_s}{l}.$

For all other reactions,

$$R_m = 6 \frac{M_m}{l}.$$

Thus for load covering the first span of seven spans, we find from the known moment, given in preceding Art., for the fourth support,

$$R_4 = \frac{6 \times 56}{11644} wl = \frac{336}{11644} wl.$$

For load over third span from left,

$$R_4 = \frac{6 \times 616}{11644} wl + \frac{1}{4} wl = \frac{6607}{11644} wl.$$

For load on sixth span,

$$R_4 = \frac{45}{11644} wl \times 6 = \frac{270}{11644} wl.$$

Hence, total reaction at fourth support for first case of loading is $\frac{7213}{11644} wl$. In the same way we can find the reactions at the first, second, and third supports, for the second case of loading, as shown in Fig. 87, and then can easily find the *shear* at any support, as D, by taking the algebraic sum of all the reactions and loads between that support and the end.

We can now, therefore, find the shear and moment at D, and thus determine the strains in the span D E for both cases of loading, as given in our tabulation, Art. 127.

142. Triangle for Reactions—Single Span loaded.—If we calculate from our formulæ the reactions at supports of loaded span, for a number of spans, we can tabulate the results, as on next page, in a triangle, where each number is four times the preceding minus the one preceding that, all in the same diagonal column.

REACTIONS AT SUPPORTS—LIVE LOAD OVER SINGLE SPAN.

Coefficients of wl given in triangle.

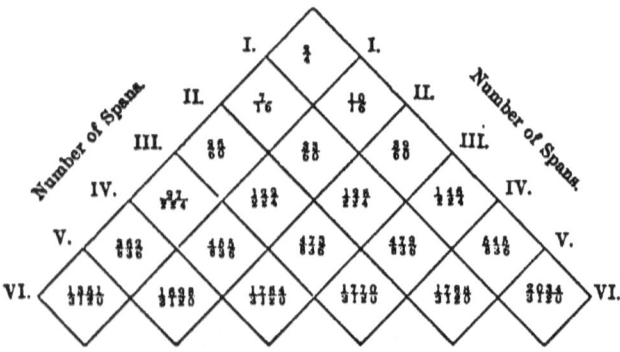

This triangle, similar to the preceding one for moments, gives the moments at the *left* support of the loaded span, when we count from left to right. Counting the other way, we have the reactions at the *right* support of the loaded span.

Thus for six spans, fourth span from left loaded, we count four from left in horizontal line for VI., and find $\dfrac{1770}{3120} wl$ for reaction at *left* support. For reaction at *right*, we count *four* also from *right* end, and find $\dfrac{1764}{3120} wl$.

143. Reactions in unloaded Spans—Load over one Span only—Table.—The formulæ of Art. 141 for the reaction at any *unloaded* support are sufficiently simple and easy to apply; still we may, if thought preferable, also draw up tables for these, to be used in connection with the triangle of the preceding Art. The following tables give the coefficients of wl for the reactions *not* adjacent to the loaded span. The denominator of the fraction is to be taken from the triangle above; the tables referred to give *only* the numerators.

REACTIONS AT UNLOADED SPANS.

Supports counted from left.

1	2	3	4	5	6	Loaded span from right.
1	6	24	90	336	1254	I′.
3	18	72	270	1008		II′.
11	66	264	990			III′.
41	246	984				IV′.
153	918					V′.
571						VI′.
2131						VII′.

Supports counted from right.

Loaded span from left.	6′	5′	4′	3′	2′	1′
I.	1254	336	90	24	6	1
II.		1008	270	72	18	3
III.			990	264	66	11
IV.				984	246	41
V.					918	153
VI.						571
VII.						2131

Tables give the numerators of the coefficients of $w\,l$. Denominators from triangle on page opposite.

These tables may be carried out to any desired extent by the law of the Clapeyronian numbers in the vertical columns.

As an example of their use, take seven spans load in fifth from left, that is, in third from right. (Make sketch.) From the triangle we take the common denominator 11644. Then from first table in the horizontal column of III.′ we have for left end

$$R_1 = \frac{11}{11644}w\,l,\ R_2 = -\frac{66}{11644}w\,l,\ R_3 = \frac{264}{11644}w\,l,\ R_4 = -\frac{990}{11644}.$$

For supports right of loaded span, we must take the second table, and look in horizontal column for V. We thus obtain

$$R_8 = \frac{153}{11644}w\,l,\ R_7 = -\frac{918}{11644}w\,l.$$

We can now, therefore, either by our tables or formulæ, or both, find the moments, reactions, and shearing force at any support for both cases of loading given in Fig. 87. The reader will do well to take the example of Art. 127, and find the moment and *shear* at **D** for both cases, and thus check our results as given in Art. 127. Thus, from the data there given, we find for the 1st case + 788.2 ft. tons, and for the 2d case — 382.5 ft. tons for the moments and 14.55 tons for shear.

144. Concentrated Load in any Span—Moments at Supports.—It only remains to consider a *concentrated* load at any point. If the formulæ for this case do not prove to be too complex or intricate for practice, we may consider the case, so far as equal spans are concerned, as fully solved.

We have seen that the "theorem of three moments," so far as uniform loads are concerned, enables us to solve the case thoroughly. It is more especially as regards concentrated or partial loads that the opinion widely prevails as to the impossibility of obtaining practically useful formulæ; and this, notwithstanding that it has been shown by *Bresse*, *Winkler*, *Weyrauch*, and many others, that the theorem of three moments can be *extended* to include concentrated loads also.

The Theorem as thus extended is as follows:

$$M_{m-1} l_{m-1} + 2 M_m \left[l_{m-1} + l_m \right] + M_{m+1} l_m =$$

$$\frac{P_{m-1} a'}{l_{m-1}} \left(l_{m-1}^2 - a'^2 \right) + \frac{P_m a}{l_m} \left(2 l_m^2 - 3 a l_m + a^2 \right),$$

where, by our notation (Fig. 89, Art. 130), a' and a are the distances of P_{m-1}, P_m, from the nearest left supports.*

By the aid of this theorem, we are able to deduce the following formulæ:

For moments left of r, and including support r, that is

when $m < r + 1$, $\quad M_m = - c_m \dfrac{A\, c_{s-r+2} + A'\, c_{s-r+1}}{c_{s+1}}$.

For moments right of $r + 1$, including support $r + 1$, or

when $m > r$, $\quad M_m = - c_{s-m+2} \dfrac{A\, c_r + A'\, c_{r+1}}{c_{s+1}}$.

In these formulæ, c represents, as above, the Clapeyronian number, and A A' stand for the following expressions:

$$A = P\, l\, (2\, k - 3\, k^2 + k^3) \quad A' = P\, l\, (k - k^3),$$

k being the fraction $\dfrac{a}{l}$, or the ratio of the distance of the weight P from the left support, to the length of span.

145. Illustration of Application of above Formulæ.—These formulæ are by no means difficult of application. Let

---

* For demonstration of this Theorem, see Supplement to this chapter.

us take the example of Art. 127 (Fig. 88), where $P = 40$ tons, $l = 80$ ft., and a becomes 10, 30, 50 and 70 ft. respectively. First, as regards the expressions $A\ A'$:

These become in the present case $3200\ (2\ k - 3\ k^2 + k^3)$ and $3200\ (k - k^3)$ respectively, where k has the values $\tfrac{1}{8}, \tfrac{3}{8}, \tfrac{5}{8},$ and $\tfrac{7}{8}$ successively. Now as the denominator is in each term always the same, in the first 8, in the second 64, in the third 512, and only the numerators of the values of k vary for the different positions of P, we may put these values of A and A' in the forms

$$A = 3200 \left(\frac{2\ h}{8} - \frac{3\ h^2}{64} + \frac{h^3}{512}\right),\quad A' = 3200 \left(\frac{h}{8} - \frac{h^3}{512}\right),$$

or

$$A = 800\ h - 150\ h^2 + 6.2305\ h^3,$$
$$A' = 400\ h - 6.2305\ h^3,$$

where h has successively the values 1, 3, 5 and 7, for P_1, P_2, P_3 and P_4 respectively. These are then the practical formulæ for substitution in the present case.

We can now apply the formulæ for M above. Thus, suppose for seven spans we have P_3 in the fourth, as shown by Fig. 88, and wish the moment due to P_3 at the fourth support D. Then $s = 7$, $r = 4$, and $m = 4$, and we have

$$M_4 = - c_4\ \frac{A\ c_5 + A'\ c_4}{c_3},$$

or, referring to Art. 136 for the Clapeyronian numbers,

$$M_4 = - 15\ \frac{-56\ A + 15\ A'}{2911} = \frac{840\ A - 225\ A'}{2911}.$$

Now for P_3 we have $k = \tfrac{5}{8}$, or $h = 5$, and therefore

$$A = 1028.81 \quad A' = 1221.2.$$

Hence

$$M_4 = \frac{589430.4}{2911} = 202.4\ \text{ft. tons.}$$

This divided by $10 =$ height of truss gives tension in $A\ a = 20.2$ tons, nearly what we have already found in our tabulation, Art. 127.

In like manner we may easily find the moment at D due to

every weight, or by giving the proper value to m in our formulæ, we may find the moment at *any* support we please.

The moments at the supports of the loaded span being found, the moments at the other supports may be obtained according to the rule given in Art. 140 for uniform live load over single span.

146. Triangle of Moments.—The reader may also by the aid of the formulæ above form a triangle similar to those already given, containing the coefficients of $P\,l$ for the moments at the supports of the loaded span.

Thus for two spans, for moment at left support, we should obtain 0 and $\frac{1}{4}\,[2\,k - 3\,k^2 + k^3]\,P\,l$, and this last value will run down the right diagonal column without change, except in its coefficient $\frac{1}{4}$, which will become successively $\frac{4}{15}$, $\frac{15}{56}$, $\frac{56}{209}$, for three, four and five spans respectively. For three spans we shall have, 0, $\frac{1}{15}\,[7\,k - 12\,k^2 + 5\,k^3]\,P\,l$, and, as above, $\frac{4}{15}\,[2\,k - 3\,k^2 + k^3]\,P\,l$. The second of these will run down the second diagonal column from the right without change, except in its coefficient, which will be $\frac{4}{56}$, $\frac{15}{209}$, etc., for five and six spans.

So, for four spans we have

$$0,\ \tfrac{1}{56}\,[26\,k - 45\,k^2 + 19\,k^3]\,P\,l,\quad \tfrac{4}{56}\,[7\,k - 12\,k^2 + 5\,k^3]\,P\,l,$$
$$\tfrac{15}{56}\,[2\,k - 3\,k^2 + k^3]\,P\,l,$$

for moments at left, for load in 1st, 2d, 3d and 4th span respectively. The second of these runs down the *third* diagonal column from the right, changing coefficient as above.

If the triangle be now drawn, and these expressions properly inserted, we shall observe that along the diagonal columns sloping down and to the *left*, the values of k in the parenthesis, as also the *denominators* of the outside fractions, follow the law of the Clapeyronian numbers. The numerators of these outside fractions in these columns remain unchanged. The outer left column is of course always zero.

Another triangle must be found for moments to the right of load, and then the moments at the *unloaded* supports may be found by the rule of Art. 140.

All the moments may also be found from a couple of tables formed similarly to those of that Art. It is unnecessary to give such tables here. From the above the reader can form them for himself, if desired. The formulæ for moments given above are so simple, and with a little practice so readily worked, that tables are scarcely needed.

147. Reactions at Supports for concentrated Load in single Span.—For the reactions we have the following formulæ:
1st. *Abutment reactions.*
When the end span contains the load, that is, when $r = 1$ or $r = s$,

$$R_1 = -\frac{M_2}{l} + P(1-k), \quad R_{s+1} = -\frac{M_s}{l} + P(1-k).$$

When the load is *not* in the end spans, *i.e.*, when $r > 1$ and $r < s$,

$$R_1 = -\frac{M_2}{l}, \quad R_{s+1} = -\frac{M_s}{l}.$$

2d. *Reactions at supports of the loaded span itself* (not end span),

$$R_r = 6\frac{M_r}{l} + P(1 - 3k + 3k^2 - k^3),$$

$$R_{r+1} = 6\frac{M_{r+1}}{l} + P k^3.$$

3d. *For all other reactions,*

$$R_m = 6\frac{M_m}{l}.$$

The above formulæ, in view of what has been said in Art. 145, are sufficiently simple to need no illustration.

For load in fourth span of seven spans we find easily for the reaction at left support,

$$R_4 = \frac{P}{2911}\left[2911 - 3k - 6387 k^2 + 3479 k^3\right].$$

This can be put in working order as explained in Art. 145, and the reader can check the results which we have given in the example of Art. 127 for himself.

A triangle and two subsidiary tables for the reactions at the supports of loaded spans may be formed similarly to the triangle and tables of Arts. 142 and 143. We leave this for the reader to accomplish for himself, if thought desirable.

148. Shear at Supports of loaded Span.—We are now in possession of all the formulæ necessary for the complete solution of a girder over any number of supports, *all spans equal*. For any desired span, we can find the maximum positive and negative moments by the cases of Fig. 80, as also the moments due to various positions of the weight **P**. We can also find the reactions at all the supports due to these cases. From the reactions and known forces, we can then easily find the algebraic sum, or *shear*, at any support. The moment and shear at any support due to any case of loading are, as we have seen, the quantities required for calculation.

Now it is not necessary to find all the reactions in order to obtain the shear. The moments at the supports being known, we can find the shear directly.

Thus, for concentrated load in a span l_r (Fig. 89) we have for any point x

$$\mathbf{M}_r - \mathbf{S}_r x + \mathbf{P}(l_r - a) - m = 0,$$

where $\mathbf{S}_r$ is the shear at the left of the loaded span, and m is the moment at any point. We see at once that, to determine this moment, it is the shear that we wish, and not the reaction. For a uniform load we have similarly,

$$\mathbf{M}_r - \mathbf{S}_r x + \frac{w\,x^2}{2} - m = 0.$$

If in both these equations we make $x = l_r$, m becomes $\mathbf{M}_{r+1}$, and we have

$$\mathbf{S}_r = \frac{\mathbf{M}_r - \mathbf{M}_{r+1}}{l_r} + q,$$

where $q = \mathbf{P}(1 - k)$ for a concentrated load, and $q = \dfrac{w\,l_r}{2}$ for uniform load.

In an unloaded span at the left support, or when $m < r$, q disappears, and we have

$$\text{when } m < r, \quad \mathbf{S}_m = \frac{\mathbf{M}_m - \mathbf{M}_{m+1}}{l_m}.$$

For the shear at the *right* support of the loaded span we have simply $\mathbf{S}_r - \mathbf{P}$ or $\mathbf{S}_r - w\,l$, and hence

$$\mathbf{S}'_{r+1} = \frac{\mathbf{M}_{r+1} - \mathbf{M}_r}{l_r} + q',$$

where $q' = Pk$ for concentrated load, and $q' = \dfrac{w\,l_r}{2}$ for uniform load. For any other span at the right support

$$S'_m = \frac{M_m - M_{m-1}}{l_{m-1}}.$$

Thus, S_m and S'_m are the shears at any support just to right and left of that support respectively. The reaction at any support is then $R_m = S_m + S'_m$.

The moments, then, at two successive supports being known, we can readily find the shear at any support, and these two, moment and shear, we repeat, are the quantities required for calculation. The reactions, and the tables for the reactions above, are *only* useful as enabling us to find the shear. It is this last, together with the moment at the support, which gives us the moment m at any point of the span in question, as is evident from the above equations. It is only in the case of the simple girder that the reactions at the ends are the same as the shears. In the continuous girder *only the latter should be used*, except for ends of end spans, where the two are identical. We have only to remember, then, that the shear at any support is the *algebraic sum* of all the reactions and loads from that support to the nearest extreme end, and then, knowing these reactions and loads, the determination of the shear is easy.

We might give tables for shears directly, as above, for reactions; but this is unnecessary. Having taken the reactions from our tables already given, and found the moments either by our formulæ or tables, we can then find the shears both by means of the reactions and also directly from the moments themselves, and thus check at once the accuracy of our determination of both. From what has already been given, the reader can easily construct tables of shears similar to those already given for reactions for himself, if desired.

149. Recapitulation of Formulæ—Continuous Girder over any Number of Level Supports, all Spans equal.*— *For notation, see Art.* 130, *Fig.* 89.

---

* These formulæ are given in similar form in Winkler's "Der Lehre von der Elasticitaet und Festigkeit," Art. 144, p. 122. They were also independently deduced by Mr. Merriman, to whose kindness we are indebted for much of this chapter. The above methods of tabulation were communicated by him, and are given in no treatise upon the subject.

1st. *Moments at supports.*

When $m < r+1$, $\mathbf{M}_m = -c_m \dfrac{\mathbf{A}\, c_{s-r+2} + \mathbf{A}'\, c_{s-r+1}}{c_{s+1}}$;

when $m > r$, $\mathbf{M}_m = -c_{s-m+2} \dfrac{\mathbf{A}\, c_r + \mathbf{A}'\, c_{r+1}}{c_{s+1}}$;

in which $k = \dfrac{a}{l}$, and $\mathbf{A} = P\, l\, (2k - 3k^2 + k^3)$, $\mathbf{A}' = P\, l\, (k - k^3)$ for concentrated loads, and $\mathbf{A} = \mathbf{A}' = \tfrac{1}{4} w\, l^3$ for a uniform load over any one span.

2d. *Shear at the supports.*

In the loaded span, to the right of the left support,
$$S_r = \frac{\mathbf{M}_r - \mathbf{M}_{r+1}}{l_r} + q.$$

To the left of the right support,
$$S'_{r+1} = \frac{\mathbf{M}_{r+1} - \mathbf{M}_r}{l_r} + q'.$$

In the *unloaded* spans, to the right of the left support,
$$S_m = \frac{\mathbf{M}_m - \mathbf{M}_{m+1}}{l_m}.$$

To the left of the right support,
$$S'_m = \frac{\mathbf{M}_m - \mathbf{M}_{m-1}}{l_{m-1}}.$$

For the reaction at any support, $\mathbf{R}_m = S'_m + S_m$.

3d. *Reactions.*

(*a*) Abutment reactions:

when $r = 1$, $\mathbf{R}_1 = -\dfrac{\mathbf{M}_2}{l} + q$; when $r = s$, $\mathbf{R}_{s+1} = -\dfrac{\mathbf{M}_s}{l} + q'$;

when $r > 1$ and $< s$, $\mathbf{R}_1 = -\dfrac{\mathbf{M}_2}{l}$, $\mathbf{R}_{s+1} = -\dfrac{\mathbf{M}_s}{l}$.

(*b*) Reactions at supports adjacent to loaded span (when this span is *not* an end span):

$$\mathbf{R}_r = 6\, \frac{\mathbf{M}_r}{l} - \frac{\mathbf{A}}{l} + q,$$

$$\mathbf{R}_{r+1} = 6\, \frac{\mathbf{M}_{r-1}}{l} - \frac{\mathbf{A}'}{l} + q'.$$

(c) All other reactions:

$$R_m = 6 \frac{M_m}{l}.$$

Where for concentrated load,

$$A = Pl(2k - 3k^2 + k^3) \qquad A' = Pl(k - k^3)$$
$$q = P(1-k) \qquad q' = Pk.$$

For uniform load in single span,

$$A = A' = \tfrac{1}{4} w l^2, \quad q = q' = \tfrac{1}{2} w l, \quad k \text{ being always } \frac{a}{l}, \quad \text{and}$$

$c_1 = 0$	$c_4 = 15$	$c_7 = -780$	$c_{10} = 40545$
$c_2 = 1$	$c_5 = -56$	$c_8 = 2911$	$c_{11} = -151316$
$c_3 = -4$	$c_6 = 209$	$c_9 = -10864$	$c_{12} = 564719$, etc.

We give also, for sake of completeness, although not needed for calculation, the formulæ

FOR UNIFORM LOAD OVER ENTIRE LENGTH OF GIRDER.[*]

Moment at any support,

$$M_m = 2 A b_m \frac{b_{m-1} b_{s+2} - b_{m+1} b_s}{b_{s+2} - b_s};$$

Reactions at abutments,

$$R_1 = R_{s+1} = q \frac{b_{s+2}}{b_{s+2} - b_s};$$

Reactions at other supports,

$$R_m = 6 \frac{M_m}{l} + q;$$

where, as before, $A = \tfrac{1}{4} w l^2$, $q = \tfrac{1}{2} w l$,
and the numbers indicated by b are as follows:

$b_1 = 0$	$b_4 = -3$	$b_8 = -41$	$b_{12} = -571$
$b_2 = 1$	$b_5 = -4$	$b_9 = -56$	$b_{13} = -780$
$b_3 = 1$	$b_6 = +11$	$b_{10} = +153$	etc.
	$b_7 = +15$	$b_{11} = +209$	

---

[*] The above equations were first given by Mr. Merriman, in the *Jour. Franklin Institute*, April, 1875.

These numbers change signs by pairs alternately, and every other one follows the law of the Clapeyronian numbers. In fact those with the odd indices *are* those numbers, and the even ones, commencing with 0, 1, 3, follow the same law.

The above comprises *all* the formulæ thus far given, in a shape very convenient for reference. The reader who has followed attentively our explanation of their use, needs nothing more to solve the case of equal spans completely. The expressions, however, for the *reactions* are *unnecessary*. As we have seen from Art. 148, we need only the *moments* and *shears* at any support in practical calculations. The practical formulæ necessary and sufficient for *any* case will be found in the next Art.

150. Girder continuous over any Number of Level Supports; Symmetrical with respect to the Centre, and with two variable end Spans nl and pl on each side.—[Fig. 89.] *

Moments at Supports:

when $m < r + 1$, $M_m = \dfrac{c_m}{l} \dfrac{A\, c_{s-r+2} + A'\, c_{s-r+1}}{p\, c_{s-1} + 2(n+p)\, c_s}$;

when $m > r$, $M_m = \dfrac{c_{s-m+2}}{l} \dfrac{A\, c_r + A'\, c_{r+1}}{p\, c_{s-1} + 2(n+p)\, c_s}$.

Shear at Supports—loaded span,

$$S_r = \dfrac{M_r - M_{r+1}}{l_r} + q,$$

$$S'_{r+1} = \dfrac{M_{r+1} - M_r}{l_r} + q';$$

*un*loaded spans,

$$S_m = \dfrac{M_m - M_{m+1}}{l_m},$$

$$S'_m = \dfrac{M_m - M_{m-1}}{l_{m-1}}.$$

For uniform load,

$$A = A' = \tfrac{1}{4} w\, l_r^3; \quad q = q' = \tfrac{1}{2} w\, l_r.$$

---

* *Jour. Franklin Institute*, March and April, 1875.

CHAP. XIII.]   ANALYTICAL FORMULÆ.   229

For concentrated load,

$$A = P\,l_r^2\,(2k - 3k^2 + k^3), \quad A' = P\,l_r^2\,(k - k^3),$$

and $\quad q = P(1 - k), \quad q' = P\,k, \quad k$ being always $\dfrac{a}{l_r}$.

The quantities denoted by c are also as follows :

$$c_1 = 0, \quad c_2 = 1,$$

$$c_3 = \frac{-2p - 2n}{p},$$

$$c_4 = \frac{p(4 + 3p) + n(4 + 4p)}{p},$$

$$c_5 = \frac{-p(14 + 12p) - n(14 + 16p)}{p},$$

$$c_6 = \frac{p(52 + 45p) + n(52 + 60p)}{p},$$

$$c_7 = \frac{-p(194 + 168p) - n(194 + 224p)}{p},$$

$$c_8 = \frac{p(724 + 617p) + n(724 + 836p)}{p}, \text{ etc.,}$$

following the law of the Clapeyronian numbers.

151. Application of the above Formulæ. — The formulæ of the preceding Art. comprise in a most compact form all the formulæ hitherto given, and are all that is necessary for the complete solution of any practical case.

Thus, by making $p =$ unity and retaining only n, we have the case of a girder with variable end spans $n\,l$, of different length from the others, which latter are all equal and represented by l. The reader will find no difficulty in using the above. For any particular case, when w or P and l, k, n and p are given, A, A', q and q' can be easily found, and the problem is solved. If n and p be both unity, we have the formulæ for all spans equal. The expressions for M_m will then reduce to those already given in Art. 144. Thus, in Art. 145 we have already found for seven equal spans, $l = 80$ ft., load $P = 40$ distant 50 ft. from left; the moment $M_4 = 202.4$. Now, from our formulæ above, we find for M_5 making $m = 5$, $s = 7$, $r = 4$; $M_5 = 272.8$.

Then by our formulæ for shear, $S_4 = +14.12$, or nearly

what we have assumed in Art. 145. We may also find the same shear by finding the algebraic sum of the reactions at A B C and D from the formulæ of Art. 147. This is more tedious, and, as we see, unnecessary. The moments can be easily found, and then the shear obtained directly from these.

We must bear in mind that l_1 always denotes the span the load is upon, whether $n\,l, p\,l$, or l, while l_m is any span in general, according to the value of m.

152. Continuous Girder with fixed ends.—It is worthy of remark that if n be made *zero* in the formulæ of Art. 149, we have a girder *with fastened ends* and variable end spans $p\,l$. If in addition p is unity, then all the spans become equal. We must, however, remember that when we thus make $n = 0$, the number of spans is $s - 2$ instead of s, as before, and the end spans are $p\,l$; the end supports are also 2 and s instead of 1 and $s + 1$.

153. Examples.—As illustrations of the use of the formulæ of Art. 150, we give a few examples.

Ex. 1. *A beam of one span is fixed horizontally at the ends. What are the end moments and reactions for a concentrated weight distant* k l *from the left end?*

Here the two outer spans of three spans are supposed zero. Therefore, $s - 2 = 1$, and $s = 3$. The left end is 2 instead of 1, and the right end 3. Hence, $r = 2, p = 0$, and $n = 1$ in the formulæ of Art. 150. We have, then,

$$c_1 = 0, \quad c_2 = 1, \quad c_3 = -2, \quad c_4 = 4, \quad \text{and hence,}$$

$$\text{for } m = 2, \quad M_2 = \frac{c_2}{l} \frac{A\, c_3 + A'\, c_2}{c_2 + 2\, c_3};$$

$$\text{for } m = 3, \quad M_3 = \frac{c_2}{l} \frac{A\, c_2 + A'\, c_3}{c_2 + 2\, c_3};$$

or, inserting the values of c above,

$$M_2 = \frac{1}{3\,l}(2\,A - A'), \quad M_3 = -\frac{1}{3\,l}(A - 2\,A').$$

For a concentrated load, $A = P\,l^2\,(2\,k - 3\,k^2 + k^3)$, and $A' = P\,l^2\,(k - k^3)$. Hence, $M_2 = P\,l\,(k - 2\,k^2 + k^3)$, and $M_3 = P\,l\,(k^2 - k^3)$.

For the reaction at the left end, which is in this case the same as the shear, we have

$$S_2 = \frac{M_2 - M_3}{l} + P(1-k) \quad \text{or} \quad S_2 = P(1 - 3k^2 + 2k^3),$$

$$S_3 = \frac{M_3 - M_2}{l} + Pk \quad \text{or} \quad S_3 = P(3k^2 - 2k^3).$$

For a load anywhere, we have simply to give the proper value to k, and we have at once the reactions and moments. Thus, for a load at $\frac{1}{4}$ the span from the left, $k = \frac{1}{4}$, and

$$S_2 = \tfrac{27}{32} P, \quad S_3 = \tfrac{5}{32} P; \quad M_2 = \tfrac{9}{64} Pl, \quad M_3 = \tfrac{3}{64} Pl.$$

For a load in centre, $k = \frac{1}{2}$, and

$$S_2 = S_3 = \tfrac{1}{2} P, \quad M_2 = M_3 = \tfrac{1}{8} Pl.$$

[Compare Supplement to Chap. VII., Arts. 16 and 17.]

Ex. 2. *For a uniform load over the same beam, what are the end moments and reactions?*

We have simply to introduce the proper values of A and A' for this case, and we have at once

$$M_2 = \tfrac{1}{12} w l^2 = M_3 \quad \text{and} \quad S_2 = S_3 = \tfrac{1}{2} w l.$$

Ex. 3. *A girder of three equal spans is "walled in" at the ends, and has a concentrated load in the first span. What are the moments, shears, and reactions at the ends and intermediate supports?*

In this case, $s - 2 = 3$, and hence $s = 5$, $r = 2$, $n = 0$, $p = 1$, and therefore

$$M_2 = \frac{c_2}{l} \frac{A c_3 + A' c_4}{c_4 + 2 c_5},$$

$$M_3 = \frac{c_4}{l} \frac{A c_2 + A' c_3}{c_4 + 2 c_5}, \quad M_4 = \frac{c_3}{l} \frac{A c_2 + A' c_3}{c_4 + 2 c_5}, \text{ etc.;}$$

also, $c_1 = 0$, $c_2 = 1$, $c_3 = -2$, $c_4 = 7$, $c_5 = -26$, etc.

Inserting these values and the values of A and A' for concentrated load, we have

$$M_2 = \frac{Pl}{45}(45k - 78k^2 + 33k^3), \quad M_3 = \frac{21}{45} Pl(k^2 - k^3),$$

$$M_4 = -\frac{2}{45}(3k^2 - 3k^3), \quad M_5 = \frac{1}{45}(3k^2 - 3k^3).$$

Observe that the moments are positive at each end of the loaded span and alternate in sign from that span, varying as the numbers 1, 2, 7, or as the numbers c [Art. 140]. A positive moment always denotes compression in lower fibre. For the shears we have, then, from Art. 150,

$$S_2 = \frac{P}{45}(45 - 99\,k^2 + 54\,k^3),\quad S'_3 = \frac{P}{45}(99\,k^2 - 54\,k^3).$$

$$S_3 = \frac{P}{45}(27\,k^2 - 27\,k^3),\qquad S'_4 = \frac{P}{45}(-27\,k^2 + 27\,k^3),$$

$$S_4 = \frac{P}{45}(-9\,k^2 + 9\,k^3),\qquad S'_5 = \frac{P}{45}(9\,k^2 - 9\,k^3).$$

For the reactions, then,

$$R_2 = S_2,\ R_3 = S'_3 + S_3 = \frac{P}{45}(126\,k^2 - 81\,k^3),$$

$$R_4 = S'_4 + S_4 = \frac{P}{45}(-36\,k^2 + 36\,k^3),\ R_5 = S'_5.$$

Observe that the reactions as also the shears are positive at the supports of the loaded span, and alternate in sign from those supports. A positive shear or reaction acts always upwards. Disregarding, then, for the present, the weight of the beam itself, it would have to be *held down* at first pier from right end.

If the weight is in the centre of first span from left, $k = \frac{1}{2}$, and

$$M_2 = \frac{57}{360}\,Pl,\ M_3 = \frac{21}{360}\,Pl,\ M_4 = -\frac{12}{360}\,Pl,\ M_5 = \frac{6}{360}\,Pl.$$

$$R_2 = S_2 = \frac{216}{360}\,P,\ R_3 = \frac{171}{360}\,P,\ R_4 = -\frac{36}{360}\,P,\ R_5 = \frac{9}{360}\,P.$$

The reactions add up to P, as they should.

If $P = 100$ tons, and $l = 15$ ft., we have

$M_2 = 237.5$ ft. tons, $M_3 = 87.5$, $M_4 = -50$, $M_5 = 25$ ft. tons.
$R_2 = 60$ tons $\quad R_3 = 47.5,\quad R_4 = -10,\quad R_5 = 2.5$ tons;
$S_2 = 60$ tons, $\quad S'_3 = 40,\quad S_3 = 7.5,\quad S'_4 = -7.5,$
$\quad\quad\quad\quad\quad\ S_4 = -2.5,\ S'_5 = 2.5.$

Ex. 4. *A beam of five spans, free at ends; centre and adjacent spans 100 ft., end spans each 75 ft., has a uniform load extending over the whole of the second span from left. What are the moments at the ends and supports?*

CHAP. XIII.]   ANALYTICAL FORMULÆ.   233

Here $s = 5$, $n = \frac{3}{4}$, $p = 1$, $r = 2$; therefore, from Art. 150,

$$M_2 = \frac{c_2}{l}\frac{A\, c_5 + A'\, c_4}{c_4 + \frac{1}{2} c_5},\quad M_3 = \frac{c_4}{l}\frac{A\, c_2 + A'\, c_3}{c_4 + \frac{1}{2} c_5}, \text{ etc.;}$$

and $c_1 = 0$, $c_2 = 1$, $c_3 = -\frac{1}{2}$, $c_4 = 13$, $c_5 = -48.5$.

Since, then, $A = A' = \dfrac{w\, l^2}{4}$ for uniform load, we have

$$M_1 = 0,\ M_2 = \tfrac{13}{627} w\, l^2,\ M_3 = \tfrac{65}{1254} w\, l^2,\ M_4 = -\tfrac{35}{2508} w\, l^2,$$
$$M_5 = \tfrac{5}{1254} w\, l^2,\ M_6 = 0.$$

If the load is two tons per ft., $w\, l^2 = 20{,}000$, and
$M_1 = 0$, $M_2 = 414.7$, $M_3 = 1036.6$, $M_4 = -279.5$, $M_5 = 79.7$.

Find the shears and reactions at each support.

Ex. 5. *A beam of four equal spans, has the second span from left covered with full load. What is the moment and shear at left of load ?*

Ans. $M_2 = \tfrac{13}{224} w\, l^2$, $\quad S_2 = \tfrac{111}{224} w\, l$.

What at right of load?

Ans. $M_3 = \tfrac{12}{224} w\, l$, $\quad S'_3 = \tfrac{113}{224} w\, l$.

What are the formulæ for concentrated load?

Ans.
$$M_2 = \frac{1}{56} [26\, k - 45\, k^2 + 19\, k^3]\, P\, l,$$

$$M_3 = \frac{4}{56} [7\, k - 12\, k^2 + 5\, k^3]\, P\, l,$$

$$S_2 = \frac{P}{56} [56 - 58\, k + 3\, k^2 - k^3],$$

$$S'_3 = [58\, k - 3\, k^2 + k^3]\, \frac{P}{56}.$$

Examples might be multiplied indefinitely.

The above is sufficient to show the comprehensiveness of our formulæ, and the ease with which results may be obtained, which, by the usual methods, would require long and intricate mathematical discussions. The points of inflection and the deflection may also in any case be easily determined, and general equations similar to the above deduced, but, as we have seen, the above are sufficient for full and complete calculation.

154. Tables for Moments.—From the formulæ of Art. 150 we can easily find the moments for both uniform and concentrated load in a single span for various numbers of spans. If these results are tabulated we shall obtain tables from which the moments may be at once taken. The formulæ for the shears are so easy when for any case the moments are known, that it is unnecessary to give tables for these.

The reader will do well to make himself perfectly familiar with the formulæ by calculating the moments for various cases, and comparing with the following tables. We give the practical case of variable end spans nl and equal intermediate spans l.

TABLE FOR MOMENTS—UNIFORM LOAD OVER ANY SINGLE SPAN.

Coefficients of wl^2 from table. End spans nl.

Supports counted from left.

1	2	3	4	5	6	
0	n^2	$(2+2n)n^2$	$(7\ 8n)n^2$	$(26+30\,n)n^2$	$(97+112\,n)n^2$	I.
0	$1+2n$	$(1+2n)(2+2n)$	$(1+2n)(7+8n)$	$(1+2n)(26+30\,n)$	etc.	II.
0	$5+6n$	$(5+6n)(2+2n)$	$(5+6n)(7+8n)$	etc.		III.
0	$19+22n$	$(19+22n)(2+2n)$	$(19+22n)(7+8n)$			IV.
0	$71+82n$	$(71+82n)(2+2n)$	etc.			V.
0	$265+304n$	etc.				VI.

Loaded span from right.

Number of spans.	One-fourth of Denominator.
1	
2	
3	$3+8n+4n^2$
4	$12+23n+16n^2$
5	$45+104n+60n^2$
6	$168+383n+224n^2$

The above table can be easily extended to include any number of spans. It is precisely the same as the table of Art. 140, and, in fact, includes that table. We have only to make $n = 1$ and we have at once the table for equal spans. Suppose we take five spans, load in second from right. From the smaller table we have at once for the denominator of the coefficient of wl^2, $4(45 + 104\,n + 60\,n^2)$. Then from the other table we have at support 1 from left $M_1 = 0$,

at support 2, $$M_2 = \frac{(1+2n)\,wl^2}{4(45+104n+60n^2)},$$

at support 3, $\quad M_3 = \dfrac{(1+2n)(2+2n)\,w\,l^2}{4\,(45+104\,n+60\,n^2)}$, etc.

If the load were in second span from *left*, and supports to right of load were required, we have simply to count the supports the other way in the table.

Thus, $M_6 = 0$, $M_5 = \dfrac{(1+2n)\,w\,l^2}{4\,(45+104\,n+60\,n^2)}$, or same as M_2 in first case, etc.

TABLE FOR MOMENTS—CONCENTRATED LOAD IN ANY SPAN, $k = \dfrac{a}{l_r}$.

End spans n l. Coefficients of $P\,l_r$ *from table.*

	$a = a'$	β	γ	$\theta = \theta'$	β'	γ'	
II.	3+2 n	1	2 n	1	2+2 n	3+4 n	II'.
III.	9+10 n	2+2 n	3+4 n	2+2 n	7+8 n	12+14 n	III'.
IV.	33+38 n	7+8 n	12+14 n	7+8 n	26+30 n	45+52 n	IV'.
V.	123+142 n	26+30 n	45+52 n	26+30 n	97+112 n	168+194 n	V'.
VI.	459+530 n	97+112 n	168+194 n	97+112 n	362+418 n	627+724 n	VI'.
VII.	etc.	etc.	etc.	etc.	etc.	etc.	VII'.

(Loaded span from left / Loaded span from right)

Number of spans.	Denominator Δ.
2	
3	$3+8n+4n^2$
4	$12+28n+16n^2$
5	$45+104n+60n^2$
6	$168+388n+224n^2$
7	etc.

The above tables give *only* the moments *at the supports* of the loaded span. The Roman numerals I., II., III., etc., denote the number of this span from *left*, and I'., II'., III'., the number of the loaded span from right. The expression for the moment at left support is

$$M = P\,l_r \dfrac{\theta}{\Delta}\left[\gamma'\,k - 3\,\beta'\,k^2 + a'\,k^3.\right]$$

For the *right* support,

$$M' = P\, l_r \frac{\theta'}{\Delta}\Big[\gamma k + 3\beta k^2 - a\, k^3\Big],$$

where the expressions for θ, θ', Δ, γ, γ', β, β', a, a', are to be taken from the tables and inserted.

Thus, for five spans load in second span from right, or third from left, we have at once $\Delta = 45 + 104\, n + 60\, n^2$. For the moment at the left support, to find θ, we must take horizontal line for III., and thus find $(2 + 2n)$. For γ', β' and a' we must take II'., and find, therefore, $3 + 4n$, $2 + 2n$ and $3 + 2n$. Hence, moment at left support is

$$M = M_4 = P\, l_r \frac{2+2n}{45+104\,n+60\,n^2}\Big[(3+4n)\,k - 3\,(2+2n)\,k^2 + (3+2n)\,k^3\Big].$$

For moment at right support, we must take line II'. for θ' and line III. for γ, β and a, and hence

$$M' = M_5 = \frac{P\, l_r}{45+104\,n+60\,n^2}\Big[(3+4n)\,k + 3\,(2+2n)\,k^2 - (9+10n)\,k^3\Big].$$

Since the span in question is not an end span, $l_r = l$ and $k = \dfrac{a}{l}$.

For a load in an end span, use the formulæ of Art. 150.

For a load in middle span of five spans, *i.e.*, third span from each end, we have

$$M = M_3 = \frac{P\, l\,(2+2n)}{45+104\,n+60\,n^2}\Big[(12+14n)\,k - 3\,(7+8n)\,k^2 + (9+10n)\,k^3\Big].$$

$$M' = M_4 = \frac{P\, l\,(2+2n)}{45+104\,n+60\,n^2}\Big[(3+4n)\,k + 3\,(2+2n)\,k^2 - (9+10n)\,k^3\Big].$$

When $k = \dfrac{1}{2}$, both these moments become, as they should, equal for any assumed value of n, as the reader may readily prove by insertion.

For the other moments *not* adjacent to the loaded span, the rule of Art. 140 holds good.

Thus, $M_1 = 0$, $M_2 = \dfrac{1}{\theta_r} M_r$, $M_3 = \dfrac{\theta_2}{\theta_r} M_r$, $M_4 = \dfrac{\theta_4}{\theta_r} M_r$, etc., and similarly on the other side,

$$M_s = 0, \quad M_{s-1} = \frac{1}{\theta_{r+1}} M_{r+1}, \quad M_{s-2} = \frac{\theta_{s-2}}{\theta_{r+1}} M_{r+1}, \text{ etc.}$$

We must remember always to give the proper *signs* to the moments, viz., positive for extremities of loaded span, and alternating each way from these for the others. From the formulæ of Art. 140 we can then easily find the shear at any support.

155. Continuous Girder—Level Supports—Spans all different—General Formulæ.[*]—The preceding formulæ comprise the case of one or, at most, two variable end spans. We give below the general formulæ for *all* spans different. These formulæ include all the others as special cases. Thus, if we make all spans equal, we have the formulæ of Art. 149. If end spans l_1 and l_s are made zero, and we take the number of spans equal to $s-2$, and first support 2, we have the continuous girder with *fixed ends*, in which the intermediate spans may or may not be equal, as we choose. If we make $l_1 = 0$ or $l_s = 0$ alone, and $s - 1 = $ No. of spans, we have a continuous girder fixed at *one* end only. In short, the formulæ comprise the entire case of level supports. They are as follows:

Let $s = $ number of spans, $l_r = $ length of loaded span, $k = \dfrac{a}{l_r}$, a being distance of load from left support; $l_1, l_2, l_3 \ldots l_{s-1} l_s$, the length of the various spans counting from left.

Then, when $m < r + 1$, $M_m = c_m \dfrac{A \, d_{s-r+2} + B \, d_{s-r+1}}{l_2 \, d_{s-1} + 2 \, (l_1 + l_2) \, d_s}$; when

$m > r$, $M_m = d_{s-m+2} \dfrac{A \, c_r + B \, c_{r+1}}{l_{s-1} \, c_{s-1} + 2 \, (l_s + l_{s-1}) c_s}$.[†] For the shear

at supports of loaded span, $S_r = \dfrac{M_r - M_{r+1}}{l_r} + q$, $S'_{r+1} = \dfrac{M_{r+1} - M_r}{l_r} + q'$. For *un*loaded spans, $S_m = \dfrac{M_m - M_{m+1}}{l_m}$, $S'_m = \dfrac{M_m - M_{m-1}}{l_{m-1}}$.

---

[*] These formulæ were first given by Mr. Merriman, and may be found in the *London Phil. Magazine*, Sept., 1875.

[†] We can put $l_2 \, d_{s-1} + 2 \, (l_1 + l_2) \, d_s = - d_{s+1} l_1$ and $l_{s-1} \, c_{s-1} + 2 \, (l_s + l_{s-1}) \, c_s = - c_{s+1} l_s$. These values may be used for supported ends. The above values are, however, the most general, and hold good not only for supported ends, but for *fixed* ends as well.

For the reaction at any support, $R_m = S'_m + S_m$. In which we have always $k = \dfrac{a}{l_r}$, $q = P(1-k)$, $q' = Pk$.

$A = P l_r^2 (2k - 3k^2 + k^3)$ and $B = P l_r^2 (k - k^3)$ for concentrated load; and $q = q' = \tfrac{1}{2} w l_r$, and $A = B = \tfrac{1}{4} w l_r^3$ for uniform load entirely covering any one span.

Also for c and d we have the following values:

$$c_1 = 0,$$
$$c_2 = 1,$$
$$c_3 = -\frac{2(l_1 + l_2)}{l_2},$$
$$c_4 = \frac{4(l_1 + l_2)(l_2 + l_3) - l_2^2}{l_2 l_3},$$
$$c_5 = -2 c_4 \left(\frac{l_3 + l_4}{l_4}\right) - c_3 \frac{l_3}{l_4},$$

or, generally,

$$c_m = -2 c_{m-1} \left(\frac{l_{m-2} + l_{m-1}}{l_{m-1}}\right) - c_{m-2} \left(\frac{l_{m-2}}{l_{m-1}}\right),$$

$$d_1 = 0,$$
$$d_2 = 1,$$
$$d_3 = -2 \frac{l_s + l_{s-1}}{l_{s-1}},$$
$$d_4 = \frac{4(l_s + l_{s-1})(l_{s-1} + l_{s-2}) - l_{s-1}^2}{l_{s-1} l_{s-2}},$$
$$d_5 = -2 d_4 \frac{l_{s-2} + l_{s-3}}{l_{s-3}} - d_3 \frac{l_{s-2}}{l_{s-3}};$$

or, generally,

$$d_m = -2 d_{m-1} \frac{l_{s-m+3} + l_{s-m+2}}{l_{s-m+2}} - d_{m-2} \frac{l_{s-m+3}}{l_{s-m+2}}.$$

As an illustration of the use of the above formulæ, let us take three unequal spans, load in the first. Then $s = 3$, $r = 1$, $k = \dfrac{a}{l_1}$. For moment at second support, $m = 2$, or $m > r$; hence

$$M_2 = d_3 \frac{A c_1 + B c_2}{l_2 c_2 + 2(l_3 + l_2) c_3}.$$

But $c_1 = 0$, $c_2 = 1$,

$$c_3 = -\frac{2(l_1 + l_2)}{l_2} \text{ and } d_3 = -2\left(\frac{l_3 + l_2}{l_2}\right);$$

hence, since $\mathbf{B} = \mathbf{P} l_1^2 (k - k^3)$,

$$\mathbf{M}_2 = \frac{2 \mathbf{P} l_1^2 (k - k^3)(l_3 + l_2)}{4(l_3 + l_2)(l_1 + l_2) - l_2^2}.$$

If in this we make $l_1 = l_3$, we have the extreme spans *equal*, and then

$$\mathbf{M}_2 = \frac{2 \mathbf{P} l_1^2 (k - k^3)(l_1 + l_2)}{4(l_1 + l_2)^2 - l_2^2}.$$

If we make in this, again, $l_1 = l_2$, we have for *all* spans equal

$$\mathbf{M}_2 = \frac{4 \mathbf{P} l (k - k^3)}{15},$$

just what we should have from Art. 149.

For the reaction at the end support, we have

$$\mathbf{R}_1 = \mathbf{S}_1 = \frac{\mathbf{M}_1 - \mathbf{M}_2}{l_1} + \mathbf{P}(1 - k),$$

or, since $\mathbf{M}_1 = 0$,

$$\mathbf{S}_1 = \frac{-\mathbf{M}^2}{l_1} + \mathbf{P}(1 - k) = \frac{-2 \mathbf{P} l_1 (k - k^3)(l_3 + l_2)}{4(l_3 + l_2)(l_1 + l_2) - l_2^2} + \mathbf{P}(1 - k).$$

For all spans equal, or $l_1 = l_2 = l_3 = l$, this reduces to

$$\mathbf{S}_1 = \frac{\mathbf{P}}{15}(15 - 19 k + 4 k^3),$$

as we should have found from Art. 149.

Ex. 1.—*A beam of one span is fixed horizontally at the right end; what are the reactions and the moments for concentrated load?*

Here $s - 1 = 1$ or $s = 2$, $r = 1$, $l_2 = 0$, and from the formulæ of Art. 155, $c_1 = 0$, $c_2 = 1$, and $d_1 = 0$, $d_2 = 1$, $d_3 = -2$,

$$\mathbf{M}_1 = 0, \quad \mathbf{M}_2 = d_2 \frac{\mathbf{A} c_1 + \mathbf{B} c_2}{l c_1 + 2 l c_2} = \frac{\mathbf{B} c_2}{2 l c_2},$$

or

$$\mathbf{M}_2 = \frac{\mathbf{B}}{2 l} = \frac{\mathbf{P} l}{2}(k - k^3)$$

$$\mathbf{S}_1 = -\frac{\mathbf{M}_2}{l} + \mathbf{P}(1 - k) = \frac{\mathbf{P}}{2}(2 - 3 k + k^3),$$

$$\mathbf{S'}_2 = \frac{\mathbf{P}}{2}(3 k - k^3).$$

240 CONTINUOUS GIRDER. [CHAP. XIII.

Ex. 2.—*A beam of three spans of 25, 50 and 40 feet respectively is fixed horizontally at the right end, and has a concentrated load of 10 tons at 12 feet from the third support from left. What are the moments at the supports?*

Here $l_1 = 25$, $l_2 = 50$, $l_3 = 40$, $l_4 = 0$, $P = 10$, $kl_3 = 12$, $k = 0.3$, $s - 1 = 3$, $s = 4$, $l_4 = 0$ and $r = 3$. Also, $c_1 = 0$, $c_2 = 1$, $c_3 = -3$, $c_4 = 12.25$ and $d_1 = 0$, $d_2 = 1$, $d_3 = -2$, $d_4 = 6.4$, $d_5 = -32.4$.

When, then, $m < 4$,

$$M_m = \frac{c_m}{860}(-2A + B) = -\frac{c_n P l_3^2}{860}(3k - 6k^2 + 3k^3).$$

Inserting $k = 0.3$ and the values of c,
for $m = 1$, $M_1 = 0$; $m = 2$, $M_2 = -8.20$; $m = 3$, $M_3 = 24.62$;
for $n = 4$,

$$M_4 = \frac{d_2}{860}(-3A + 12.25 B) = \frac{P l_3^2}{860}(6.25 k + 9k^2 - 15.25 k^3),$$

or $M_4 = 42.29$ ft. tons.

Find the shears. Also moments and shears for uniform load over third span.

Ex. 3.—*A beam of four spans $l_1 = 80$, $l_2 = 100$, $l_3 = 50$, $l_4 = 40$ ft., free at the ends, has a load of 10 tons in the second span at 40 ft. from left. What are the moments?*

Here $s = 4$, $c_1 = 0$, $c_2 = 1$, $c_3 = -3.6$, $c_4 = 19.6$, $c_5 = -83.7$, $d_1 = 0$, $d_2 = 1$, $d_3 = -3.6$, $d_4 = 10.3$, $d_5 = -41.85$, $r = 2$.

For $m < 3$, $M_m = -\frac{c_m}{80 d_5}(A d_4 + B d_3)$,

$m > 3$, $M_m = -\frac{d_{6-m}}{40 c_5}(A c_2 + B c_3)$.

Hence, $M_1 = 0$,

$$M_2 = \frac{10.3 A - 3.6 B}{3348} = \frac{P l_2^2}{3348}(17 k - 30.9 k^2 + 13.9 k^3) = 82.01,$$

$$M_3 = -3.6 \frac{A - 3.6 B}{3348} = \frac{3.6 P l_2^2}{3348}(1.6 k + 3 k^2 - 4.6 k^3) = 88.56,$$

$$M_4 = \frac{A - 3.6 B}{3348} = -\frac{P l_2^2}{3348}(1.6 k + 3 k^2 - 4.6 k^3) = -24.05$$

$M_5 = 0$.

Find the shears. Also find the moments and shear for uniform load over second span.

156. Thus we see that, as in Art. 150, a few short and simple formulæ, which may be written on a piece of paper the size of one's hand, are all that we need for the complete solution of any case of level supports—whether the spans be all equal or the end ones only different, or all different; whether the girder merely rest on the end supports or be fastened horizontally at one or both ends. We have only to remember that a positive moment causes tension in upper flange at support, and therefore compression in lower; inversely for negative moment. Also, that a positive shear acts upwards, and a negative shear downwards. Also, that both moment and shear are positive at supports of loaded span, and alternate in sign both ways. This is all that we need to form properly the equation of moments at any apex, and determine the quality of the strains in flanges and diagonals. We can thus solve any practical case of framed continuous girder which can ever occur with little more difficulty than in the case of a simple girder.

Thus, for the span **D E** (Fig. 87) we have only to find the moments at **D** and **E** due to every position of **P** in the span **D E**, and the corresponding shears at **D**. These once known, and, as we have seen, they can be easily obtained from our formulæ, we can find and tabulate the strains in every piece due to each weight, as shown in Art. 127. An addition of these strains gives, then, the maxima of each kind due to interior loading.

We have, then, to find, in like manner, the strains due to the two cases of *exterior* loading as represented in Fig. 87, or else due to each exterior span loaded, making a column for each span. From the columns thus obtained, we can deduce the dead load strains, and then finally the total maximum strains of each kind for every piece. [See, for illustration of the above, Art. 127.]

Thus, the whole subject is solved with the aid of but four simple formulæ, and for a problem generally considered impossible by reason of its "complexity," our results will, we trust, be found sufficiently simple and practical.

In view of the fact that the necessary formulæ for practical computations have been often given in the later works of French and German authors, although perhaps never before in so compact and available a shape as above, it is indeed surprising that they should have been so completely ignored by English and American writers.

The tables and formulæ which we have given will, we trust,

bring the subject fairly within the reach of the practical engineer, and should they be the means of calling more general attention to this important class of structures, will not, we hope, be considered as out of place in the present treatise.

For the *influence of difference of level* of the supports, as well as for *variable cross-section* and the relative economy of the continuous girder, see Arts. 17 and 18 of the Appendix.

SUPPLEMENT TO CHAPTER XIII.

DEMONSTRATION OF ANALYTICAL FORMULÆ GIVEN IN TEXT.

In the following we shall give the complete development of the general formulæ of Art. 155. As these formulæ include, as we have seen, all the others as special cases, it is sufficient to show how they are obtained in order to enable the reader to deduce all the others.

1. Conditions of Equilibrium.—In the rth span of a continuous girder, whose length is l_r (see Fig.), take a point o vertically above the rth

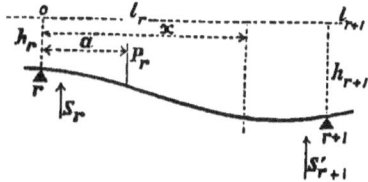

support as the origin of co-ordinates, and the horizontal line $o\,l$ as the axis of abscissas. At a distance x from the left support pass a vertical section, and between the support and this section let there be a single load P_r whose distance from the support is a.

Now all the exterior forces which act on the girder to the left of the support r we consider as replaced, without disturbing the equilibrium, by a resultant *moment* M_r and a resultant vertical shearing force S_r. This moment is equal and opposite to the moment of the internal forces at the section through the support r; while the vertical force is equal and opposite to the shear.

Not only over the support, but also at every section, the interior forces must hold the exterior ones in equilibrium, and therefore we have the conditions:

1st. The sum (*algebraic*) of all the horizontal forces must be zero.
2d. The sum (*algebraic*) of all the vertical forces must be zero.
3d. The sum (*algebraic*) of the moments of all the forces must be zero.

Thus, for the section x, we have from the third condition

$$\Sigma M = M_r - S_r\,x + P_r\,(x-a) - m = 0 \quad \ldots \quad (1)$$

where m is the moment at the section. From this we have

$$m = M_r - S_r\,x + P_r\,(x-a) \quad \ldots \ldots \ldots (2)$$

If in this we make $x = l_r$, m becomes M_{r+1}, and we thus have for the shear just to the right of the left support of the loaded span

$$S_r = \frac{M_r - M_{r+1}}{l_r} + \frac{P_r}{l_r}(l_r - a).$$

For an unloaded span the weight P disappears, and

$$S_m = \frac{M_m - M_{m+1}}{l_m}.$$

For the shear just to the left of the *right* support of loaded span,

$$S'_{r+1} = P - S_r = \frac{M_{r+1} - M_r}{l_r} + \frac{P_r a}{l_r}.$$

For unloaded span, the weight P disappears, and

$$S'_m = \frac{M_m - M_{m-1}}{l_{m-1}}.$$

S'_m is then the shear to the left of any support m, and S_m that to the right. The reaction at any support is therefore

$$R_m = S'_m + S_m.$$

These are the formulæ already given in Art. 148.

2. Equation of the Elastic Line.—We can now easily make out the equation of the elastic line for the continuous girder of constant cross-section, or constant moment of inertia.

The differential equation of the elastic line is,[*]

$$E I \frac{d^2 y}{d x^2} = m \quad \ldots \ldots \quad (3)$$

where E is the coefficient of elasticity, and I the moment of inertia.

If now we insert in (3) the value of m, as given in (2), we have

$$\frac{d^2 y}{d x^2} = \frac{M_r - S_r x + P_r (x - a)}{E I}.$$

Integrating [†] this between the limits $x = 0$ and x, and upon the condition that x cannot be less than a, the constant of integration $\frac{dy}{dx} = t_r =$ the tangent of the angle, which the tangent to the deflected curve makes with the horizontal at r; and we have, since we must take the $\int P_r (x - a)$ simultaneously between the limits $x = a$ and x for $x = 0$ and x;

$$\frac{dy}{dx} = t_r + \frac{2 M_r x - S_r x^2 + P_r (x - a)^2}{2 E I} \quad \ldots \quad (3, a)$$

If we take the origin at a distance h_r (see Fig.) above the support r, then integrating again, the constant is h_r, and we have

$$y = h_r + t_r x + \frac{3 M_r x^2 - S_r x^3 + P_r (x - a)^3}{6 E I} \quad \ldots \quad (4)$$

which is the general equation of the elastic curve. If in this we make

---

[*] See Supplement to Chapter VII., Art. 11.
[†] Notice that when $x = 0$, $a = 0$, and hence $(x - a) = 0$ also.

ART. 3.] SUPPLEMENT TO CHAP. XIII. 245

$x = l_r$, y becomes h_{r+1}. If also we put $\dfrac{a}{l_r} = k$, or $a = k\,l_r$, and insert also for S_r its value as given in (2 a), we find for t_r

$$t_r = \frac{h_{r+1} - h_r}{l_r} - \frac{1}{6\,E\,I}\left[2\,M_r\,l_r + M_{r+1}\,l_r - P_r\,l_r^2\,(2\,k - 3\,k^2 + k^3)\right]\ ..(5)$$

We see, then, that the equation of the curve is completely determined, when we know M_r and M_{r+1}, the moments at the supports. These, as we shall see in the next Art., are readily found by the remarkable "theorem of three moments," already alluded to in Art. 144.

3. Theorem of Three Moments.

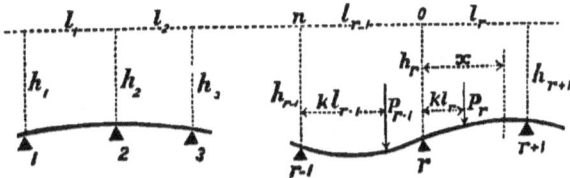

In the Fig. we have represented a portion of a continuous girder, the spans being l_1 l_2 ... l_r, etc., and the supports 1, 2 ... r, etc. Upon the spans l_{r-1} and l_r are the loads P_{r-1} and P_r, whose distances from the nearest left-hand supports are $k\,l_{r-1}$ and $k\,l_r$; k being any fraction expressing the ratio of the distance to the length of span.

The equation of the elastic line between P_r and the $r + 1^{\text{th}}$ support is given by (4), and the tangent of the angle which the curve makes with the axis of abscissas is given by (3 a). If in (3 a) we substitute for S_r its value from (2 a), and for t_r its value from (5), and make at the same time $x = l_r$, then $\dfrac{dy}{dx}$ becomes t_{r+1}, the tangent at $r + 1$, and we have

$$t_{r+1} = \frac{h_{r+1} - h_r}{l_r} + \frac{1}{6\,E\,I}\left[M_r\,l_r + 2\,M_{r+1}\,l_r - P_r\,l_r^2\,(k - k^3)\right].$$

Remove now the origin from o to n, and we may derive an expression for t_r by simply diminishing each of the indices above by unity; therefore

$$t_r = \frac{h_r - h_{r-1}}{l_{r-1}} + \frac{1}{6\,E\,I}\left[M_{r-1}\,l_{r-1} + 2\,M_r\,l_{r-1} - P_{r-1}\,l_{r-1}^2\,(k - k^3)\right].$$

Now, comparing these two equations, we may eliminate the tangents, and thus obtain

$$M_{r-1}\,l_{r-1} + 2\,M_r\,(l_{r-1} + l_r) + M_{r+1}\,l_r =$$

$$-\,6\,E\,I\left[\frac{h_r - h_{r-1}}{l_{r-1}} + \frac{h_r - h_{r+1}}{l_r}\right] + P_{r-1}\,l_{r-1}^2\,(k - k^3) + P_r\,l_r^2\,(2\,k - 3\,k^2 + k^3),$$

which is the most general form of the theorem of three moments for a girder of constant cross-section.

When the ends of the girder are merely supported, the end moments are, of course, zero. Then, for each of the piers, we may write an equation of the above form, and thus have as many equations as there are unknown moments.

4. Determination of the Moments—Supports all on level.

—When all the supports are in the same horizontal, the ordinates h_1, h_2, h_r, etc., are equal; and hence the term involving $\mathbf{E\,I}$ disappears, and we have simply

$$\mathbf{M}_{r-1}\, l_{r-1} + 2\, \mathbf{M}_r\, (l_{r-1} + l_r) + \mathbf{M}_{r+1}\, l_r =$$
$$\mathbf{P}_{r-1}\, l^2_{r-1}\, (k - k^3) + \mathbf{P}_r\, l_r^2\, (2\,k - 3\,k^2 + k^3),$$

as already given in Art. 144.

Now let s = number of spans, and let a single load $\mathbf{P}$ be placed on the rth span. [Pl. 23, Fig. 89.]

From the above theorem, since $\mathbf{M}_1$ and $\mathbf{M}_{s+1}$ are zero, we may write the following equations:

$$\left.\begin{aligned}
2\,\mathbf{M}_2\,(l_1 + l_2) + \mathbf{M}_3\,l_2 &= 0\,; \\
\mathbf{M}_2\,l_2 + 2\,\mathbf{M}_3\,(l_2 + l_3) + \mathbf{M}_4\,l_3 &= 0. \\
\cdots\cdots\cdots\cdots\cdots& \\
\cdots\cdots\cdots\cdots\cdots& \\
\mathbf{M}_{r-1}\,l_{r-1} + 2\,\mathbf{M}_r\,(l_{r-1} + l_r) + \mathbf{M}_{r+1}\,l_r &= \\
\mathbf{P}_r\,l_r^2\,(2\,k - 3\,k^2 + k^3) &= \mathbf{A}\,; \\
\mathbf{M}_r\,l_r + 2\,\mathbf{M}_{r+1}\,(l_r + l_{r+1}) + \mathbf{M}_{r+2}\,l_{r+1} &= \\
\mathbf{P}_r\,l_r^2\,(k - k^3) &= \mathbf{B}. \\
\cdots\cdots\cdots\cdots\cdots& \\
\cdots\cdots\cdots\cdots\cdots& \\
\mathbf{M}_{s-2}\,l_{s-2} + 2\,\mathbf{M}_{s-1}\,(l_{s-2} + l_{s-1}) + \mathbf{M}_s\,l_{s-1} &= 0\,; \\
\mathbf{M}_{s-1}\,l_{s-1} + 2\,\mathbf{M}_s\,(l_{s-1} + l_s) &= 0.
\end{aligned}\right\} \quad\cdot\cdot\ (6)$$

The solution of these equations can be best effected by the method of *indeterminate coefficients*, as referred to in Art. 136.

Thus we multiply the first equation by a number c_2, whose value we shall hereafter determine, so as to satisfy desired conditions. The second we multiply by c_3, the third by c_4, the rth by c_{r+1}, etc., the index of c corresponding always to that of $\mathbf{M}$ in the middle term. Having performed these multiplications, add the equations, and arrange according to the coefficients of $\mathbf{M}_2, \mathbf{M}_3$, etc. We thus have the equation

$$[2\,c_2\,(l_1 + l_2) + c_3\,l_2]\,\mathbf{M}_2 + [c_2\,l_2 + 2\,c_3\,(l_2 + l_3) + c_4\,l_3]\,\mathbf{M}_3 + \ldots$$
$$+ [c_{r-1}\,l_{r-1} + 2\,c_r\,(l_{r-1} + l_r) + c_{r+1}\,l_r]\,\mathbf{M}_r + \ldots$$
$$+ [c_{s-2}\,l_{s-2} + 2\,c_{s-1}\,(l_{s-2} + l_{s-1}) + c_s\,l_{s-1}]\,\mathbf{M}_{s-1}$$
$$+ [c_{s-1}\,l_{s-1} + 2\,c_s\,(l_{s-1} + l_s)]\,\mathbf{M}_s = \mathbf{A}\,c_r + \mathbf{B}\,c_{r+1}.$$

Now suppose we wish to determine $\mathbf{M}_s$. We have only to require that such relations shall exist among the multipliers c that all the terms in the first member of the above equation, except the last, shall disappear. We have then evidently, for the conditions which these multipliers must satisfy,

$$2\,c_2\,(l_1 + l_2) + c_3\,l_2 = 0\,;$$
$$c_2\,l_2 + 2\,c_3\,(l_2 + l_3) + c_4\,l_3 = 0\,;$$
$$\cdots\cdots\cdots\cdots\cdots$$
$$c_{r-1}\,l_{r-1} + 2\,c_r\,(l_{r-1} + l_r) + c_{r+1}\,l_r = 0\,;$$
$$\cdots\cdots\cdots\cdots\cdots$$
$$c_{s-2}\,l_{s-2} + 2\,c_{s-1}\,(l_{s-2} + l_{s-1}) + c_s\,l_{s-1} = 0\,;$$

ART. 4.]   SUPPLEMENT TO CHAP. XIII.   247

while for M_s we have at once,

$$M_s = \frac{A\, c_r + B\, c_{r+1}}{c_{s-1}\, l_{s-1} + 2\, c_s\, (l_{s-1} + l_s)} = -\frac{A\, c_r + B\, c_{r+1}}{c_{s+1}\, l_s}.$$

If, in like manner, we should multiply the *last* of equations (6) by the number d_2, the last but one by d_3, the rth by d_{s-r+1}, etc.; then add, and make all terms, except that containing M_2, equal to zero; we should have the conditions:

$$2\, d_2\, (l_s + l_{s-1}) + d_3\, l_{s-1} = 0;$$
$$d_2\, l_{s-1} + 2\, d_3\, (l_{s-1} + l_{s-2}) + d_4\, l_{s-2} = 0;$$
$$\cdots \cdots \cdots \cdots \cdots$$
$$d_{s-r+1}\, l_r + 2\, d_{s-r+2}\, (l_r + l_{r-1}) + d_{s-r+3}\, l_{r-1} = 0;$$
$$\cdots \cdots \cdots \cdots \cdots$$
$$d_{s-2}\, l_2 + 2\, d_{s-1}\, (l_2 + l_1) + d_s\, l_1 = 0;$$

while for the moment we have

$$M_2 = \frac{A\, d_{s-r+2} + B\, d_{s-r+1}}{d_{s-1}\, l_2 + 2\, d_s\, (l_2 + l_1)} = -\frac{A\, d_{s-r+2} + B\, d_{s-r+1}}{d_{s+1}\, l_1}.$$

The values of M_2 and M_s are thus given in terms of the quantities A and B and c and d.

A and B depend simply upon the load and its position in the rth span. Thus $\quad A = P\, l_r^2 (2\, k - 3\, k^2 + k^3), \quad B = P\, l_r^2 (k - k^3).$

As for the multipliers c and d, they depend only upon the lengths of the spans, and need only satisfy the conditions above. Hence, assuming $c_1 = 0$, $c_2 = 1$, and $d_1 = 0$, $d_2 = 1$, we can deduce the proper values for all the others. Thus,

$$c_1 = 0,\qquad\qquad\qquad\qquad d_1 = 0,$$
$$c_2 = 1,\qquad\qquad\qquad\qquad d_2 = 1,$$
$$c_3 = -2\frac{l_1 + l_2}{l_2},\qquad\qquad d_3 = -2\frac{l_s + l_{s-1}}{l_{s-1}},$$
$$c_4 = -2\, c_3\frac{l_2 + l_3}{l_3} - c_2\frac{l_2}{l_3},\qquad d_4 = -2\, d_3\frac{l_{s-1} + l_{s-2}}{l_{s-2}} - d_2\frac{l_{s-1}}{l_{s-2}},$$
$$c_5 = -2\, c_4\frac{l_3 + l_4}{l_4} - c_3\frac{l_3}{l_4},\qquad d_5 = -2\, d_4\frac{l_{s-2} + l_{s-3}}{l_{s-3}} - d_3\frac{l_{s-2}}{l_{s-3}},$$
$$c_6 = -2\, c_5\frac{l_4 + l_5}{l_5} - c_4\frac{l_4}{l_5},\qquad d_6 = -2\, d_5\frac{l_{s-3} + l_{s-4}}{l_{s-4}} - d_4\frac{l_{s-3}}{l_{s-4}},$$
$$\text{etc., etc.}\qquad\qquad\qquad \text{etc., etc.}$$

Now from equations (6) we see at once that $M_3 = c_3\, M_2$, $M_4 = c_4\, M_2$, etc., or, universally, when $n < r + 1$,

$$M_n = c_n\, M_2 = -\frac{c_n}{d_{s+1}\, l_1}\, (A\, d_{s-r+2} + B\, d_{s-r+1}) \quad . \quad . \quad (7)$$

Also taking the same equations in reverse order, $M_{s-1} = d_3\, M_s$, $M_{s-2} = d_4\, M_s$, etc., or, universally, when $n > r$,

$$M_n = d_{s-n+2}\, M_s = -\frac{d_{s-n+2}}{c_{s+1}\, l_s}\, (A\, c_r + B\, c_{r+1}) \quad . \quad . \quad . \quad (8)$$

Equations (7) and (8) are the general equations given in Art. 155, which, as we have seen, include the whole case of level supports.

5. Uniform Load.—For uniform load the same equations hold good. We have only to give a different value to **A** and **B**.

Thus, for several concentrated loads we should have

$$\mathbf{A} = \Sigma \, \mathbf{P} \, l_r^2 \, (2\,k - 3\,k^2 + k^3).$$

For a uniform load over the whole span l_r, let w be the load per unit of length, then

$$\Sigma \mathbf{P} = \int_0^{l_r} w\, d\,a\,; \text{ or since } a = k\,l_r, \; \Sigma \mathbf{P} = \int_0^{l_r} w\, l_r\, d\,k.$$

Inserting this in place of $\Sigma \mathbf{P}$ above, and integrating, we have

$$\mathbf{A} = \mathbf{B} = \tfrac{1}{4}\,w\,l_r^3.$$

Thus the equations of Art. 155 hold good for concentrated and uniform load in any span, for any number and any lengths of spans.

The above formulæ were first published in an article on the *Flexure of Continuous Girders*, by Mansfield Merriman, C.E., in the *London Phil. Magazine*, Sept., 1875.

6. Formulæ for the Tipper.—The expressions for the reactions in this case, already given in Art. 120, may be easily deduced. The solution is tedious by reason of lengthy reductions, but the process of deduction is simple.

The construction in this case is indicated in Fig. 83, Pl. 22. We suppose, as shown there, a weight upon the first span only. Under the action of this weight the beam deflects, and one centre support falls and the other rises *an equal amount*. Thus, if we take the level line as reference, $h_2 = -h_3$. Moreover, the reactions at these two supports must always be equal.

We have, then, as representing this state of things, $h_2 = -h_3$, and calling the supports 1, 2, 3 and 4, we have from Art. 1, since $\mathbf{M}_1 = \mathbf{M}_4 = 0$, and $l_1 = l_3$,

$$\left.\begin{aligned}
\mathbf{R}_1 = \mathbf{S}_1 = \mathbf{S}_r &= -\frac{\mathbf{M}_2}{l_1} + \mathbf{P}\,(1-k), \\
\mathbf{R}_2 = \mathbf{S'}_2 + \mathbf{S}_2 &= \frac{\mathbf{M}_2}{l_1} + \mathbf{P}\,k + \frac{\mathbf{M}_2 - \mathbf{M}_3}{l_2}, \\
\mathbf{R}_3 = \mathbf{S'}_3 + \mathbf{S}_3 &= \frac{\mathbf{M}_3 - \mathbf{M}_2}{l_2} + \frac{\mathbf{M}_3}{l_1}, \\
\mathbf{R}_4 = \mathbf{S'}_4 &= -\frac{\mathbf{M}_3}{l_1}.
\end{aligned}\right\} \quad \ldots \quad (9)$$

These reactions will evidently be known, if we can determine the moments.

Let $\mathbf{Y}_r = 6\,\mathbf{E}\,\mathbf{I} \left[\dfrac{h_r - h_{r-1}}{l_{r-1}} + \dfrac{h_r - h_{r+1}}{l_r} \right]$. Then the general equation of three moments of Art. 3 becomes, when we neglect $\mathbf{P}_r$, that is, suppose only the first span loaded:

$$\mathbf{M}_{r-1}\,l_{r-1} + 2\,\mathbf{M}_r\,(l_{r-1} + l_r) + \mathbf{M}_{r+1}\,l_r = -\mathbf{Y}_r + \mathbf{P}_{r-1}\,l_{r-1}^2\,(k - k^3).$$

This expresses a relation between the moments at three consecutive

ART. 7.]   SUPPLEMENT TO CHAP. XIII.   249

supports for load between the first two. Let $r-1=1$, or $r=2$. Then, since $M_1 = M_4 = 0$, we have

$$2 M_2 (l_1 + l_2) + M_3 l_2 = - Y_2 + P l_1^2 (k - k^3) = R \quad . \quad . \quad (10)$$

where R stands for convenience equal to the expression on right.

Let $r-1=2$, or $r=3$. Then the weight disappears, and since $l_1 = l_3$,

$$M_2 l_2 + 2 M_3 (l_2 + l_1) = - Y_3 \quad . \quad . \quad . \quad . \quad . \quad (11)$$

From (11) we have

$$M_3 = \frac{- Y_3 - 2 M_2 (l_2 + l_1)}{l_2} \quad . \quad . \quad . \quad . \quad . \quad (12)$$

But since R_2 must always equal R_3, we have from (9)

$$\frac{M_2 - M_3}{l_1} + \frac{2 M_2 - 2 M_3}{l_2} = - P k \quad . \quad . \quad . \quad . \quad (13)$$

Substituting (12) in (10), we have

$$M_2 = \frac{- R l_2 - 2 Y_3 (l_1 + l_2)}{3 l_2^2 + 8 l_2 l_1 + 4 l_1^2} \quad . \quad . \quad . \quad . \quad . \quad (14)$$

Substituting (12) in (13),

$$M_2 = \frac{l_1 l_2^2 P k - Y_3 (l_2 + 2 l_1)}{3 l_2^2 + 8 l_2 l_1 + 4 l_1^2} \quad . \quad . \quad . \quad . \quad . \quad (15)$$

From (14) and (15), we have then

$$- Y_3 - R = l_1 l_2 P k.$$

Insert in this the value of R from (10), and

$$Y_2 - Y_3 = P l_1^2 (k - k^3) + P l_1 l_2 k = P (l_1^2 k - l_1^2 k^3 + l_1 l_2 k).$$

Now in the present case $h_1 = 0$, $h_4 = 0$, and $h_3 = - h_2$, and since also $l_2 = l_1$,

$$Y_2 = 6 E I \left[- \frac{2 h_2}{l_2} - \frac{h_2}{l_1} \right]$$

and

$$Y_3 = 6 E I \left[\frac{2 h_2}{l_2} + \frac{h_2}{l_1} \right]. \quad \text{That is, } Y_2 = - Y_3.$$

Hence, from our equation above,

$$Y_2 = \frac{P}{2} \left[l_1^2 k - l_1^2 k^3 + l_1 l_2 k \right],$$

$$Y_3 = -\frac{P}{2} \left[l_1^2 k - l_1^2 k^3 + l_1 l_2 k \right].$$

Substituting these values of y_2 and y_3 in (11) and (13), we can obtain at once M_2 and M_3, which finally substituted in eq. (9), will give us the reactions as already given in Art. 120, when we put $n l$ in place of l_2.

7. In similar manner we can solve other problems. Thus—*what are the reactions for a girder continuous over three supports, the two right-hand ones resting upon an inflexible body which is pivoted at the centre ?*

This is the case of the tipper when raised at the centre so that the ends just touch, and then subjected to a load at any point of first span—the other end *not being latched down*, so that it rises freely, as though without weight of its own.

In this case we have from (9), since now $M_2 = 0$,

$$R_1 = -\frac{M_1}{l_1} + P(1-k), \quad R_2 = \frac{M_1}{l_1} + \frac{M_3}{l_2} = -Pk = Pk, \quad R_3 = -\frac{M_3}{l_2},$$

and $R_4 = 0$.

By the conditions R_2 must equal R_3, hence

$$\frac{2M_3}{l_2} + \frac{M_1}{l_1} = -Pk, \text{ or}$$

$$2l_1 M_3 + l_2 M_1 = -l_1 l_2 Pk \quad \ldots \ldots \quad (16)$$

From the equation of three moments above, we have, making $r - 1 = 2$, or $r = 3$, since then P disappears,

$$M_1 l_2 = -Y_2 \quad \ldots \ldots \ldots \quad (17)$$

or $$M_1 = -\frac{Y_2}{l_2}.$$

Substituting this value of M_1 in (16), we find

$$Y_2 = \frac{l_1 l_2^2 Pk}{2l_1 + l_2}; \text{ hence } M_1 = -\frac{l_1 l_2 Pk}{2l_1 + l_2},$$

and therefore, at once,

$$R_2 = \frac{l_1 Pk}{2l_1 + l_2} = R_3 \text{ and } R_1 = \frac{l_2 Pk}{2l_1 + l_2} + P(1-k).$$

Putting nl in place of l_2, we have

$$R_1 = \frac{nPk}{2+n} + P(1-k), \quad R_2 = R_3 = \frac{Pk}{2+n}.$$

$R_1 + 2 R_2$, it will be observed, equals P, as should be.

The conception of a beam *tipping*, as in the last two Arts., is due to Clemens Herschel (*Continuous, Revolving Drawbridges*, Boston, 1875), and the above formulæ were first deduced by him in the above work.

LITERATURE UPON THE CONTINUOUS GIRDER.

We give below, for the benefit of students and those interested in the subject, a list, chronologically arranged, of works upon the continuous girder. A glance at this list will convince the reader as to the thoroughness with which the problem has been treated.

1. REHMANN.—" Theorie der Holz-und Eisenconstructionen." Wien, 1856.—[Treats the continuous girder of constant cross-section and equal spans according to the old method of Navier; first determining the reactions at the supports. A load in any single span only is considered, either total uniformly distributed, or concentrated and acting at the centre.]

2. KÖPKE.—" Ueber die Dimensionen von Balkenlagen, besonders in Lagerhäusern." Zeitschr. des Hannov. Arch. u. Ing. Ver., 1856. [The simple and continuous girder. Attention is here first called to the advantage gained from sinking the supports.]

3. SCHEFFLER.—"Theorie der Gewölbe, Futtermauern und Eisernen Brücken." Braunschweig, 1857. [Continuous girder with total uniformly distributed load, and invariable concentrated loading. Advantage of sinking the supports.]

4. CLAPEYRON.—Calcul d'une poutre élastique reposant librement sur des appuis inégalement espèces." Comptes rendus, 1857. [Here, for the first time, the well-known Clapeyronian method is developed, by which a series of equations between the moments at the supports is first obtained. Application to total distributed loads, but varying in different spans.]

5. MOLLINOS ET PRONIER.—" Traité théoretique et practique de la construction des ponts métalliques." Paris, 1857. [Treatment of the continuous girder of constant cross-section, according to Clapeyron.]

6. GRASHOF.—" Ueber die relative Festigkeit mit Rücksicht auf deren möglichste Vergrösserung durch angemessene Unterstützung und Einmauerung der Träger bei constantem Querschnitte.". Zeitschr. des Deutsch. Ing. Ver., 1857, 1858, 1859.

7. MOHR.—" Beitrag zur Theorie der Holz-und Eisenconstructionen." Zeitschr. des Hannöv. Arch. u. Ing. Ver., 1860. [Theory of continuous girder, with reference to relative height of supports. Application to girders of two and three spans. Best sinking of supports for constant cross-section. Disadvantage of accidental changes of height of supports. Influence of breadth of piers.]

8. H.—" Continuirliche Brückenträger." Bornemann's Civil-Ingenieur, 1860. [Continuous girder of constant cross-section of three spans. Best ratio of spans, and sinking of supports.]

9. WINKLER.—" Beiträge zur Theorie der continuirlichen Brückenträger." Civil-Ingenieur, 1862. [General Theory. Determination of methods of loading causing maximum strains; and, for the first time, general rules for the same given. Best ratio of end spans.]

10. Bresse.—"Cours mécanique appliquée professé à l'école impériale des ponts et chaussées." Seconde Partie. Paris, 1862. [Analytical treatment of the continuous girder of constant cross-section. The transverse forces are not considered. The exact determination of the most dangerous methods of loading, with reference to the moments in the neighborhood of the supports, is also wanting.]

11. Albaret.—"Etude des ponts métalliques à poutres droits réposant sur plus de deux appuis." Ann. des ponts et chaussées, 1866. [Continuous girder of constant cross-section, treated after Clapeyron.]

12. Renaudot.—"Mémoire sur le calcul et le contrôle de la resistance des poutres droites à plusiers travées." Ann. des ponts et chaussées, 1866. [Continuous girder, treated according to Clapeyron.]

13. Culmann.—"Die Graphische Statik." Zürich, 1866. [Graphical treatment of simple and continuous girder of constant and variable cross-section. Moments at the supports are determined analytically.]

14. H. Schmidt.—"Ueber die Bestimmung der ausseren auf ein Brückensystem wirkenden Kräfte." Förster's Bauz., 1866. [Data for the amount of live load for Railroad and Way Bridges. Determination of the equivalent uniformly distributed load. Data for dead weight and wind pressure.]

15. Grashof.—"Die Festigkeits Lehre." 1866. [General analytical treatment of the girder without special reference to bridges. Continuous girder of uniform strength.]

16. Winkler.—"Die Lehre von der Elasticitaet und Festigkeit." Prag., 1867.—[General analytical theory of the continuous girder of constant and variable cross-section. Application to total uniformly distributed loading. Influence of difference of height of supports.]

17. Fränkel.—"Ueber die ungünstigste Stellung eines Systems von Einzellasten auf Trägern über eine und über zwei Oeffnungen, speciell auf Trägern von Drehscheiben." Bornemann's Civil-Ingenieur, 1868.

18. Mohr.—"Beitrag zur Theorie der Holz- und Eisenconstructionen." Zeitschr. des Hannov. Arch. u. Ing. Ver., 1868. [Here, for the first time, the elastic line is regarded as an equilibrium curve, and the graphical treatment of the continuous girder founded.]

19. H. Schmidt.—"Betrachtungen über Brückenträger, welche auf zwei und mehr Stützpunkte frei aufliegen, sowie über den Einfluss der ungleichen Höhenlage der Stützpunkte." Förster's Bauz., 1868.

20. Collignon.—"Cours de mécanique appliquée aux constructions." Paris, 1869. [Continuous girder of constant cross-section and uniform load.]

21. Laissle and Schübler.—"Der Bau der Brückenträger mit besonderer Rücksicht auf Eisenconstructionen." III. Aufl., I. Theil. Stuttgart, 1869. [Treatment of the continuous girder, according to Clapeyron.]

22. Leygue.—"Etude sur les surcharges à considérer dans les calculs des tabliers metallique d'après les conditions générales d'exploitation des chemins de fer." Paris, 1871.

23. Lipricii.—"Theorie des continuirlichen Trägers constanten Querschnittes. Elementare Darstellung der von Clapeyron und Mohr begrün-

deten Analytischen und Graphischen Methoden und ihres Zusammenhanges." Förster's Bauz., 1871, also separate reprint. [The geometrical constructions are deduced from the analytical formulæ.]

24. SEEFEHLNER, G.—"A tobbnyugpontú vasrácstartókról—A magyar mérrök és épitèsz—egylet közlönye," 1871 [Hungarian].

25. RITTER, W.—"Die elastische Linie und ihre Anwendung auf den continuirlichen Balken. Ein Beitrag zur graphischen Statik." Zürich, 1871. [This and the preceding work treat the continuous girder after the Culmann-Mohr method.]

26. OTT.—"Vorträge über Baumechanik," II. Theil. Prag, 1872. [Analytical determination of the shearing forces and moments for the simple and continuous girder of constant cross-section and level supports.]

27. WEYRAUCH, J. I.—"Allgemeine Theorie und Berechnung der continuirlichen und einfachen Träger." Leipzig, 1873. [A work well deserving to close the list. Gives the general theory for constant and variable cross-section for any number of spans from 1 to ∞, and for all kinds of regular or irregularly distributed and concentrated loads. The formulæ are general, and for given loading free from integrals. Difference of level of supports; most unfavorable position of load; exact theory of the fixed and movable inflexion and influence points, etc. Examples illustrating use of formulæ, and complete calculations of girders.]

This last work leaves but little to be desired in thoroughness and comprehensiveness.

It will be observed that England and America have contributed but little to the literature of the subject. Indeed the standard works of both countries show scarcely a trace of the influence of the labors of French and German mathematicians in this field. The only works which, so far as we are aware, can be mentioned in this connection are as follows:

RANKINE, W. J. M.—"Civil Engineering." 1870. [Very brief and incomplete.]

HUMBER.—"Strains in Girders." American Ed. New York: Van Nostrand. 1870. [Graphical constructions, holding good only under the supposition that the end spans are so proportioned that the girder may be considered as fixed at the intermediate supports, for full load.]

STONEY, B.—"Theory of Strains." London, 1873. [Very brief notice of the subject. Points of inflection are found for full load, and the flanges then *cut* at these points.]

HEPPEL, J. M.—Phil. Mag. (London), Vol. 40, p. 446.

Also Minutes of the Proceed. of the Inst. Civ. Eng. [Excellent papers, which might well have been followed up.]

In the latter publication also:

BELL, W.—Vol. 32, p. 171.
STONEY, E. W.—Vol. 29, p. 382.
BARTON, JAMES.—Vol. 14, p. 443.

In American literature:

FRIZELL.—"Theory of Continuous Beams." Jour. Frank. Inst., 1872. [Development of the subject according to *Scheffler*. See 3.]

GREENE, CHAS. E.—"Graphical Method for the Analysis of Bridge Trusses." Van Nostrand. 1875. [Force and equilibrium polygons are used, but the moments at the supports are found by an original method of approximation, or *balancing of moment areas*.]

HERSCHEL, CLEMENS.—"Continuous, Revolving Drawbridges." Boston, 1875. [The formulæ of *Weyrauch* are made use of. The case of "secondary central span" is for the first time investigated, and the appropriate formulæ given. The fact that the live load reactions for supports out of level are unchanged, provided the dead load reactions are zero, is also for the first time clearly stated. The draw span is thoroughly treated, and the idea of weighing off the reactions at the piers of a continuous girder suggested.]

MERRIMAN, MANSFIELD.—"Upon the Moments and Reactions of the Continuous Girder"—Journal of the Franklin Inst. for March and April, 1875; Van Nostrand's Eng. Mag., July, 1875, August and Sept., 1876; London Phil. Mag., Sept., 1875, "On the Theory and Calculation of Continuous Bridges," D. Van Nostrand, New York, 1876, 16mo, pp. 130; as well as the formulæ contained in Chapter XII. of this work. [By the aid of the properties of the Clapeyronian numbers, Mr. Merriman has deduced new and general formulæ eminently suited for practical use. Also relations are deduced from which tables for moments and reactions may be drawn up to any desired extent by simple additions and subtractions, independently of the general formulæ. (See Chap. XII.) The simple girder appears as a special case of the continuous girder. The formulæ are, in respect to simplicity and ease of application, superior to any heretofore given.]

PART III.
APPLICATION OF THE GRAPHICAL METHOD TO THE ARCH.

CHAPTER XIV.

THE BRACED ARCH.

157. Different kinds of Braced Arches.—Just as in girders, we may distinguish between the solid beam, or "plate girder," and the open work, or framed girder; so, regarding the arch as a bent beam, we may distinguish the braced arch and the solid arch, or arch proper. The strains in the various pieces composing the braced arch may be easily found by the method of Arts. 8–15, or by calculation by the method of moments of Art. 14 for any loading, *if only all the outer forces acting upon the arch are known :* that is, so soon as, in addition to the load, we know also the *reactions* at the abutments, or the horizontal thrust and vertical reactions at the points of support, and the moments, if any, which exist at these points.

We may distinguish three classes of braced arches : viz., 1st. Arch hinged at both crown and springing; 2d. Arch hinged at spring line only—continuous at crown; 3d. Arch continuous at crown and *fixed* at abutments.

158. Arch hinged at both Crown and Abutments.—This form of construction [Pl. 23, Fig. 90], owing to the hinges at crown and abutments, affords for live load but little of the advantage of a true arch. It is, in fact, an arch only in form, but in principle is more nearly analogous to a simple triangular truss of two rafters, these rafters being curved and braced ; the thrust being taken by the abutments, instead of resisted by a tie line **A B**.

The case presents no especial difficulty, and may be easily calculated or diagramed, provided that not more than two pieces, the strains in which are unknown, meet at any apex. Thus, in Pl. 23, Fig. 90, the resultant at the abutment due to any weight **P** being known, it may be directly resolved into the two pieces which meet there. The strains in these two pieces being thus found, those in two others in equilibrium with each of them may be obtained. In Art. 13 we have already illustrated the method of procedure for such a case, as also the method of finding graphically the resultant at crown and abutments due to any position of the weight.

Thus the resultant at the crown for the unloaded half must, for equilibrium, pass through the hinge at **B** also. Its direction is thus constant for all positions of **P** upon the other half. The resultant for the other half must then pass through a and the hinge at **A** (Fig. 90).

We have then simply to draw a **B**, prolong **P** to intersection a, and draw a **A**. A a and **B** a are the directions of the resultant at **A** and **B**, and by resolving **P** along these lines, we may find the vertical reaction $\mathbf{V} = a\,b$ and the horizontal thrust $\mathbf{H} = c\,b$.

We can thus easily find the reactions at the abutments in intensity and direction, and following these reactions through the structure, as illustrated in Arts. 8–13, Chap. I., can determine the strains upon all the pieces for any position of the weight. A tabulation of the strains for each weight will then give us the strains for uniform load as well as live load, as already explained in the preceding chapter, Art. 156.

There must be only two pieces meeting at the abutments. Thus the pieces in Fig. 90, represented by broken lines, can serve only to support a superstructure, or transmit load to the arch, and have no influence upon the strains in the other pieces.

If the span $\mathbf{A\,B} = 2\,a$, the rise of the arch is h, and the distance of the weight **P** from the crown is x, positive to the left; then taking moments about the end **B**, we have

$$2\,\mathbf{V}\,a = \mathbf{P}\,(a+x), \text{ or } \mathbf{V} = \frac{\mathbf{P}\,(a+x)}{2\,a}.$$

Similarly, taking moments about the crown,

$$-\mathbf{V}\,a + \mathbf{H}\,h = -\mathbf{P}\,x, \quad \text{or} \quad \mathbf{H} = \frac{\mathbf{V}\,a - \mathbf{P}\,x}{h} = \frac{\mathbf{P}\,(a-x)}{2\,h}.$$

The same formulæ apply for a weight upon the other half, for **V** and **H** at the other end.

The values of **V** and **H** can easily be found from these formulæ, and the strains then calculated by moments, thus checking the diagrams. If these reactions are found for the given dimensions of the *centre line*, we may, if we choose, suppose the depth of the arch to vary above and below the centre line equally, from the crown to ends. The lever arms of the pieces, and hence their strains, will be different, but **V** and **H** are the same as before. Thus, whatever the shape of the arch, we can easily find the strains both by diagram and calculation. If we draw a line through **A** and the hinge at crown, we may easily prove that the greatest vertical ordinate between this line and the arch is

$$y = \frac{r}{2a}\sqrt{4a^2 + 4h^2} - r = \frac{r}{a}\sqrt{a^2 + h^2} - r,$$

where r is the radius.

Now if the depth d of the arch is made greater than this ordinate, it may be shown that both flanges will always be in compression. This condition serves, then, to determine the proper depth of circular arch, which should *not be less* than

$$\frac{r}{a}\sqrt{a^2 + h^2} - r.$$

It is unnecessary to give here an example.* The method is so simple that the reader will find no difficulty in applying the principles above to any case. He will do well to calculate or diagram the strains in an arch similar to that shown in the Fig. for comparison with the two cases which follow.

159. Arch hinged at Abutments—continuous at Crown. —If we suppose the hinge at the crown removed—those at the abutments being, however, retained—then, for any position of the weight, the resultant at each end must for equilibrium pass, as before, through the end hinges. In the preceding case, a, for load on left half, was always to be found at intersection of weight with the line through **B** and hinge at crown, and was therefore fully determined. Now, however, a, the common intersection of weight and resultant abutment pressures, has a different position, and hence the resultants and horizontal and vertical reactions are different.

If we can find or know the *locus* or curve in which this point a must always lie, we can easily find, as before, the resultants or reactions by simply prolonging the line of direction of the weight till it meets this *locus*, and then drawing from the point

---

* See Note to this Chap. in Appendix.

of intersection lines to **A** and **B**, and resolving **P** in these directions.

The equation of this locus can be found analytically without much difficulty.

1st. PARABOLIC ARC.—Thus, for a *parabolic arc*, we have *

$$y = \frac{32\ a^2\ h}{5\ (5\ a^2 - x^2)}.$$

Where [Pl. 23, Fig. 91] a is the half span, and h the rise of the arc; x the distance of the weight from the crown, and y the ordinate **N** d of the *locus c d e i k*.

For a given arc, then—that is, a and h given—we have only to substitute different values for x, as $x = 0, 0.1, 0.2$, etc., of the span, and we can easily find the corresponding ordinates y, and thus construct the locus $c\,d\,e\,i\,k$. It is then easy to find the reactions at **A** and **B** for any position of **P**, as above indicated.

The vertical reaction at the abutment may also be easily found by moments—thus,

$$\mathbf{V}_1 \times 2\,a = \mathbf{P}\,(a + x),\ \ \text{or}\ \ \mathbf{V}_1 = \frac{\mathbf{P}}{2\,a}(a + x).$$

The horizontal thrust is

$$\mathbf{H}_1 = \frac{5}{64}\,\mathbf{P}\,\frac{(5\,a^2 - x^2)(a^2 - x^2)}{a^3\,h}.*$$

These values, though not needed for the construction above, may be of use, and are therefore given. In the following tables we give the values of **H** and y for different values of x:

x	**H**	y	x	**H**	y
0	0.3906	1.280	0.5	0.2783	1.347
0.1	0.3859	1.283	0.6	0.2320	1.379
0.2	0.3706	1.290	0.7	0.1797	1.415
0.3	0.3490	1.304	0.8	0.1226	1.468
0.4	0.3176	1.322	0.9	0.0622	1.527
0.5	0.2783	1.347	1.0	0	1.620
$.a$	$.\mathbf{P}\frac{a}{h}$	$.h$	$.a$	$.\mathbf{P}\frac{a}{h}$	$.h$

* For the demonstration of the analytical results made use of in this chapter, we refer the reader to *Die Lehre von der Elasticität und Festigkeit*, by E. Winkler. Prag, 1867. See also the Supplement to this chapter.

From the table, a and h and **P** being known, **H** and y can be found for the successive positions of **P** at 0.1, 0.2, etc., of a, or the half span, by multiplying $\mathbf{P}\dfrac{a}{h}$ by the tabular number for **H**, and h by the tabular number for y.

2d. CIRCULAR ARC.—For a circular arc we have for the equation of the locus $c\,d\,e\,i\,k$ [Fig. 91],

$$y = \frac{1 + \mathbf{B}\kappa}{1 - \mathbf{A}\kappa} y_0,$$

where $\kappa = \dfrac{\mathbf{I}}{\mathbf{A}\,r^2}$, **I** being the moment of inertia of the *constant* cross-section, **A** its area, and r the radius of the circle: also where

$$y_0 = r\,\frac{(\sin^2 a - \sin^2 \beta)(a - 3\sin a \cos a + 2a \cos^2 a)}{\sin a [\sin^2 a - \sin^2 \beta + 2\cos a (\cos \beta - \cos a) - 2\cos a (a \sin a - \beta \sin \beta)]},$$

a being the angle subtended at centre by the half span, and

$$\mathbf{A} = \frac{2 \cos a\,(a \sin a - \beta \sin \beta)}{\sin^2 a - \sin^2 \beta + 2\cos a (\cos \beta - \cos a - a \sin a + \beta \sin \beta)},$$

$$\mathbf{B} = \frac{2 a \cos^2 a}{2(a - 3 \sin a \cos a + 2 a \cos^2 a)},$$

or, approximately,

$$\mathbf{A} = \frac{24}{5\,a^2 - \beta^2} = \frac{6\,a^2}{5\,h^2\left[1 - \dfrac{1}{5}\left(\dfrac{\beta}{a}\right)^2\right]}$$

$$\mathbf{B} = \frac{15}{4\,a^4} = \frac{15}{64}\frac{a^4}{h^4},$$

where β is the angle from crown to weight.

$\dfrac{\mathbf{I}}{\mathbf{A}}$ = the square of the radius of gyration, or, approximately, the square of the half depth, hence $\kappa = \dfrac{d^2}{4\,r^2}$ approximately.

For the exact values of **A** and **B**, we have the following table:

	β	a = 0	a=10°	a=20°	a=30°	a=40°	a=50°	a=60°	a=90°	
A	0	1.20	1.19	1.17	1.14	1.08	1.00	0.88	0	$\dfrac{a^2}{h^2}$
	0.2	1.21	1.20	1.18	1.15	1.10	1.01	0.90	0	
	0.4	1.24	1.24	1.21	1.18	1.13	1.05	0.94	0	
	0.6	1.29	1.29	1.27	1.24	1.20	1.13	1.02	0	
	0.8	1.38	1.38	1.36	1.34	1.30	1.24	1.18	0	
	1.0	1.50	1.50	1.49	1.47	1.45	1.41	1.36	0	
B	a	0.234	0.233	0.221	0.203	0.178	0.146	0.107	0	$\dfrac{a^4}{h^4}$

For the values of y_0 we have the following table:

β	a=0	a=10°	a=20°	a=30°	a=40°	a=50°	a=60°	a=90°
0	1.280	1.282	1.288	1.300	1.316	1.340	1.375	1.571
0.2	1.290	1.292	1.298	1.309	1.327	1.348	1.380	1.571
0.4	1.322	1.324	1.329	1.340	1.354	1.374	1.403	1.571
0.6	1.370	1.380	1.385	1.393	1.405	1.421	1.443	1.571
0.8	1.468	1.469	1.471	1.476	1.483	1.490	1.504	1.571
1.0	1.600	1.600	1.599	1.597	1.594	1.591	1.588	1.571
.a				.h				

It will be seen that for the *semi-circle* the locus *is a straight line*, for which $y = \tfrac{1}{2}\pi r = 1.5708\,r$. Thus, for any given case—that is, **I, A** and r given—we can easily calculate κ. Then from our tables, for given value of a, we can find **A, B** and y_0 for values of β of 0, $\tfrac{2}{10}$ths, $\tfrac{4}{10}$ths, etc., of a. These values inserted in the equation for y above, will enable us to plot the curve or locus $c\,d\,e\,i\,k$, which being once known, the rest is easy. We have thus by a union of analytical results with our graphical method a very easy and practical solution of this important case. We may, if we choose, only use our method to determine the horizontal thrust and vertical reaction as shown by the Fig. 91, and then calculate the strains by the method of moments. The availability and ease of the method here given, as compared with calculation, will be seen from a consideration of the analytical formulæ for the horizontal thrust and vertical reaction at **A**. Thus, for the vertical reaction, we have, as before, simply

$$V_1 = \frac{P}{2a}(a + x).$$

For the horizontal thrust, however, we have the following very clumsy formula:

$$H = P \frac{\sin^2 \alpha - \sin^2 \beta + 2 \cos \alpha (\cos \beta - \cos \alpha) - 2(1+\kappa) \cos \alpha (\alpha \sin \alpha - \beta \sin \beta)}{2[\alpha - 3 \sin \alpha \cos \alpha + 2(1+\kappa) \alpha \cos^2 \alpha]}$$

For the semi-circle, this reduces to

$$H = P \frac{\cos^2 \beta}{\pi},$$

κ being, as before, $= \frac{I}{A r^2}$; where **A** is the area and **I** the moment of inertia of the cross-section, r the radius of the arch, and the angles α and β, as represented in Fig. 91, viz., the angle of the half span, and the angle to the load, subtended by x. The first of the above formulæ is sufficiently simple, and by it we may *check* the accuracy of our construction. Thus having plotted the curve $c\,d\,e\,i\,k$ by the aid of our expression for y and the tables above for any position of **P** required, we draw d **A** d **B**, and resolve **P** along these lines, thus finding **V** and **H** [Fig. 91]. We can then calculate **V** from the formulæ above, viz., $V = \frac{P}{2a}(a + x)$. If this calculated value agrees with that found by diagram, we may have confidence that the curve is properly plotted, and hence that the value of **H** is also correct. Thus, with very little calculation and great ease, rapidity and accuracy, we can find the reactions at the end **A** for any given position of **P** in any given case. These reactions once known, we can easily find the strains either by diagram, as illustrated in Chap. l., or by calculation by the method of moments of Art. 14.

160. Arch fixed at Abutments—continuous at Crown. This is by far the most important case of braced arch, as by the continuity of the crown and fixity of ends we obtain all the advantage possible due to the combined strength and elasticity of the arch. It is also the most difficult case of solution, as the formulæ obtained by a mathematical investigation are complex, and give rise to tedious and laborious computations in practice. A method combining simple analytical results with graphical construction similar to the preceding, will, however, obviate these difficulties, and bring the subject fairly within the reach of the practical engineer.

In the present case, as before, the common intersection of the weight and the reactions lies in a curve, the equation of which may be found, and the curve itself thus plotted for any given case.

But this curve, or *locus*, **I L K** [Pl. 24, Fig. 92] being constructed, in order to find the directions of the reactions *which now no longer pass through the ends of the arc* **A** and **B**, it is necessary to find and construct also the *curve enveloped by these reactions* for every position of **P**; that is, the curve to which these reactions are tangent. If, then, these two curves are constructed, we have only to draw through **L** [Fig. 92] lines tangent to this enveloped curve, and we have at once the reactions in proper direction, and by resolving **P** along these lines, can easily find their intensities, and therefore **V** and **H**, as before.

1*st*. PARABOLIC ARC.

For a parabolic arc we have for the locus **I L K**, $y = \frac{1}{4} h$; that is, the *locus is a straight line at $\frac{1}{4}$th the rise of the arch above the crown;* since we *now* take y as the ordinate to the locus measured above the horizontal tangent at the *crown*. The origin is, therefore, at the crown instead of at the centre of the half span, as in the previous case.

For the second curve, or curve enveloped by the reactions, we have,* taking v as the abscissa and w as the ordinate of any point [Fig. 92], $v = \dfrac{2 a^2}{3 a + x}$, $w = \dfrac{(23 a^2 + 20 a x + 5 x^2) h}{15 (a + x)(3 a + x)}$,

where, as before, a is the half span, h the rise, and x the distance of the weight from crown. For $x = 0$, $v = \frac{2}{3} a$, and $w = \frac{23}{45} h$. For $x = a$, $v = \frac{1}{2} a$, and $w = \frac{4}{5} h$. For $x = -a$, $v = a$, and $w = -\infty$. Eliminating x from both equations, we have
$$\dfrac{5 a^2 - 5 a v + 2 v^2}{15 a (a - v)} h.$$

Hence the curve enveloped by the reactions is on each side an

---

* For the proof of all the expressions assumed, see the Supplement to this chapter.

CHAP. XIV.]   THE BRACED ARCH.   263

hyperbola, which has for asymptotes the vertical through the abutment and a straight line which cuts the axis of symmetry of the arch at the point b [Pl. 24, Fig. 93], $\frac{1}{3} h$ under the crown, the tangent at the crown at $\frac{2}{3} a$ from the crown, and the chord of the arc at $- 6 a$ from the centre. The centre of the hyperbola is at e, $\frac{1}{15} h$ below the horizontal through the crown. The two hyperbolas osculate at the point $\frac{1}{3} a$ vertically below the crown. [See Fig. 93.]

As an aid to the construction of these hyperbolas, we give the following table:

x	v	w	x	v	w
−	+	+	+	+	+
1	1.0000	∞	0	0.6667	0.5111
0.9	0.9524	2.7721	0.1	0.6452	0.4897
0.8	0.9091	1.5455	0.2	0.6249	0.4722
0.7	0.8695	1.1065	0.3	0.6061	0.4577
0.6	0.8334	0.9999	0.4	0.5882	0.4463
0.5	0.8000	0.7600	0.5	0.5714	0.4349
0.4	0.7693	0.6756	0.6	0.5555	0.4258
0.3	0.7407	0.6155	0.7	0.5405	0.4160
0.2	0.7144	0.5714	0.8	0.5263	0.4102
0.1	0.6897	0.5377	0.9	0.5128	0.4053
0	0.6667	0.5111	1.0	0.5000	0.4000
.a	.a	.h	.a	.a	.h

From the table it is easy to construct the hyperbola for any given case. We have, of course, a perfectly similar hyperbola for the other half, its centre e being similarly situated with respect to the crown, to the right of c. We have then simply to draw a line through the intersection m of the weight **P** [Fig. 93] with the line ik, at $\frac{1}{3} h$ above c, tangent to the hyperbola, and we have at once the direction of the resultant. This tangent may be drawn by eye, or geometrically constructed if desired.* A similar tangent to the hyperbola on the other side determines the direction of the other reaction. We can then resolve **P** in these two directions, and find at once **V** and **H**. The problem, then, so far as a parabolic arc is concerned, is sufficiently simple and easy of solution. We have only to draw a straight line and two easily constructed curves. The formulæ for **V** and **H** and moment at crown M_0 are for this case also simple, and may be used for checking our results. They are:

$$M_0 = -\tfrac{1}{32} P \frac{(a-x)^2 (3 a^2 - 10 a x - 5 x^2)}{a^3},$$

$$H = 1\tfrac{5}{32} P \frac{(a^2 - x^2)^2}{a^3 h}, \quad V = \tfrac{1}{4} P \frac{(a-x)^2 (2 a + x)}{a^3},$$

where, as before, a is the half span, h the rise, and x the distance of weight **P** from crown. A negative moment always indicates tension in lower or inner flange.

2*d*. CIRCULAR ARC.

In this case we have for the locus **I L K** [Fig. 92], for small central angles a, the equation:

$$y = \tfrac{1}{4} h \left[1 - 30 \frac{(a^2 - 2 a x - x^2) g^2}{(a+x)^2 h^2} \right],$$

a, h, and x being as above, and $g^2 = \dfrac{I}{A}$ = the square of radius of gyration; **A** being the area and **I** the moment of inertia of cross-section.

[*Note.*—In *all* the cases hitherto considered, or which we shall consider hereafter, the cross-section is assumed *constant.*]

According to the exact formula, which is too complicated to make it desirable to be given here, we have the following

---

* For the construction of a tangent to a conic section, see Appendix, Fig. 4.

CHAP. XIV.]   THE BRACED ARCH.   265

TABLE, FOR VALUE OF y.

β	$a = 0°$	$a = 30°$	$a = 60°$	$a = 90°$
0	0.200	0.211	0.252	0.329
0.2	0.200	0.210	0.246	0.312
0.4	0.200	0.207	0.236	0.280
0.6	0.200	0.202	0.217	0.228
0.8	0.200	0.197	0.200	0.158
1.0	0.200	0.188	0.151	0.082
a	h			

Instead of determining the curves enveloped by the reactions, the expressions for which are in this case somewhat complicated, it will be found preferable to find the distances $c_1 c_2$ of the *intersections* ϕ and ψ of the reactions [see Fig. 92] with the verticals through the centres of gravity of the end cross-sections. For small central angles a, we have

$$c_1 = -\frac{2h}{15(a+x)}\left[a - 5x - \frac{45\,\mathrm{I}\,a}{\mathrm{A}\,h^2}\right],$$

$$c_2 = -\frac{2h}{15(a+x)}\left[a + 5x - \frac{45\,\mathrm{I}\,a}{\mathrm{A}\,h^2}\right],$$

where a, h, A and x, have the same signification as above.

Since I divided by A equals the square of the radius of gyration $= g^2$, we have

$$c_1 = -\frac{2h}{15(a+x)}\left[a - 5x - \frac{45\,g^2 a}{h^2}\right],$$

$$c_2 = -\frac{2h}{15(a+x)}\left[a + 5x - \frac{45\,g^2 a}{h^2}\right].$$

For braced arches when the material is nearly all in the flanges, the material in the bracing being very small, we may call the radius of gyration half the depth of the arch measured upon the radius from centre to centre of flanges; or representing this depth by d,

$$c_1 = -\frac{2h}{15(a+x)}\left[a - 5x - \frac{45\,d^2 a}{4\,h^2}\right],$$

$$c_2 = -\frac{2h}{15(a+x)}\left[a + 5x - \frac{45\,d^2 a}{4\,h^2}\right].$$

[A negative result indicates that the distance is to be laid off *below* the centre of cross-section.] These formulæ are easy of application, and sufficiently exact for arches whose rise is small compared to the span; when $\dfrac{h}{2\,a}$ is, say, not greater than $\frac{1}{10}$.

All the above formulæ are for *constant cross-sections*. Exact formulæ for *variable* cross-section give results but little *less*, and are much more complicated. The effect of using the above formulæ is therefore, merely, to *increase* slightly the coefficient of safety.

161. We are now able to determine readily and accurately the strains in the various pieces of braced arches hinged at crown and abutments, and hinged at abutments only. We have only to construct in each case the reactions at the abutments, as explained in Arts. 158 and 159, Figs. 90 and 91, and then, by the method already detailed in Arts. 8–13, we can follow these reactions through the structure, and thus find the strains in each piece due to every position of the load. We may also, having found the reactions for given position of weight, calculate the strain in each piece by moments.

For the case of the arch continuous at the crown and *fixed at the abutments*, we must remember that we have also a *moment* at each end tending to cause either tension or compression in the inner flanges according as it is negative or positive. The case is precisely analogous to the continuous girder, or girder fixed at ends. As in that case [see Fig. 77, Art. 111] the moment at one end, as **B**, was the product of **H** into the vertical distance **B D**, so here the moment at **A** (Figs. 92 and 93) is the product of **H** into c_1, found by the formulæ above. This moment can, then, be easily found when c_1 and **H** are known. We can then lay it off, according to the directions of Art. 125, for "passing from one span to another of a continuous girder," and thus commence our diagram of strains; or we can calculate the strains by the method of moments.

162. Illustration of Method of Solution.—As an illustration, take a portion of a braced arch, as represented in Pl. 24, Fig. 94. We have first to plot the upper curve or locus of m for *the given dimensions of the centre line* of the arch. This curve once plotted, then, for any position of the weight, we have only to prolong **P** to m, and draw a line from m to the end of centre line if the arch is hinged at ends, or to ϕ at a distance c_1, above or below the end of centre line if the arch is fixed at ends; c_1 being easily found from our formulæ above. In similar manner, we draw a line from m to the other end, or c_2. Now these two lines are the *resultants* of the outer forces **P**, and by simply resolving **P** in these directions, we have at once **V** and **H**, while the moment at the end $\mathbf{M}_1 = -\mathbf{H}\,c_1$, positive if it tends to cause compression in lower flange, or since c_1 is negative down, if it acts *below* the end.

We can now easily find the strain in any flange, as **D**, *whether the arch vary in depth or not*, provided only it is symmetrical with respect to its centre line. Thus for **D**, take the opposite apex a as the centre of moments. The moment of **H** with reference to a, as shown in the Fig., tends to cause tension in **D**, while that of **V** causes compression. We have then, representing tension by minus,

$$\text{strain in } \mathbf{D} = \frac{\text{moment of } \mathbf{V} - \text{moment of } \mathbf{H}}{\text{lever arm of } \mathbf{D}},$$

all with reference to a. If the result is minus, it indicates thus tension, if plus, compression; if it is zero, the two moments are equal, and at a, therefore, no moment exists; hence a must be a point of inflection. Note that **H** and **V** must be taken as acting at ϕ, Fig. 94. We can also evidently take them as acting at the centre of the end cross-section, if we *take into account the moment* $\mathbf{H}\,c_1$.

In similar manner, for **C** we take b as centre of moments, and then, since **H** now causes compression in **C** and **V** tension, we have for **V** and **H**, acting at ϕ,

$$\text{strain in } \mathbf{C} = \frac{\text{moment of } \mathbf{H} - \text{moment of } \mathbf{V}}{\text{lever arm of } \mathbf{C}}.$$

For **V** and **H**, considered as acting at the end of centre line, we have
$$\mathbf{C} = \frac{\text{moment of } \mathbf{H} + \mathbf{H}\,c_1 - \text{moment of } \mathbf{V}}{\text{lever arm of}},$$

taking c_1 without regard to its sign, but simply to the kind of strain it tends to cause in the piece in question. Properly, since when H is below c_1 is negative, we should have $-H c_1$ for moment causing compression in C.

Thus we may proceed till we pass **P**, and then the moment of **P**, with its proper sign, as producing tension or compression in the piece in question, must *also* be taken into account, or we may instead take the moments of **V** and **H** at the other end, that is, the same side of the weight as the piece itself.

The diagonals may be similarly found by moments. It will, however, be best to determine them by diagram, one of the flanges being first calculated (in this case the first upper flange), as explained in Art. 125. They may also be calculated from the *resultant shear* at any apex. Thus, for diagonal 3 find the vertical components of the previously determined strains in **D** and **C**. These vertical components, together with the vertical component of the strain in 3, must for equilibrium be equal and opposite to the *total* shear at b.

Calling this shear **F**, and a, β and γ the inclinations of **D** and 3, we have for the strain in 3,

$$S_3 = (F - S_1 \sin a - S_2 \sin \beta) \cos \gamma.$$

If either of the vertical components of the strains in **D** or **C** acts opposite to the shear **F**, it must, of course, be subtracted; if in the same direction, added to **F**. For the ready determination of the proper signs, see Appendix, Art. 16 (4).

The moment $H c_1$ is the moment at the fixed end, and is constant throughout the arch for any one position of the load. It causes tension in outer and compression in inner flanges, provided, as in the Fig., ϕ fall *below* the centre of the end section. This moment is increased (or diminished if ϕ is above) by the varying moment of **H** for each apex.

The above method of determining the strains in the braced arch, though not strictly graphical, but rather a combination of analytical and graphical methods, offers such a ready solution of this important and difficult case, that we have not thought it out of place to notice it somewhat in detail. We consider it by far the simplest and easiest method which has yet appeared.

163. Analytical Formulæ for V and H.—A comparison of our method with the long and involved analytical expressions to which the theory of flexure conducts us, will render its advantages still more apparent.

CHAP. XIV.] THE BRACED ARCH. 269

Thus, for a load of w per unit of horizontal length, reaching from left end to a point whose angle from vertical through crown is β (Fig. 92), a being the angle subtended by the half span, we have *

$$H = \frac{w R}{k}\left[k_1 \sin \beta - k_2 \mu + k_3 + \tfrac{1}{6}\sin^3 \beta\right],$$

where R is radius of arch, and

$$k = a + \sin a \cos a - \frac{2 \sin a}{a}, \quad k_1 = \frac{\sin a \cos a}{a} + \frac{\sin^2 a}{2},$$

$$k_2 = \frac{\sin a}{4 a}, \quad k_3 = \frac{\sin a}{4 a}\left[a - \sin a \cos a\right], \text{ and}$$

$$\mu = \beta + 2 \beta \sin^2 \beta + 3 \sin \beta \cos \beta.$$

For V we have

$$V = w R\left[\frac{\cos a \sin^2 \beta}{2(a - \sin a \cos a)} - \frac{\sin \beta}{2} - \frac{K}{a - \sin a \cos a} + \frac{3 \cos a - \cos^3 a}{6(a - \sin a \cos a)} + \sin a\right],$$

where

$$K = \frac{\cos \beta}{2} + \frac{\beta \sin \beta}{2} - \frac{\cos^3 \beta}{6}.$$

For a concentrated load P for any point [Fig. 92], we have †

$$V = \frac{P}{4 a^3}\left[2 a^3 - 3 a^2 \beta + \beta^3 - \tfrac{7}{30}(a^2 - \beta^2)^2 \beta + \ldots\right],$$

or, more correctly,

$$V = P \frac{a - \beta - \sin a \cos a - \sin \beta \cos \beta + 2 \cos a \sin \beta}{2(a - \sin a \cos a)}.$$

For the semi-circle, this becomes

$$V = P \frac{\pi - 2\beta - 2 \sin \beta \cos \beta}{2 \pi}.$$

For H we have

$$H = P \frac{2 \sin a \left[\cos \beta - \cos a + (1 + \kappa)\beta \sin \beta\right] - (1 + \kappa) a (\sin^2 a + \sin^2 \beta)}{2\left[(1 + \kappa) a (a + \sin a \cos a) - 2 \sin^2 a\right]}.$$

---

* Taken from Capt. Eads' *Report to the Illinois and St. Louis Bridge Co.*, May, 1868.
† *Die Lehre von der Elasticität und Festigkeit.* Winkler. Prag. 1867.

where $\kappa = \dfrac{I}{A\,r^2}$; I being the moment of inertia, and **A** area of the cross-section, and r the radius of circle.

These formulæ, it will be observed, involve much labor in any particular case. Where the number of weights is large, the computation is tedious in the extreme. A method which shall give accurate results and avoid such formulæ as the above is certainly very desirable, and such we believe to be the method which we have given.

For the analytical investigation of arches, and the demonstration of the formulæ for the curves of which we have made use, the reader may consult *Die Lehre von der Elasticitaet und Festigkeit*, by *Dr. E. Winkler*, to which we have already referred, and which contains a thorough discussion of the whole subject. The tables which we have given, as well as the formulæ for y, c_1 and c_2, will, it is hoped, give the method here presented a practical value, and render the solution of any particular case easy and rapid.

164. For a *solid* or plate girder arch of given cross-section, we may also determine the proper proportions by finding, as above, the moment **M** of the exterior forces at any point.

Then
$$\mathbf{M} = \frac{\mathbf{T}\,\mathbf{I}}{t},$$

where **T** is the strain per unit of area in any fibre distant t from the axis, and **I** the moment of inertia of the cross-section. Thus, for a rectangular cross-section $\mathbf{I} = \tfrac{1}{12}\,b\,d^3$, where b is the breadth and d the depth.

Hence $\mathbf{M} = \tfrac{1}{6}\,\mathbf{T}\,b\,d^2$,

if we take $t = \dfrac{d}{2}$.

The strain, then, per unit of outer fibre will be
$$\mathbf{T} = \frac{6\,\mathbf{M}}{b\,d^2}.$$

The safe working strain should not exceed for iron 5 tons per sq. inch for tension and 4 tons for compression, and therefore d being assumed, we can easily proportion b so as to satisfy this condition.

As examples of braced arches, such as we have considered, viz., continuous at crown and fixed at abutments, we may mention the *Bridge over the Mississippi River at St. Louis, by Capt. Eads;* one over the *Elbe* near *Hamburg* on the *Paris-Hamburg R. R.*, in which, however, the outward thrust of the arch is balanced by a precisely similar *inverted* braced arch, or suspension system. Thus the piers have to support a vertical reaction only, and the necessity of large and expensive abutments of masonry for resisting the horizontal thrust is obviated.

The strains in the *inverted* arch of this character are found in a precisely similar manner. The only difference is that the reactions, and therefore the vertical and horizontal components, act now in a direction opposite to the direction for the upright arch, and the strains, though the same in amount, are of reverse character in each piece.

The bridge over the *Rhine* at *Coblenz* is an illustration of the braced arch pivoted at the abutments only.

Examples of the solid or cast-iron arch of all kinds are common.

165. Strains due to Temperature.—In the first class of braced arches, viz., pivoted at both abutments and crown, there are evidently no strains due to changes of temperature. The arch can accommodate itself to any change of length by rising at the crown and turning at the abutments, and no strains are induced.

We represent by ϵ the coefficient of linear expansion for one degree [*about* 0.000012 *for iron, for every degree centigrade*], and by t, the temperature above or below the mean temperature t_0, for which no strain exists.

Then for arch pivoted at abutments only, we have for the increase of thrust,*

$$H = \frac{2\,\mathbf{E}\,\mathbf{I}\,\epsilon\,t \sin a}{r^2 (a - 3 \sin a \cos a + 2\,a \cos^2 a) + 2\,\kappa\,r^2\,a \cos^2 a},$$

where $\mathbf{E}$ is modulus of elasticity, $\mathbf{I}$ moment of inertia of cross-section, and $\kappa = \dfrac{\mathbf{I}}{\mathbf{A}\,r^2}$; $\mathbf{A}$ being area of cross-section, r radius of arch, a the angle of half span, or, approximately,

$$H = \frac{15\,\mathbf{E}\,\mathbf{I}\,\epsilon\,t}{r^2 (2\,a^4 + 15\,\kappa)} = \frac{15\,\mathbf{E}\,\mathbf{I}\,\mathbf{A}\,\epsilon\,t}{8\,\mathbf{A}\,h^2 + 15\,\mathbf{I}},$$

where h is the rise of arch.

------

* *Lehre von der Elasticität.* Winkler. Also Supplement to this chapter, Art. 26.

For the moment at any point, then, due to change of temperature, we have
$$M = H\,r\,(\cos\beta - \cos a),$$
β being the angle from vertical to that point.

This moment, if positive, causes tension in outer and compression in inner flanges, and we can, as before, easily find the corresponding strains either by diagram or calculation.

For an arch fixed at ends and continuous at crown, we have
$$H = \frac{2\,\mathbf{E\,I}\,e\,t\,(1+\kappa)\,a\,\sin a}{r^2\,[(1+\kappa)\,(a^2 + a\,\sin a\,\cos a) - 2\sin^2 a]},$$
or, approximately,
$$H = \frac{45\,\mathbf{E\,I}\,e\,t}{r^2\,(a^4 + 45\,\kappa)} = \frac{45\,\mathbf{E\,I\,A}\,e\,t}{4\,\mathbf{A}\,h^2 + 45\,\mathbf{I}}.$$

But this thrust does not act at the abutment, since, if it did, there would be no moment there. It must be considered as acting at a distance for *rise* of temperature, *below the crown* of
$$e_0 = \frac{(1+\kappa)\,a - \sin a}{(1+\kappa)\,a}\,r,$$
or, at a distance *above* the end abutment of $h - e_0$.

Approximately, we have
$$e_0 = \frac{a^2 + 6\,\kappa}{6}\,r = \frac{(\mathbf{A}\,a^2 + 6\,\mathbf{I})\,h}{3\,\mathbf{A}\,a^2},$$
a being the half span.

This thrust and its point of application being known, we can easily find the moment, and hence the strains at any point. We see that the horizontal thrust is about six times as great as for the case of an arch pivoted at both ends.

The constant moment acting at the abutment, which may be considered as acting at every point, is
$$M_A = -\left(\frac{\sin a}{a} - \cos a\right) H\,r;\ *$$
it acts to cause compression in outer and tension in inner flanges at abutment. If we find this moment, we can then consider H as acting at the end, and then we have for the moment at any point
$$M_x = H\,y - M_A;$$

---

* Capt. Eads' *Report to the Illinois and St. Louis Bridge Co.*, May, 1868.

a positive result, giving tension, a negative, compression in the outer flanges.

166. Effects of Temperature.—We are now able to solve accurately and thoroughly any class of braced arch, both for variable loading and changes of temperature, and here the following remarks upon the latter subject may not be without interest. We quote from *Culmann—Die Graphische Statik*, p. 487:

"The question arises whether the fears which the additional strains, due to variations of temperature, have given rise to, are well founded. Before the construction of the *Arcole Bridge* in *Paris* the Engineer *Oudry* made various experiments with a rib of about the same span as the bridge itself, of which the following seems decisive as regards the present question. By driving in the wedges upon which the rib rested above and below, he could raise and lower the crown much more than the distance due to variation of temperature without diminishing its supporting capacity. . . . . . *Oudry*, having thus assured himself of the harmlessness of temperature variations, decided upon broad and firm bearing surfaces.

"Interesting observations have also been made upon the changes of form of the cast-iron arch of 60 metres span over the *Rhone* at *Tarascon*, published in the *Annales des Ponts et Chemins*, 1854, from which, however, it only appeared that the changes of form followed slowly the temperature; that they were less than the received coefficients would have led us to expect, and were nowhere found to be prejudicial.

"Since, then, this question appears to have been settled more than ten years ago, may we not fear that those who still wish to pivot *iron* may some day seize upon the idea of pivoting *stone* arches also!

"Stone, as is well known, expands not much less than iron for equal changes of temperature, and, moreover, its modulus of elasticity is much less. The expanded stone arch cannot accommodate itself to the given span, therefore, as easily as the iron arch, and it would then be clearly more advantageous to *pivot the stone arch!* As, however, such a clumsy contrivance would give no great impression of *stability*, we feel justified in recommending a broad and solid bearing surface for all arches."

As the opinion of an eminent engineer, the above may not be without interest. We would only add that, according to the accepted formulæ for temperature strains already given, the results are of more importance than the above remarks would indicate. As will be seen in the Appendix, the temperature strains in the braced arch, fixed at ends and continuous at crown, are very considerable, and, if the formulæ are accepted as correct, can by no means be disregarded. By comparison of our numerical results for the three cases of braced arch there given, it appears that the one hinged at crown, and springing, is by far the best form of construction, but it must be remembered that a different proportion of span to height and depth may considerably affect this conclusion. Upon this point we refer the reader to Art. 28 of the Appendix.

With the above, we conclude our discussion of braced arches, or arches whose weight is not so great that the effect of the live load can be disregarded, and pass on to the stone arch, or arch proper.*

---

* See Appendix, Art. 17, for a practical application of the principles of this chapter.

SUPPLEMENT TO CHAPTER XIV.

DEMONSTRATION OF ANALYTICAL FORMULÆ GIVEN IN TEXT.

In order to complete our discussion of the braced arch, we shall now give the analytical development of the formulæ of which we have made use in the preceding chapter. We do this the more readily, as in no book of easy access to the student are these formulæ made out. In the work of *Winkler*, already referred to in the text, will be found a very thorough discussion of the subject. We shall confine ourselves at present to the case of a single concentrated load.

CHAPTER I.

GENERAL CONSIDERATIONS AND FORMULÆ FOR FLEXURE.

1. Fundamental Equations.—The resultant of all the forces acting upon a curved piece in a common plane may be decomposed into a force normal to the piece **N**, and into a compressive or tensile force in the direction of the axis or of the tangent to the axis **G**; and this latter force, if taking effect above or below the axis, acts to bend the piece, and gives rise to a moment **M** as well as to a compressive or tensile force **G**. These forces cause corresponding *strains*. Thus, if **P** is the tangential *strain* per unit of area da, then

$$\int P\, da = G \quad \ldots \ldots \ldots (1)$$

while, if v is the distance of any fibre from the axis,

$$\int P v\, da = M \quad \ldots \ldots \ldots (2)$$

(*a*) *Coefficient of elasticity.*

Let the length of a piece be s, its area of cross-section **A**, and, as above the force acting upon this area **G**. Then $\frac{G}{A}$ will be the force per unit of area. Let the displacement [elongation or compression] produced by this force $\frac{G}{A}$ be Δs; the sign Δ indicating and reading "elongation." Now experiment shows that within narrow limits, *i.e.*, within the *elastic limits*, the elongation or compression is directly as the force of extension or compression. Supposing that this held true always for all values of Δs, then, since a force $\dfrac{G}{A}$ produces a displacement Δs, the force necessary to produce a displacement s, will be $\dfrac{s}{\Delta s}$ *times* as great. Calling this force E, we have

$$E = \frac{G\,s}{A\,\Delta s}.$$

The force E, then, is the force *which would be necessary to produce a displacement* s *equal to the original length*, if the law of proportionality of the displacement to the force always held good for all values of Δs. This value

$$E = \frac{G\,s}{A\,\Delta s} \quad \ldots \ldots \ldots \quad (3)$$

we call the *coefficient of elasticity*.

From (3) we easily obtain for the force in the direction of the tangent to the axis

$$G = E A \frac{\Delta s}{s} \quad \ldots \ldots \ldots \quad (4)$$

and for the relative displacement

$$\frac{\Delta s}{s} = \frac{G}{E A} \quad \ldots \ldots \ldots \quad (5)$$

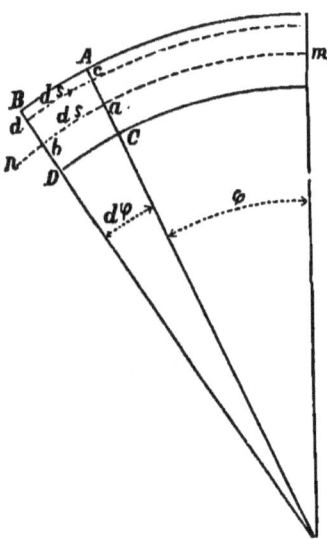

(*b*) *Fibre strain* **P**, *and moment* **M**.

As seen in eq. (1), the longitudinal *strain* upon an element of cross-section da is called **P**. In a curved piece conceive two cross-sections, as shown in the Fig., as **A C B D** perpendicular to the axis of the piece $m\, n$. Let these sections be infinitely near; then, the distance $b\, a$ *upon the axis* is $d\, s$. Let $d\, s_v$ be the length of any fibre, as $d\, c$, *before* the change of form. Then, after deformation, its length is $= d\, s_v + \Delta\, d\, s_v$. But if $d\, \phi$ is the small angle between the normals, $d\, s_v = d\, s + v\, d\, \phi$, where v is the distance $a\, c$ of any fibre from the centre of gravity of the cross-section. After deformation, $d\, s$ becomes $d\, s + \Delta\, d\, s$, and $d\, \phi$ becomes $d\, \phi + \Delta\, d\, \phi$, and $d\, s_v$ becomes $d\, s_v + \Delta\, d\, s_v$. Hence the length of any fibre *after* deformation is $\quad d\, s_v + \Delta\, d\, s_v = d\, s + \Delta\, d\, s + v\, (d\, \phi + \Delta\, d\, \phi)$.

Subtracting this from the eq. for $d\, s_v$ above, we have

$$\Delta\, d\, s_v = \Delta\, d\, s + v\, \Delta\, d\, \phi.$$

Therefore the ratio of the change of length to the original length of fibre

is $\quad\dfrac{\Delta\, d\, s_v}{d\, s_v} = \dfrac{\Delta\, d\, s + v\, \Delta\, d\, \phi}{d\, s + v\, d\, \phi}.$

If r is the radius of curvature, then $r\, d\, \phi = d\, s$, $\dfrac{d\, \phi}{d\, s} = \dfrac{1}{r}$; hence

$$\frac{\Delta\, d\, s_v}{d\, s_v} = \left[\frac{\Delta\, d\, s}{d\, s} + v\, \frac{\Delta\, d\, \phi}{d\, s}\right] \frac{r}{r + v} \quad \ldots \ldots (6)$$

From eq. (1) we have the strain on a fibre $\mathbf{P} = \dfrac{\mathbf{G}}{\mathbf{A}}$.

From eq. (4), $\mathbf{G} = \mathbf{E}\, \mathbf{A}\, \dfrac{\Delta\, s}{s}$. Hence $\mathbf{P} = \mathbf{E}\, \dfrac{\Delta\, s}{s}$. In the present case $\dfrac{\Delta\, s}{s}$ is given by (6); therefore

$$\mathbf{P} = \mathbf{E} \left[\frac{\Delta\, d\, s}{d\, s} + v\, \frac{\Delta\, d\, \phi}{d\, s}\right] \frac{r}{r + v} \quad \ldots \ldots (7)$$

Since now from (1) $\mathbf{G} = \displaystyle\int \mathbf{P}\, d\, a$, we have from (7)

$$\frac{\mathbf{G}}{\mathbf{E}} = r\, \frac{\Delta\, d\, s}{d\, s} \int \frac{d\, a}{r + v} + r\, \frac{\Delta\, d\, \phi}{d\, s} \int \frac{v\, d\, a}{r + v}.$$

Since from (2) $\mathbf{M} = \displaystyle\int \mathbf{P}\, v\, d\, a$, we have again from (7)

$$\frac{\mathbf{M}}{\mathbf{E}} = r\, \frac{\Delta\, d\, s}{d\, s} \int \frac{v\, d\, a}{r + v} + r\, \frac{\Delta\, d\, \phi}{d\, s} \int \frac{v^2\, d\, a}{r + v}.$$

But $\displaystyle\int \frac{r\, v^2\, d\, a}{r + v}$ is, when v is very small compared to r, equal to $\displaystyle\int v^2\, d\, a$, which is the *moment of inertia* of the cross-section $\mathbf{I}$. Also,

$$r \int \frac{d\, a}{r + v} = \int d\, a - \int \frac{v\, d\, a}{r + v},$$

or, $r\int\dfrac{d\,a}{r+v}=\int d\,a-\dfrac{1}{r}\int v\,d\,a+\dfrac{1}{r}\int\dfrac{v^2\,d\,a}{r+v}=A+\dfrac{I}{r^2}$, because since

v is measured from the centre of gravity, $\int v\,d\,a=0$.

Again, $\quad r\int\dfrac{v\,d\,a}{r+v}=\int v\,d\,a-\int\dfrac{v^2\,d\,a}{r+v}=-\dfrac{I}{r}.$

Therefore the insertion of these values of the integrals in the equations for $\dfrac{G}{E}$ and $\dfrac{M}{E}$ above gives

$$\left.\begin{array}{l}\dfrac{G}{E}=\dfrac{\Delta d\,s}{d\,s}A-\left[\dfrac{\Delta d\,\phi}{d\,s}-\dfrac{\Delta d\,s}{r\,d\,s}\right]\dfrac{I}{r}\\[6pt]\dfrac{M}{E\,r}=\left[\dfrac{\Delta d\,\phi}{d\,s}-\dfrac{\Delta d\,s}{r\,d\,s}\right]\dfrac{I}{r}.\end{array}\right\}$$

Inserting the second in the first of the above equations, $\dfrac{G}{E}=\dfrac{\Delta d\,s}{d\,s}A-\dfrac{M}{E\,r}$, and hence

$$\dfrac{\Delta d\,s}{d\,s}=\dfrac{G}{E\,A}+\dfrac{M}{E\,A\,r}\quad\ldots\ldots\quad(8)$$

Inserting this in the second equation above,

$$\dfrac{\Delta d\,\phi}{d\,s}=\dfrac{M}{E\,I}+\dfrac{M}{E\,A\,r^2}+\dfrac{G}{E\,A\,r}\quad\ldots\ldots\quad(9)$$

(c) *Change of length and position of axis.*

From (8) we have at once for the elongation of axis,

$$\Delta s=\dfrac{1}{E}\int\left(\dfrac{G}{A}+\dfrac{M}{A\,r}\right)d\,s\quad\ldots\ldots\quad(10)$$

or, when v is very small compared to r,

$$\Delta s=\dfrac{1}{E}\int\dfrac{G}{A}\,d\,s\,.\quad\ldots\ldots\ldots\quad(11)$$

From (9) we have for the change of direction of the tangent to the axis

$$\Delta\phi=\dfrac{1}{E}\int\left(\dfrac{M}{I}+\dfrac{M}{A\,r^2}+\dfrac{G}{A\,r}\right)d\,s,\text{ or from (10) for }r$$

constant, that is, for a circle,

$$\Delta\phi=\dfrac{1}{E}\int\dfrac{M}{I}\,d\,s+\dfrac{\Delta s}{r}\quad\ldots\ldots\quad(12)$$

When v is very small compared to r, we have

$$\Delta\phi=\dfrac{1}{E}\int\dfrac{M}{I}\,d\,s\quad\ldots\ldots\ldots\quad(13)$$

If the piece had been *originally straight*, $d \Delta \phi$ would be equal to $d \phi$, and $\frac{d \Delta \phi}{d s} = \frac{d \phi}{r d \phi} = \frac{1}{r}$; hence from (13) we have $\mathbf{M} = \frac{\mathbf{E I}}{r}$.

From the calculus we have the radius of curvature

$$r = \frac{\left[1 + \left(\frac{dy}{dx}\right)^2\right]^{\frac{3}{2}}}{\frac{d^2 y}{dx^2}}, \text{ or, approximately, } r = \frac{dx^2}{d^2 y};$$

hence
$$\mathbf{M} = \mathbf{E I} \frac{d^2 y}{dx^2} \quad \ldots \ldots \quad (13\ b)$$

This is the equation assumed in the Supplement to Chap. XIII., Art. 2.

2. Displacement of any point.—We indicate the horizontal displacement of any point of the axis along the axis of x by Δx, along y by Δy. The corresponding changes of dx, dy, and ds are Δdx, Δdy, Δds. The total horizontal displacement is then $dx + \Delta dx = (ds + \Delta ds) \cos (\phi + \Delta \phi)$. The total vertical displacement is $dy + \Delta dy = (ds + \Delta ds) \sin (\phi + \Delta \phi)$. Hence, since $\Delta dx = d \Delta x$, $\Delta dy = d \Delta y$,

$$d \Delta x = (ds + \Delta ds) \cos (\phi + \Delta \phi) - dx,$$
$$d \Delta y = (ds + \Delta ds) \sin (\phi + \Delta \phi) - dy.$$

By Trigonometry, $\cos (\phi + \Delta \phi) = \cos \phi \cos \Delta \phi - \sin \phi \sin \Delta \phi$, $\sin (\phi + \Delta \phi) = \sin \phi \cos \Delta \phi + \cos \phi \sin \Delta \phi$, or if $\cos \Delta \phi = 1$,

$$\sin \Delta \phi = \Delta \phi, \cos \phi = \frac{dx}{ds}, \sin \phi = \frac{dy}{ds}.$$

$$\cos (\phi + \Delta \phi) = \frac{dx}{ds} - \Delta \phi \frac{dy}{ds},$$

$$\sin (\phi + \Delta \phi) = \frac{dy}{ds} + \Delta \phi \frac{dx}{ds}.$$

Substituting these in the equations above,

$$d \Delta x = (dx - \Delta \phi\, dy)\left(1 + \frac{\Delta ds}{ds}\right) - dx,$$
$$d \Delta y = (dy + \Delta \phi\, dx)\left(1 + \frac{\Delta ds}{ds}\right) - dy,$$

or, removing the parentheses, and neglecting quantities of the second order with respect to $\Delta \phi$ and $\frac{\Delta ds}{ds}$,

$$\left.\begin{array}{l} d \Delta x = - \Delta \phi\, dy + \dfrac{\Delta ds}{ds} dx \\[1em] d \Delta y = + \Delta \phi\, dx + \dfrac{\Delta ds}{ds} dy \end{array}\right\} \quad \ldots \ldots \quad (14)$$

When the radius of curvature is very great with reference to the thickness of the beam, and the relative change of length $\frac{\Delta ds}{ds}$ is disregarded, we have simply

$$d\Delta x = -\Delta\phi\, dy,$$
$$d\Delta y = \Delta\phi\, dx.$$

But from (12) $\Delta\phi$ is equal to $\int \frac{M\, ds}{EI} + \frac{\Delta ds}{r}$;

or, in the above case, $\Delta\phi = \int \frac{M\, ds}{EI}$, hence, for v very small with reference to r,

$$\left. \begin{array}{l} d\Delta x = -\displaystyle\int \frac{M\, ds}{EI} dy \\[2mm] d\Delta y = \displaystyle\int \frac{M\, ds}{EI} dx \end{array} \right\} \quad \ldots \quad (15)$$

We shall have frequent occasion to refer to formulæ (8), (12), (13), (14) and (15) in the following discussion.

CHAPTER II.

HINGED ARCH IN GENERAL.

3. Notation—The outer forces in general.—We suppose the ends of the arch to be hinged at the abutments at the centre of gravity of the end cross-sections. Then the end reactions must pass through these points. These end reactions and the loads constitute, then, the outer forces. For equilibrium, then, the horizontal components of these reactions must be equal. Each of these components we call the *horizontal thrust*.
We use the following notation [Pl. 23, Fig. 91]:
R and **R'**, the reactions at ends **A** and **B**.
V and **V'**, their vertical components.
H, their horizontal component, or the thrust.
a, the half span.
h, the rise of arch.
α, the half central angle, if arch is circular.
The origin of co-ordinates we take *at crown*, x horizontal, y vertical. The angle of radius of curvature at any point with y, or of tangent to curve at any point with x, we call ϕ.

THE OUTER FORCES IN GENERAL.

Suppose a single load **P** to act at any point **E**. Let its horizontal distance from crown be z, the corresponding central angle **E O C** be β.
Then the conditions of equilibrium are:

$$V + V' = P,$$
$$Va - V'a - Pz = 0.$$

From these last, we have

$$\left. \begin{array}{l} V = P \dfrac{a+z}{2a} \\[1em] V' = P \dfrac{a-z}{2a} \end{array} \right\} \quad \ldots \ldots (16)$$

For a circular arc, since $a = r \sin \alpha$, $z = r \sin \beta$,

$$V = P \frac{\sin \alpha + \sin \beta}{2 \sin \alpha}, \quad V' = P \frac{\sin \alpha - \sin \beta}{2 \sin \alpha} \quad \ldots (17)$$

We distinguish three segments in the arch [Fig. 91], viz., **A E**, or from end to the load; **E C**, or from load to crown; **C B**, or from crown to right end. Quantities referring to the second we indicate by primes, to the third by double primes. Thus for the tangential component of the resultant at any point within **A E**, we put **G**; for the normal component, **N**. In **E C**, then, we have **G'** and **N'**; in **C B**, **G'** and **N'**. We have then

282   SUPPLEMENT TO CHAP. XIV.   [CHAP. II.

$$\begin{aligned}\mathbf{G} &= -\mathbf{H}\cos\phi - \mathbf{V}\sin\phi, \quad \mathbf{G}' = -\mathbf{H}\cos\phi + \mathbf{V}'\sin\phi \\ \mathbf{N} &= -\mathbf{H}\sin\phi + \mathbf{V}\cos\phi, \quad \mathbf{N}' = -\mathbf{H}\sin\phi - \mathbf{V}'\cos\phi\end{aligned} \bigg\} \quad \ldots (18)$$

$$\mathbf{M} = \mathbf{H}(h-y) - \mathbf{V}(a-x), \quad \mathbf{M}' = \mathbf{H}(h-y) - \mathbf{V}'(a+x) \ldots (19)$$

In the case of a circular arc, $a = r\sin\alpha$, $h = r(1-\cos\alpha)$, $x = r\sin\phi$, $y = r(1-\cos\phi)$, and hence

$$\begin{aligned}\mathbf{M} &= \mathbf{H}r(\cos\phi - \cos\alpha) - \mathbf{V}r(\sin\alpha - \sin\phi) \\ \mathbf{M}' &= \mathbf{H}r(\cos\phi - \cos\alpha) - \mathbf{V}'r(\sin\alpha + \sin\phi)\end{aligned} \bigg\} \quad \ldots (20)$$

4. Intersection Line.—We call the locus of d [Pl. 23, Fig. 91], or the curve $c\,d\,e\,i\,k$, the *intersection line*.

If now there are three hinges, one at crown and one at each abutment, then the resultant for each half must pass through the crown O. If, therefore, for the crown ($x=0$, $y=0$), we make in (19) $\mathbf{M}'=0$, we have $\mathbf{H}=\mathbf{V}'\dfrac{a}{h}$, or inserting the value of $\mathbf{V}'$ from (16),

$$\mathbf{H} = \frac{\mathbf{P}(a-z)}{2h} \quad \ldots \ldots \ldots (21)$$

If the load lies to the left of O, then only the resultant $\mathbf{R}'$ acts upon the right half, and must pass, as above, through the crown O. The point d lies then always upon B O or A O prolonged. Hence, *the intersection lines are two straight lines, which pass through the crown and ends*.

5. Parabolic Arc—concentrated Load.—For a parabola we have $y = \dfrac{h}{a^2}x^2$; hence, $dy = \dfrac{2h}{a^2}x\,dx$, and, approximately, $d\,s = d\,x$.

(a) *Change of direction of tangents.*

Inserting this value of y in equations (18) for $\mathbf{M}$ and $\mathbf{M}'$, we have from equation (13), Art. 1, since $r\,d\phi = d\,s = d\,x$ for the change of direction of the tangent at any point before and after flexure,

$$\mathbf{E}\mathbf{I}\,d\,\Delta\phi = \left[\mathbf{H}h\left(1 - \frac{x^2}{a^2}\right) - \mathbf{V}(a-x)\right]dx.$$

Integrating this, we have for the three segments A E, E O and O B,

$$\begin{aligned}\mathbf{E}\mathbf{I}\,\Delta\phi &= \mathbf{H}\,h\,x\left(1 - \frac{x^2}{3a^2}\right) - \mathbf{V}\,a\,x\left(1 - \frac{x}{2a}\right) + \mathbf{A} \\ \mathbf{E}\mathbf{I}\,\Delta\phi' &= \mathbf{H}\,h\,x\left(1 - \frac{x^2}{3a^2}\right) - \mathbf{V}'\,a\,x\left(1 + \frac{x}{2a}\right) + \mathbf{A}' \\ \mathbf{E}\mathbf{I}\,\Delta\phi'' &= \mathbf{H}\,h\,x\left(1 - \frac{x^2}{3a^2}\right) - \mathbf{V}'\,a\,x\left(1 + \frac{x}{2a}\right) + \mathbf{A}''\end{aligned} \bigg\} \quad \ldots (22)$$

where $\mathbf{A}$, $\mathbf{A}'$, $\mathbf{A}''$ are constants of integration, to be determined by the assignment of the proper limits. Thus, if we make $x = z$, the two first of equations (22) are equal; hence,

$$-\mathbf{V}\,a\,z\left(1 - \frac{z}{2a}\right) + \mathbf{A} = -\mathbf{V}'\,a\,z\left(1 + \frac{z}{2a}\right) + \mathbf{A}',$$

and accordingly $A - A' = (V - V') a z - \frac{1}{2}(V + V') z$. Since $V + V' = P$, and from (16) $V - V' = P\frac{z}{a}$, we have,

$$A - A' = \tfrac{1}{2} P z^2 \quad \ldots \ldots \quad (I.)$$

(b) *Horizontal displacement.*

Inserting the value of dy for the parabola, viz., $a y = \frac{2h}{a^2} x\, d x$, and the value of M from (19), and inserting in this last the value of y, viz., $y = \frac{h}{a^2} x^2$, we have from equation (15), Art. 2,

$$E I d \Delta x = -\frac{2h}{a^2}\left[H h x\left(1 - \frac{x^2}{8 a^2}\right) - V a x\left(1 - \frac{x}{2a}\right) + A\right] x\, d x.$$

Integrating this, we have for the three divisions of the arch, as before,

$$\left.\begin{aligned}
E I \Delta x &= -\frac{2h}{a^2}\left[\tfrac{1}{2} H h x^2\left(1 - \frac{x^2}{5 a^2}\right) - \tfrac{1}{3} V a x^2\left(1 - \frac{3x}{8a}\right) + \tfrac{1}{2} A x^2 + B\right] \\
E I \Delta x' &= -\frac{2h}{a^2}\left[\tfrac{1}{2} H h x^2\left(1 - \frac{x^2}{5 a^2}\right) - \tfrac{1}{3} V' a x^2\left(1 + \frac{3x}{8a}\right) + \tfrac{1}{2} A' x^2 + B'\right] \\
E I \Delta x'' &= -\frac{2h}{a^2}\left[\tfrac{1}{2} H h x^2\left(1 - \frac{x^2}{5 a^2}\right) - \tfrac{1}{3} V' a x^2\left(1 + \frac{3x}{8a}\right) + \tfrac{1}{2} A'' x^2 + B''\right]
\end{aligned}\right\} \ldots(23)$$

For the point E, or $x = z$, we have from the two first of these equations, $B - B' = \frac{1}{3}(V - V') a z^3 - \frac{1}{6}(V + V') z^4 - \frac{1}{2}(A - A') z^2$, that is,

$$B - B' = -\tfrac{1}{12} P z^4 \quad \ldots \ldots \quad (II.)$$

For the crown, $x = 0$, and the second and third equations are equal, hence

$$B' = B' \quad \ldots \ldots \quad (III.)$$

For the left end, that is, for $x = a$, since the end of the arch must not slip, we must have $x' - x = 0$. So also for the right end, for $x = -a$. Therefore, from the first and third equations, putting B' for B'', we have

$$\tfrac{1}{15} H a^3 h - \tfrac{5}{24} V a^4 + \tfrac{1}{2} A a^2 + B = 0,$$
$$-\tfrac{4}{15} H a^3 h + \tfrac{5}{24} V' a^4 + \tfrac{1}{2} A' a^2 + B' = 0.$$

By the addition and subtraction of these equations, we have, since $V + V' = P$, and $(V - V') a = P z$,

$$\tfrac{5}{24} P a^3 z - \tfrac{1}{2}(A + A') a^2 - (B + B') = 0 \quad \ldots \quad (IV.)$$

$$\tfrac{8}{15} H a^3 h - \tfrac{1}{24} P(5 a^4 + z^4) + \tfrac{1}{2}(A - A') a^2 = 0 \quad \ldots \quad (V.)$$

We might, in a precisely similar manner, form three equations similar to (23) for the *vertical* displacement Δy. This would introduce three more constants and four more equations of condition between them. By the nine equations I. to IX. thus obtained, these constants may be then determined in terms of the known quantities H, h, P, a and z, and thus the change of shape at any point may be fully determined.

The complete discussion, as indicated, is unnecessary for the purpose we have in view, and we shall not, therefore, pursue it further. We have already all the general formulæ of which we shall need to make use in the discussion of the parabolic arch.

6. Circular Arc—concentrated Load.—In a perfectly similar manner we may make out analogous formulæ for the circular arch. Thus, referring to equation (8), Art. 1, and inserting for **G** and **M** their values as given in (18) and (19), Art. 3, we have for the force in the direction of the axis (see eq. 4),

$$\left. \begin{array}{l} \mathbf{E A} \dfrac{\Delta ds}{ds} = -\mathbf{H}\cos\alpha - \mathbf{V}\sin\alpha \\[6pt] \mathbf{E A} \dfrac{\Delta ds'}{ds} = -\mathbf{H}\cos\alpha - \mathbf{V}'\sin\alpha \end{array} \right\} \quad \ldots \quad (24)$$

Putting for **H**, **V** and **V**' their values from (21), Art. 4, and (17), Art. 3, we have,

$$\left. \begin{array}{l} \mathbf{E A} \dfrac{\Delta ds}{ds} = -\mathbf{P}\dfrac{\sin\alpha + \sin\beta - 2\cos\alpha\sin\beta}{2(1-\cos\alpha)} \\[6pt] \mathbf{E A} \dfrac{\Delta ds'}{ds} = -\mathbf{P}\dfrac{\sin\alpha - \sin\beta}{2(1-\cos\alpha)} \end{array} \right\} \quad \ldots \quad (25)$$

(a) Change of direction of tangents.
Referring to equation (12), Art. 1, we have, since $r\,d\phi = ds$ and $r\phi = s$,

$$\Delta\phi = \int \frac{\mathbf{M}}{\mathbf{E I}} r\,d\phi + \frac{\Delta ds}{ds}\phi.$$

The value of **M** is given in (20), of $\dfrac{\Delta ds}{ds}$ in (24). Inserting these values, we have

$$\Delta\phi = \frac{r^2}{\mathbf{E I}} \int \Big[\mathbf{H}(\cos\phi - \cos\alpha) - \mathbf{V}(\sin\alpha - \sin\phi) \Big]$$
$$- \frac{1}{\mathbf{E A}}(\mathbf{H}\cos\alpha + \mathbf{V}\sin\alpha)\phi.$$

Performing the integration,* and putting, for brevity, $\kappa = \dfrac{\mathbf{I}}{\mathbf{A} r^2}$, we have for the three segments of the arc, as before,

$$\left. \begin{array}{l} \mathbf{E I}\Delta\phi = r^2\Big[\mathbf{H}(\sin\phi - \phi\cos\alpha) - \mathbf{V}(\phi\sin\alpha + \cos\phi)\Big] - \kappa r^2(\mathbf{H}\cos\alpha + \mathbf{V}\sin\alpha)\phi + \mathbf{A} \\ \mathbf{E I}\Delta\phi' = r^2\Big[\mathbf{H}(\sin\phi - \phi\cos\alpha) - \mathbf{V}'(\phi\sin\alpha - \cos\phi)\Big] - \kappa r^2(\mathbf{H}\cos\alpha + \mathbf{V}'\sin\alpha)\phi + \mathbf{A}' \\ \mathbf{E I}\Delta\phi'' = r^2\Big[\mathbf{H}(\sin\phi - \phi\cos\alpha) - \mathbf{V}'(\phi\sin\alpha - \cos\phi)\Big] - \kappa r^2(\mathbf{H}\cos\alpha + \mathbf{V}'\sin\alpha)\phi + \mathbf{A}'' \end{array} \right\} 26$$

For the point **E**, $\phi = \beta$, and the first and second equations become simultaneous. Hence, after reduction,

$$\mathbf{A} - \mathbf{A}' = (\mathbf{V} - \mathbf{V}')\, r^2 \beta\sin\alpha + (\mathbf{V} + \mathbf{V}')\, r^2 \cos\beta + \kappa(\mathbf{V}-\mathbf{V}')\, r^2\beta\sin\alpha.$$

But $\mathbf{V} + \mathbf{V}' = \mathbf{P}$, and from (6) $\mathbf{V} - \mathbf{V}' = \mathbf{P}\dfrac{\sin\beta}{\sin\alpha}$, hence

$$\mathbf{A} - \mathbf{A}' = \mathbf{P}\, r^2\Big[\cos\beta + (1+\kappa)\beta\sin\beta\Big] \quad \ldots \quad (\mathrm{I.})$$

---

* See Art. 7, following.

(b) *Horizontal displacement.*

According to eq. (14), Art. 2, since $x = r \sin \phi$, $y = r(1 - \cos \phi)$, $dx = r \cos \phi \, d\phi$, $dy = r \sin \phi \, d\phi$, we have

$$EI \Delta x = r \int \Big[H r^2 (\sin \phi - \phi \cos a) - V r^2 (\phi \sin a + \cos \phi),$$

$$- \kappa r^2 (H \cos a + V \sin a) \phi + A \Big] \sin \phi \, d\phi,$$

$$- \kappa r^2 \int (H \cos a + V \sin a) \cos \phi \, d\phi.$$

The integration gives us the three equations,

$$EI \Delta x = - r^3 \Big[H(\tfrac{1}{2}\phi - \tfrac{1}{2}\sin\phi\cos\phi - \cos a \sin\phi + \phi \cos a \cos\phi)$$
$$- V (\sin a \sin\phi - \phi \sin a \cos\phi + \tfrac{1}{2}\sin^2\phi) \Big]$$
$$- \kappa r^3 (H \cos a + V' \sin a) \phi \cos \phi - A r (1 - \cos \phi) + B.$$

$$EI \Delta x' = - r^3 \Big[H(\tfrac{1}{2}\phi - \tfrac{1}{2}\sin\phi\cos\phi - \cos a \sin\phi + \phi \cos a \cos\phi)$$
$$- V' (\sin a \sin\phi - \phi \sin a \cos\phi - \tfrac{1}{2}\sin^2\phi) \Big]$$
$$- \kappa r^3 (H \cos a + V' \sin a) \phi \cos \phi - A' r (1 - \cos \phi) + B'.$$

$$EI \Delta x'' = - r^3 \Big[H(\tfrac{1}{2}\phi - \tfrac{1}{2}\sin\phi\cos\phi - \cos a \sin\phi + \phi \cos a \cos\phi)$$
$$- V' (\sin a \sin\phi - \phi \sin a \cos\phi - \tfrac{1}{2}\sin^2\phi) \Big]$$
$$- \kappa r^3 (H \cos a + V' \sin a) \phi \cos \phi - A' r (1 - \cos \phi) + B''.$$

For $\phi = \beta$, that is at the load, Δx must equal $\Delta x'$.
Hence, after reduction,

$$B - B' = - \tfrac{1}{2} P r^3 (2 + \sin^2\beta - 2\cos\beta - 2\beta \sin\beta) + \kappa P r^3 \beta \sin \beta \quad . . \text{ (II.)}$$

For the crown $\phi = 0$, and $\Delta x' = \Delta x''$; hence

$$B' = B'' \quad . . . . \text{ (III.)}$$

For the left end, $\phi = a$ and $\Delta x = 0$; for the right end, $\phi = -a$ and $\Delta x'' = 0$; that is,

$$- \tfrac{1}{2} H r^3 (a - 3 \sin a \cos a + 2 a \cos^2 a) + \tfrac{1}{2} V r^3 (3 \sin^2 a - 2 a \sin a \cos a)$$
$$- \kappa r^3 (H \cos a + V \sin a) a \cos a - A r (1 - \cos a) + B = 0, \text{ and}$$
$$+ \tfrac{1}{2} H r^3 (a - 3 \sin a \cos a + 2 a \cos^2 a) - \tfrac{1}{2} V' r^3 (3 \sin^2 a - 2 a \sin a \cos a)$$
$$+ \kappa r^3 (H \cos a + V' \sin a) a \cos a - A' r (1 - \cos a) + B'' = 0.$$

The subtraction and addition of these equation gives, after reduction,

$$H r^2 (a - 3 \sin a \cos a + 2 a \cos^2 a)$$
$$- \tfrac{1}{2} P r^2 (3 \sin^2 a - 2 a \sin a \cos a - 2 - \sin^2 \beta + 2 \cos \beta + 2 \beta \sin \beta)$$
$$+ \kappa r^2 \Big[2 H a \cos^2 a + P (a \sin a \cos a - \beta \sin \beta) \Big]$$
$$+ (A - A')(1 - \cos a) = 0 \quad . . . . . \text{ (IV.)}$$

and
$$\tfrac{1}{2} P r^2 \sin \beta (3 \sin a - 2 a \cos a)$$
$$- (A + A') r (1 - \cos a) - \kappa P r^2 a \cos a \sin \beta + B + B'' = 0 \quad . . \text{ (V.)}$$

Here, as before, we shall leave the discussion, as we have already all the equations of which we shall make use.

7. Integrals used in the above Discussion.—For convenience of reference, we here group together the known integrals employed in the preceding discussion.

$$\int \sin x \, dx = -\cos x, \quad \int \cos x \, dx = \sin x.$$

$$\int \sin^2 x \, dx = \tfrac{1}{2} x - \tfrac{1}{2} \sin x \cos x, \quad \int \cos^2 x \, dx = \tfrac{1}{2} x + \tfrac{1}{2} \sin x \cos x.$$

$$\int \sin x \cos x \, dx = \tfrac{1}{2} \sin^2 x.$$

$$\int \sin^3 x \, dx = -\tfrac{1}{3} \cos x \, (2 + \sin^2 x), \quad \int \cos^3 x \, dx = \tfrac{1}{3} \sin x \, (2 + \cos^2 x).$$

$$\int \sin^2 x \cos x \, dx = \tfrac{1}{3} \sin^3 x, \quad \int \sin x \cos^2 x \, dx = -\tfrac{1}{3} \cos^3 x.$$

$$\int x \sin x \, dx = \sin x - x \cos x, \quad \int x \cos x \, dx = \cos x + x \sin x.$$

$$\int x \sin^2 x \, dx = \tfrac{1}{4} x^2 + \tfrac{1}{4} \sin^2 x - \tfrac{1}{2} x \sin x \cos x.$$

$$\int x \cos^2 x \, dx = \tfrac{1}{4} x^2 - \tfrac{1}{4} \sin^2 x + \tfrac{1}{2} x \sin x \cos x.$$

$$\int x \sin x \cos x \, dx = \tfrac{1}{4} (2 x \sin^2 x - x + \sin x \cos x).$$

CHAPTER III.

ARCH HINGED AT ABUTMENTS ONLY — CONTINUOUS AT CROWN.

A. PARABOLIC ARC—CONSTANT CROSS-SECTION—CONCENTRATED LOAD.

8. Horizontal Thrust.—We can apply here directly the results of Art 5. Thus, in equations (22) for $x = 0$, $\Delta \phi' = 0$, and $\Delta \phi'' = 0$, hence $A' = A''$. If then in eq. (V.) of that Art. we put A' for A'', and then for $A - A'$ its value from (I.), we have at once

$$H = \tfrac{5}{64} P \frac{5 a^4 - 6 a^2 z^2 + z^4}{a^3 h} = \tfrac{5}{64} P \frac{(5 a^2 - z^2)(a^2 - z^2)}{a^3 h} \quad \ldots (27)$$

This is the formula which we have given in Art. 159 of the text, without demonstration. The thrust is greatest when the load is at the crown. We have then $z = 0$ and $H = \tfrac{25}{64} P \dfrac{a}{h}$. The value of V is given in Art. 3.

9. Intersection Curve.—Denote the ordinate of the curve $c\,d\,e\,i\,k$ (Pl. 23, Fig. 91), taken above the line AB by y. Then we see from the Fig. that $y = AN$ tang. $dAN = (a - z)\dfrac{V}{H}$. The value of H is given above, that of V has already been found in Art. 3, eq. (16), viz.,

$$V = P \frac{a + z}{2a}.$$ Hence we have, after reduction,

$$y = \frac{32\, a^2 h}{5\,(5\, a^2 - z^2)} \quad \ldots \ldots (28)$$

which is the equation already given in Art. 159, and from (18) and (19) the values of the table in that Art. have been calculated.

The above values of H and V are simple and of easy application, not involving much calculation in any special case. Hence we can readily compute H and V, and thus check the accuracy of our method of construction given in Chap. XIV.

B. CIRCULAR ARC—CONSTANT CROSS-SECTION—CONCENTRATED LOAD.

10. Horizontal Thrust.—Here we can apply directly the results of Art. 6. Thus, inserting in eq. (IV.) of that Art. $A - A'$ for $A - A''$, and taking the value of $A - A'$ from (I.), we have an equation for the determination of H. This, after reduction, becomes

$$H = P \frac{\sin^2 a - \sin^2 \beta + 2 \cos a\, (\cos \beta - \cos a) - 2(1 + \kappa) \cos a\, (a \sin a - \beta \sin \beta)}{2\,[a - 3 \sin a \cos a + 2(1 + \kappa)\, a \cos^2 a]},$$

which is the equation given in Art. 159 (2) of the text.

For the semi-circle, $a = 90°$, $\sin a = 1$, and $\cos a = 0$, and this becomes

$$H = P \frac{\cos^2 \beta}{\pi}.$$

If we put

$A_1 = \sin^2 \alpha - \sin^2 \beta + 2 \cos \alpha (\cos \beta - \cos \alpha - \alpha \sin \alpha + \beta \sin \beta)$,
$A_2 = 2 \cos \alpha (\alpha \sin \alpha - \beta \sin \beta)$,
$B_1 = 2 (\alpha - 3 \sin \alpha \cos \alpha + 2 \alpha \cos^2 \alpha)$,
$B_2 = 2 \alpha \cos^2 \alpha$,

we have

$$H = P \frac{A_1 - A_2 k}{B_1 + B_2} = P \frac{1 - \frac{A_2}{A_1} \kappa}{1 + \frac{B_2}{B_1} \kappa} \left(\frac{A_1}{B_1}\right),$$

or, if we put $A = \frac{A_2}{A_1}$, $B = \frac{B_2}{B_1}$, and $H_0 = P \frac{A_1}{B_1}$, we have

$$H = H_0 \frac{1 - A \kappa}{1 + B \kappa} \quad \ldots \ldots \ldots (29)$$

But $H_0 = P \frac{A_1}{B_1}$ is the value of H from the formula above, when the terms containing k are disregarded.

We have also, by series (see Art. 20, following),

$A_1 = \tfrac{1}{7} (\alpha^2 - \beta^2) [(5 \alpha^2 - \beta^2) - \tfrac{1}{30} (49 \alpha^4 + 34 \alpha^2 \beta^2 - 11 \beta^4) + \ldots]$
$A_2 = 2 (\alpha^2 - \beta^2) [1 - \tfrac{1}{6} (4 \alpha^2 + \beta^2) + \ldots]$
$B_1 = \tfrac{1}{15} \alpha^5 (1 - \tfrac{4}{21} \alpha^2 + \ldots)$ $B_2 = 2 \alpha (1 - \alpha^2 + \ldots)$

Approximately, therefore, since for h small with respect to a, the tangent may be taken for the arc, and hence $\frac{2r}{ra} = \frac{a}{h}$, or $a = \frac{a}{2h}$, we have

$$A = \frac{24}{5 \alpha^2 - \beta^2} = \frac{6 a^2}{5 h^2 \left[1 - \frac{1}{5}\left(\frac{\beta}{\alpha}\right)^2\right]} \qquad B = \frac{15}{4 \alpha^4} = \frac{15 a^4}{64 h^4}$$

Hence, when rise is small compared with span, we have the approximate expression

$$H = H_0 \frac{1 - \frac{24}{5 \alpha^2 - \beta^2} \kappa}{1 + \frac{15}{4 \alpha^4} \kappa} = H_0 \frac{1 - \frac{6 a^2}{5 h^2 \left[1 - \frac{1}{5}\left(\frac{\beta}{\alpha}\right)^2\right]} \kappa}{1 + \frac{15 a^4}{64 h^4} \kappa}$$

By means of a table calculated for H_0, for various values of α and $\beta = 0$, 0.2, 0.4, etc., of α, the thrust can be readily found in any case from the above formula.

We give in the following Tables the values of H_r, A and B, calculated from the exact formulæ. The formula for H above is thus made of easy practical application, without tedious calculation, and the results given by the method of Chap. XIV. may easily be checked.

The value of V is given in Art. 3.

TABLE FOR H_0.

β	a = 0	a = 10°	a = 20°	a = 30°	a = 40°	a = 50°	a = 60°	a = 90°
0	0.391	0.391	0.388	0.385	0.380	0.373	0.364	0.318
0.2	0.372	0.372	0.369	0.365	0.359	0.352	0.342	0.288
0.4	0.318	0.317	0.315	0.309	0.301	0.292	0.278	0.208
0.6	0.232	0.231	0.228	0.223	0.213	0.202	0.187	0.110
0.8	0.123	0.122	0.119	0.115	0.108	0.099	0.086	0.030
1	0	0	0	0	0	0	0	0
a	Coefficients of $P\dfrac{a}{h}$.							

Thus, for $\beta = 0$, $\tfrac{2}{10}$, $\tfrac{4}{10}$, etc., of a, the numbers in the table give the coefficients of $P\dfrac{a}{h}$ for $a = 0$, 10°, 20°, etc.

For the values of **A** and **B**, we have the following

TABLE FOR **A** AND **B**.

	$\dfrac{\beta}{a}$	a = 0	a = 10°	a = 20°	a = 30°	a = 40°	a = 50°	a = 60°	a = 90°	
	0	1.20	1.19	1.17	1.14	1.08	1.00	0.88	0	Coefficients of $\dfrac{a^2}{h^2}$
	0.2	1.21	1.20	1.18	1.15	1.10	1.01	0.90	0	
A	0.4	1.24	1.24	1.21	1.18	1.13	1.05	0.94	0	
	0.6	1.29	1.29	1.27	1.24	1.20	1.13	1.02	0	
	0.8	1.38	1.38	1.36	1.34	1.30	1.24	1.18	0	
	1	1.50	1.50	1.49	1.47	1.45	1.41	1.36	0	
B		0.234	0.233	0.221	0.203	0.178	0.146	0.107	0	$\dfrac{a^4}{h^4}$

Thus, for various values of $\dfrac{\beta}{a}$, we have the coefficients of $\dfrac{a^2}{h^2}$, which give **A** for $a = 10°$, 20°, etc., and for the values of a have the coefficients of $\dfrac{a^4}{h^4}$, which give **B**.

11. Intersection Curve.—Indicating, as before, by y the ordinate

Nd of the curve $cdeik$ [Fig. 91], we have, as before, $y = AN$ tang.

$dAN = (a - z)\dfrac{V}{H}$, or since $a = r \sin a$, $z = r \sin \beta$, $V = P\dfrac{\sin a + \sin \beta}{2 \sin a}$,

from eq. (17),

$$y = \frac{\sin^2 a - \sin^2 \beta}{2 \sin a} \cdot \frac{P}{H} r.$$

Inserting the value of H above, we have

$$y = r\frac{\sin^2 a - \sin^2 \beta}{2 \sin a}\left[\frac{1 + B\kappa}{1 - A\kappa}\right]\frac{B_1}{A_1},$$

or if $y_i = r\left(\dfrac{\sin^2 a - \sin^2 \beta}{2 \sin a}\right)\dfrac{B_1}{A_1}$; that is, if y_i is the value of y when k is neglected,

$$y = \frac{1 + B\kappa}{1 - A\kappa} y_i \quad \ldots \ldots (30)$$

which is the value of y given in Art. 159 (2) of the text. In that Art. we have already tabulated the values of A and B, as also of y_i for various values of a and β.

For $\beta = a$, that is, for the end ordinate, our expression for y reduces to $\dfrac{0}{0}$. In this case, by differentiating numerator and denominator, we have

$$y = r\frac{a - 3 \sin a \cos a + 2(1 + \kappa) a \cos^2 a}{\sin a - a \cos a - \kappa (\sin a + a \cos a)}.$$

For the semi-circle, $a = 90° = \dfrac{\pi}{2}$, $\sin a = 1$, $\cos a = 0$, and hence $y = \tfrac{1}{2} \pi r = 1.5708\, r$. Hence, *for the semi-circle the intersection curve becomes a horizontal straight line at $0.5708\, r$ above the crown.* In all cases for small central angle a, κ may be disregarded.

The above results are sufficient to enable us to either diagram or calculate the strains in every piece for any given position of load.

CHAPTER IV.

ARCH FIXED AT ENDS.

12. Introduction.— In the previous case, the end reactions pass always through the ends. If, however, the ends are "walled in," so that the end cross-sections remain unchanged in position, and cannot turn, these reactions pass then no longer through the centres of the end cross-sections. In the first case, the moments at the ends are zero; now, however, we have end moments to be determined, viz., M_1 and M_2, left and right. For their determination we have the condition that the tangents to the curve at the ends must always remain invariable in direction, or for the ends, $\Delta \phi = 0$.

In the arch above with hinges at ends, we have always considered a portion lying between the end and any point. In the present case, however, we shall consider the portion between the *crown* and any point. Both methods lead, of course, to the same results, but the latter, in the present case, is somewhat simpler.

Accordingly, we conceive the arch cut through at the crown [Pl. 24, Fig. 93]. The total resultant force exerted upon the one-half by the other, we decompose into a vertical force **V** at the crown, and a horizontal force **H**. The distance ck of this last from the centre of gravity of the section at crown is e_0, and hence the moment at crown is $M_0 = -H e_0$.

13. Concentrated Load—General Formulæ.—Let a weight **P** act at any point; then representing, as before, by primes, quantities relating to the portion between the load and right end, we have, as in (18),

$$G = -H \cos \phi - (P - V) \sin \phi, \quad G' = -H \cos \phi + V \sin \phi$$
$$N = -H \sin \phi + (P - V) \cos \phi, \quad N' = -H \sin \phi - V \sin \phi \quad \}\ ..\ (31)$$

Also, $M = -H(e_0 + y) - Vx + P(x - z), \quad M' = -H(e_0 + y) - Vx$,

or, since $-H e_0 = M_0 =$ moment at crown,

$$M = M_0 - H y - V x + P(x - z), \quad M' = M_0 - H y - V x\ ..\ (32)$$

(a) *Intersection curve.*

The two reactions, **R** and **R'**, intersect, as before, in a point **L** (Fig. 92), which must lie upon **P** prolonged, as otherwise **R**, **R'** and **P** could not be in equilibrium. The *locus* of the point **L** we call, as before, the *intersection curve*. The equation of this curve can be easily found when **V**, **H** and M_0 are known.

The force acting upon the portion **B E** (Fig. 92) is the resultant of **V** and **H**. The component **H** acts at the point of intersection o of this resultant **L** ψ with the vertical through **C**. The vertical distance of this point o from c is, as above, e_0; its horizontal distance from **P** is z. Then $z \cot L \psi k$ is the vertical distance of this point from **L**, and $e_0 + z \cot$

$L \psi k = e_1 + s \dfrac{V}{H} = y$, where, as in the Fig., e_1 is negative. In any case, e_1 is given by $e_1 = - \dfrac{M_1}{H}$; hence, for the intersection curve,

$$y = \frac{V s - M_1}{H} \quad \ldots \ldots \quad (33)$$

(b) *Direction curves and segments.*

The *direction* of the resultants **R** and **R'** can be determined in two ways First, by the points of intersection ϕ and ψ with the verticals through the centres of the end cross-sections; second, by means of the *curves enveloped* by these resultants for every position of **P**. We call the first distances $A \phi = c_1$, $B \psi = c_2$, the *direction segments*, and the enveloped curves the *direction curves*.

Taking c_1 and c_2 as positive when laid off upwards above the ends, we have $M_1 = - H c_1$ $M_2 = - H c_2$; therefore

$$c_1 = - \frac{M_1}{H} \quad c_2 = - \frac{M_2}{H} \quad \ldots \ldots \quad (34)$$

We may also easily determine the equation of the direction curves. Let the co-ordinates with reference to the crown of any point be v and w (Fig. 92). If the load **P** is now moved through an indefinitely small distance, the new resultant cuts the former in a point of the curve required. These two resultants intersect the vertical through **C** in two points. Let the distances of these points from **C** be c and $c + d c$, and let γ and $\gamma + d \gamma$ be the angles of the resultants with the vertical. Then $v = (w + c) \tan \gamma$, $v = (w + c + d c) \tan (\gamma + d \gamma)$.

Eliminating v,

$$w = - \frac{(c + d c) \tan (\gamma + d \gamma) - c \tan \gamma}{\tan (\gamma + d \gamma) - \tan \gamma} = - \frac{d (c \tan \gamma)}{d \tan \gamma}.$$

From the first of the equations above we have then

$$v = - \frac{d c}{d \tan \gamma} \tan^2 \gamma.$$

But $\tan \gamma = \dfrac{H}{V}$, $c = - \dfrac{M_1}{H}$, $c \tan \gamma = - \dfrac{M_1}{V}$, hence

$$\left.\begin{array}{l} v = \dfrac{d \left(\dfrac{M_1}{H}\right)}{d \left(\dfrac{H}{V}\right)} \left(\dfrac{H}{V}\right)^2 = \dfrac{H d M_1 - M_1 d H}{V d H - H d V} \\[2ex] w = \dfrac{d \left(\dfrac{M_1}{V}\right)}{d \left(\dfrac{H}{V}\right)} = \dfrac{V d M_1 - M_1 d V}{V d H - H d V} \end{array}\right\} \quad \ldots \quad (35)$$

Thus we see that in any case we have only to determine **H**, **V** and **M₁**, and we can then from (33) and (34) or (35) determine at once the intersection curve, and the direction segments or curves. These are all we need for our method of construction as given in Chap. XIV.; once given, we can then easily construct H, V and M_t or M_1 for any position of weight.

A. PARABOLIC ARC—CONSTANT CROSS-SECTION—CONCENTRATED LOAD.

14. Determination of H, V and M_o.—We put $y = h\dfrac{x^2}{a^2}$,

$dy = 2\dfrac{hx}{a^2}dx$, $ds = dx$, as before. Then from the values of M given in (32) we have, according to equation (13), Art. 1, after inserting the values of y and ds above, and integrating,

$$E I \Delta \phi = x\left[M_o - \tfrac{1}{3}\dfrac{Hh}{a^2}x^2 - \tfrac{1}{2}Vx + \tfrac{1}{2}P(x - 2z)\right] + A,$$

and

$$E I \Delta \phi' = x\left[M_o - \tfrac{1}{3}\dfrac{Hh}{a^2}x^2 - \tfrac{1}{2}Vx\right] + A'.$$

For $x = z$, $\Delta\phi = \Delta\phi'$, hence $\tfrac{1}{2}P(z - 2z)z + A = A'$, or

$$A - A' = \tfrac{1}{2}Pz^2 \quad \ldots \ldots \ldots \text{(I.)}$$

For $x = a$, $\Delta\phi = 0$, and for $x = -a$, $\Delta\phi' = 0$, hence

$$0 = a\left[M_o - \tfrac{1}{3}Hh - \tfrac{1}{2}Va + \tfrac{1}{2}P(a - 2z)\right] + A,$$

$$0 = -a\left[M_o - \tfrac{1}{3}Hh + \tfrac{1}{2}Va\right] + A',$$

and by addition and subtraction,

$$A + A' = Va^2 - \tfrac{1}{2}P(a - 2z)a \quad \ldots \ldots \text{(II.)}$$

$$2M_o a - \tfrac{2}{3}Hah + \tfrac{1}{2}P(a - z)^2 = 0 \quad \ldots \text{(III.)}$$

From I. and II. we have

$$\left.\begin{array}{l}A = \tfrac{1}{2}Va^2 - \tfrac{1}{4}P(a^2 - 2az - z^2) \\ A' = \tfrac{1}{2}Va^2 - \tfrac{1}{4}P(a^2 - 2az + z^2)\end{array}\right\} \ldots \ldots (a)$$

For the horizontal and vertical displacement of any point, we have from 15), Art. 2, after integration,

$$E I \Delta z = -\dfrac{2h}{a^2}\left[\tfrac{1}{3}M_o x^3 - \dfrac{Hh}{15a^2}x^5 - \tfrac{1}{8}Vx^4 + \tfrac{1}{2}P(\tfrac{1}{4}x^4 - \tfrac{2}{3}zx^3) + \tfrac{1}{2}Ax^2 + B\right],$$

$$E I \Delta z' = -\dfrac{2h}{a^2}\left[\tfrac{1}{3}M_o x^3 - \dfrac{Hh}{15a^2}x^5 - \tfrac{1}{8}Vx^4 + \tfrac{1}{2}A'x^2 + B'\right],$$

and

$$E I \Delta y = \tfrac{1}{2}M_o x^2 - \dfrac{Hh}{12a^2}x^4 - \tfrac{1}{6}(V-P)x^3 - \tfrac{1}{2}Pzx^2 + Ax + C,$$

$$E I \Delta y' = \tfrac{1}{2}M_o x^2 - \dfrac{Hh}{12a^2}x^4 - \tfrac{1}{6}Vx^3 + A'x + C'.$$

For $z = z$, $\Delta z = \Delta z'$, and $\Delta y = \Delta y'$, hence
$B - B' = \frac{1}{14} P z^4 - \frac{1}{3}(A - A')z^3$, and $C - C' = \frac{1}{3} P z^3 - (A - A')z$, or

$$B - B' = -\tfrac{1}{14} P z^4 \quad \ldots \ldots \quad (IV.)$$

$$C - C' = -\tfrac{1}{6} P z^3 \quad \ldots \ldots \quad (V.)$$

For $z = a$, $\Delta z = 0$, $\Delta y = 0$, and for $z = -a$, $\Delta z' = 0$, $\Delta y' = 0$, hence,

$0 = +\tfrac{1}{3} M_0 a^3 - \tfrac{1}{12} H h a^2 - \tfrac{1}{3} V a^4 + \tfrac{1}{24} P a^5 (3a - 8z) + \tfrac{1}{2} A a^2 + B$,

$0 = -\tfrac{1}{3} M_0 a^3 + \tfrac{1}{12} H h a^2 - \tfrac{1}{3} V a^4 + \tfrac{1}{2} A' a^2 + B'$,

$0 = +\tfrac{1}{2} M_0 a^2 - \tfrac{1}{12} H h a^2 - \tfrac{1}{2}(V - P) a^4 - \tfrac{1}{2} P z a^3 + A a + C$,

$0 = +\tfrac{1}{2} M_0 a^2 - \tfrac{1}{12} H h a^2 + \tfrac{1}{6} V a^3 - A' a + C'$.

The addition and subtraction of the two first and two last of these equations gives, when we put for $A + A'$, $A - A'$, $B - B'$, $C - C'$, their values above:

$$B + B' = -\tfrac{1}{4} V a^4 + \tfrac{1}{14} P a^3 (3a - 4z) \quad \ldots \quad (VI.)$$

$$2 M_0 a^3 - \tfrac{2}{3} H h a^2 + \tfrac{1}{3} P(3 a^4 - 8 a^3 z + 6 a^2 z^2 - z^4) = 0 \ldots (VII.)$$

$$C + C' = - M_0 a^2 + \tfrac{1}{6} H h a^2 - \tfrac{1}{6} P a (a^3 - 3 a z + 3 z^2) \ldots (VIII.)$$

$$\tfrac{2}{3} V a^3 - \tfrac{1}{6} P(2 a^3 - 3 a^2 z + z^3) = 0 \quad \ldots \ldots \quad (IX)$$

Equations III. and VII. contain only H and M_0 unknown. Their solution gives

$$H = \tfrac{15}{32} P \frac{(a^2 - z^2)^2}{a^3 h} \quad \ldots \ldots \quad (36)$$

$$M_0 = -\tfrac{1}{32} P \frac{(a-z)^2 (3 a^2 - 10 a z - 5 z^2)}{a^2} \quad \ldots \quad (37)$$

From IX. we find directly

$$V = \tfrac{1}{4} P \frac{(a - z)^2 (2a + z)}{a^3}. \quad \ldots \ldots \quad (38)$$

These are the equations given in Art. 160. From them we have the following table:

z	H	V	M_0	z	H	V	M_0
0	0.4688	0.5000	− 0.09375	0.5	0.2637	0.1563	+ 0.02539
0.1	0.4594	0.4252	− 0.04936	0.6	0.1920	0.1040	+ 0.02400
0.2	0.4320	0.3520	− 0.01606	0.7	0.1219	0.0607	+ 0.01814
0.3	0.3882	0.2818	+ 0.00689	0.8	0.0607	0.0280	+ 0.01025
0.4	0.3308	0.2160	+ 0.02025	0.9	0.0169	0.0078	+ 0.00314
0.5	0.2637	0.1562	+ 0.02539	1.0	0.	0.	0.
.a	.$P\frac{a}{h}$	.P	.Pa	.a	.$P\frac{a}{h}$	.P	.Pa

From (34) and (35) we can now find the intersection and direction curves. The preceding table gives us sufficient data for complete calculation by moments according to Art. 162. The intersection and direction curves will, as already explained, enable us to find the above quantities graphically.

15. Intersection Curve.—From (33) we have $y = \dfrac{\mathbf{V}z - \mathbf{M}_0}{\mathbf{H}}$, or inserting the values of $\mathbf{H}$, $\mathbf{V}$ and $\mathbf{M}_0$ above, and reducing,

$$\mathbf{H}y = \tfrac{1}{16}\frac{\mathbf{P}(a-z)^2}{a^3}(a+z)^2 = \tfrac{1}{8}\mathbf{H}h, \text{ hence } y = \tfrac{1}{8}h.$$

For the parabolic arch with fixed ends, then, *the intersection curve becomes a straight horizontal line, $\tfrac{1}{8}h$ above the crown.*

16. Direction Curve.—From (36), (37) and (38) we have

$$\frac{d\mathbf{H}}{dz} = -\frac{15\mathbf{P}(a^2 - z^2)z}{8a^3h}, \quad \frac{d\mathbf{V}}{dz} = -\frac{3\mathbf{P}(a^2 - z^2)}{4a^3},$$

$$\frac{d\mathbf{M}_0}{dz} = \frac{\mathbf{P}(a-z)(4a^2 - 5az - 5z^2)}{8a^3}.$$

Inserting these in (35), as also the values of $\mathbf{H}$, $\mathbf{V}$ and $\mathbf{M}_0$ themselves, and reducing, we have

$$\left.\begin{array}{l} v = \dfrac{2a^2}{3a+z} \\[2mm] w = \dfrac{(23a^2 + 20az + 5z^2)h}{15(a+z)(3a+z)} \end{array}\right\} \quad \ldots \ldots (39)$$

For $z = 0$, $v = \tfrac{2}{3}a$, $w = \tfrac{23}{45}h$. For $z = a$, $v = \tfrac{1}{2}a$, $w = \tfrac{2}{5}h$.
For $z = -a$, $v = a$, $w = -\infty$. Eliminating z, we have

$$w = \frac{5a^2 - 5av + 2v^2}{15a(a-v)}h \quad \ldots \ldots (40)$$

This is the equation of an hyperbola. Hence, for the parabolic arc with fixed ends, *the direction curve is upon each side of the crown an hyperbola.* This hyperbola is described in Art. 160 of the text, (Fig. 93), and a table to facilitate its construction is there given.

B. CIRCULAR ARC—CONSTANT CROSS-SECTION—CONCENTRATED LOAD.

17. Fundamental Equations.—From eq. (32) we have, since $z = r \sin\phi$, $y = r(1 - \cos\phi)$, $z = r \sin\beta$,

$$\left.\begin{array}{l} \mathbf{M} = \mathbf{M}_0 - \mathbf{H}r(1 - \cos\phi) + (\mathbf{P} - \mathbf{V})r\sin\phi - \mathbf{P}r\sin\beta \\ \mathbf{M}' = \mathbf{M}_0 - \mathbf{H}r(1 - \cos\phi) - \mathbf{V}r\sin\phi \end{array}\right\} \ldots (41)$$

The expressions for $\mathbf{G}$, Art. 13, eq. (12), apply here directly.
Therefore, from eq. (8), Art. 1, we have

$$\left.\begin{array}{l} \mathbf{E}\mathbf{A}\dfrac{\Delta ds}{ds} = \dfrac{\mathbf{M}_0}{r} - \mathbf{H} - \mathbf{G}\sin\beta \\[2mm] \mathbf{E}\mathbf{A}\dfrac{\Delta ds'}{ds} = \dfrac{\mathbf{M}_0}{r} - \mathbf{H} \end{array}\right\} \ldots (42)$$

Hence from (12), Art. 1, since $ds = r\,d\phi$,

$$d\Delta\phi = \frac{r}{EI}\Big[M_0 - Hr(1-\cos\phi) + (P-V)r\sin\phi - Pr\sin\beta\Big]d\phi + \frac{\Delta ds}{ds}d\phi,$$

$$d\Delta\phi' = \frac{r}{EI}\Big[M_0 - Hr(1-\cos\phi) - Vr\sin\phi\Big]d\phi + \frac{\Delta ds'}{ds}d\phi.$$

Substituting the values of $\dfrac{\Delta ds}{ds}$ above, integrating, and putting, as before, for brevity, $\kappa = \dfrac{I}{A r^2}$, we have

$$EI\Delta\phi = r\Big[M_0\phi - Hr(\phi-\sin\phi) - (P-V)r\cos\phi - Pr\phi\sin\beta\Big]$$
$$+ \kappa r(M_0 - Hr - Gr\sin\beta)\phi + A.$$

$$EI\Delta\phi' = r\Big[M_0\phi - Hr(\phi-\sin\phi) + Vr\cos\phi\Big] + \kappa r(M_0 - Hr)\phi + A'.$$

For $\phi = \beta$, $\Delta\phi = \Delta\phi'$, and we obtain

$$A - A' = Pr^2\Big[\cos\beta + (1+\kappa)\beta\sin\beta\Big] \quad \ldots \quad (I.)$$

For $\phi = a$, $\Delta\phi = 0$, and for $\phi = -a$, $\Delta\phi' = 0$. Adding and subtracting the equations thus obtained, and eliminating $A - A'$, we have

$$A + A' = Pr^2\Big[\cos a + (1+\kappa)a\sin\beta\Big] - 2Vr^2\cos a \quad \ldots \text{(II.)}$$

$$2M_0 a - 2Hr(a - \sin a) - Pr\Big[\cos a - \cos\beta + (a-\beta)\sin\beta\Big]$$
$$+ \kappa\Big[2M_0 a - 2Hra - Pr(a-\beta)\sin\beta\Big] = 0 \quad \ldots \text{(III.)}$$

From eq. (14), Art. 2, we have, as before, after integrating, for the horizontal displacement,

$$EI\Delta z = -M_0 r^2(\sin\phi - \phi\cos\phi) + \tfrac{1}{2}Hr^2(2\sin\phi - 2\phi\cos\phi - \phi + \sin\phi\cos\phi)$$
$$-\tfrac{1}{2}Vr^2\sin^2\phi + \tfrac{1}{2}Pr^2(\sin^2\phi + 2\sin\beta\sin\phi - 2\phi\sin\beta\cos\phi)$$
$$+ \kappa r^2(M_0 - Hr - Pr\sin\beta)\phi\cos\phi + Ar\cos\phi + B.$$

$$EI\Delta z' = -M_0 r^2(\sin\phi - \phi\cos\phi) + \tfrac{1}{2}Hr^2(2\sin\phi - 2\phi\cos\phi - \phi + \sin\phi\cos\phi)$$
$$-\tfrac{1}{2}Vr^2\sin^2\phi + \kappa r^2(M_0 - Hr)\phi\cos\phi + A'r\cos\phi + B'.$$

For $\phi = \beta$, $\Delta z = \Delta z'$, hence

$$B - B' = -\tfrac{1}{2}Pr(2 + \sin^2\beta) \quad \ldots \quad \text{(IV.)}$$

Further, for $\phi = a$, $\Delta z = 0$, and for $\phi = -a$, $\Delta z' = 0$. Hence, by adding and subtracting,

$$B + B' = Vr(2 - \sin^2 a) - \tfrac{1}{2}Pr(2 - \sin^2 a + 2\sin a\sin\beta)\ldots\text{(V.)}$$

$$2M_0(\sin a - a\cos a) - Hr(2\sin a - 2a\cos a - a + \sin a\cos a)$$
$$+ \tfrac{1}{2}Pr\Big[2 - \sin^2 a + \sin^2\beta - 2\cos(a-\beta) + 2(a-\beta)\cos a\sin\beta\Big]$$
$$- \kappa\Big[2(M_0 - Hra\cos a - Pr(a-\beta)\cos a\sin\beta\Big] = 0\ldots\text{(VI.)}$$

Multiplying III. by $\cos a$, and adding to VI., we have
$$2\mathbf{M}_0 \sin a - \mathbf{H}r(2\sin a - \sin a\cos a - a) + \tfrac{1}{2}\mathbf{P}r(\sin a - \sin \beta)^2 = 0.$$
In similar manner, we have from eq. (14), Art. 2, for the vertical displacement

$\mathbf{E}\mathbf{I}\Delta y = \mathbf{M}_0 r^2 (\cos\phi + \phi\sin\phi) - \tfrac{1}{2}\mathbf{H}r^3(2\cos\phi + 2\phi\sin\phi - \sin^2\phi)$
$+ \tfrac{1}{2}\mathbf{V}r^3(\phi + \sin\phi\cos\phi) - \tfrac{1}{3}\mathbf{P}r^3(\phi + \sin\phi\cos\phi + 2\sin\beta\cos\phi + 2\phi\sin\beta\sin\phi)$
$\quad + \kappa r^2 (\mathbf{M}_0 - \mathbf{H}r - \mathbf{P}r\sin\beta)\phi\sin\phi + \mathbf{A}r\sin\phi + \mathbf{C}.$

$\mathbf{E}\mathbf{I}\Delta y' = \mathbf{M}_0 r^2(\cos\phi + \phi\sin\phi) - \tfrac{1}{2}\mathbf{H}r^3(2\cos\phi + 2\phi\sin\phi - \sin^2\phi)$
$+ \tfrac{1}{2}\mathbf{V}r^3(\phi + \sin\phi\cos\phi) + \kappa r^2(\mathbf{M}_0 - \mathbf{H}r)\phi\sin\phi + \mathbf{A}'r\sin\phi + \mathbf{C}'.$

For $\phi = \beta$, $\Delta y = \Delta y'$, hence
$$\mathbf{C} - \mathbf{C}' = \tfrac{1}{2}\mathbf{P}r^3(\beta + \sin\beta\cos\beta) \quad \dots \quad \text{(VIII.)}$$
Finally, for $\phi = a$, $\Delta y = 0$. For $\phi = -a$, $\Delta y' = 0$, and hence
$\mathbf{C} + \mathbf{C}' = -2\mathbf{M}_0 r^2(\cos a + a\sin a) + \mathbf{H}r^3(2\cos a + 2a\sin a - \sin^2 a)$
$\quad + \tfrac{1}{2}\mathbf{P}r^3\Big[a + \sin a\cos a - 2\sin(a-\beta) + 2(a-\beta)\sin a\sin\beta\Big]$
$\quad - 2\mathbf{H}r^2(\mathbf{M}_0 - \mathbf{H}r)a\sin a + \kappa\mathbf{P}r^3(a-\beta)\sin a\sin\beta \quad \dots \text{(IX.)}$

$\mathbf{V}r(a - \sin a\cos a) = \tfrac{1}{2}\mathbf{P}r(a - \beta - \sin a\cos a - \sin\beta\cos\beta + 2\cos a\sin\beta)\dots\text{(X.)}$

18. Determination of H, V and $\mathbf{M}_0$.
(a) *Vertical Reaction.*
The vertical force $\mathbf{V}$ is given directly by eq. X. Thus
$$\mathbf{V} = \mathbf{P}\frac{a - \beta - \sin a\cos a - \sin\beta\cos\beta + 2\cos a\sin\beta}{2(a - \sin a\cos a)} \dots (43)$$
an expression independent of κ.
Transforming by means of series, we have, approximately,
$$\mathbf{V} = \frac{\mathbf{P}}{4a^3}\left[2a^3 - 3a^2\beta + \beta^3 - \tfrac{1}{10}(a^2 - \beta^2)^2\beta + \dots\right] \dots (44)$$
For the semi-circle
$$\mathbf{V} = \mathbf{P}\frac{\pi - 2\beta - 2\sin\beta\cos\beta}{2\pi} \quad \dots \quad (45)$$
From the exact formula (43) we have the following table:

β	$a=0$	$a=10°$	$a=20°$	$a=30°$	$a=40°$	$a=50°$	$a=60°$	$a=90°$
0	0.5	0.5	0.5	0.5	0.5	0.5	0.5	0.5
0.2	0.3520	0.3515	0.3500	0.3475	0.3439	0.3392	0.3332	0.3005
0.4	0.2160	0.2152	0.2130	0.2092	0.2037	0.1966	0.1876	0.1486
0.6	0.1040	0.1033	0.1014	0.0981	0.0934	0.0874	0.0799	0.0486
0.8	0.0280	0.0277	0.0269	0.0255	0.0238	0.0211	0.0182	0.0065
1	0	0	0	0	0	0	0	0
.a				$\mathbf{P}$				

(b) *Horizontal thrust.*

Eliminating M_1 from III. and VI. we obtain, after reduction,

$$H = P \frac{2 \sin \alpha \left[\cos \beta - \cos \alpha + (1+\kappa)\beta \sin \beta \right] - (1+\kappa)\alpha(\sin^2 \alpha + \sin^2 \beta)}{2 \left[(1+\kappa)\alpha(\alpha + \sin \alpha \cos \alpha) - 2\sin^2 \alpha \right]} \quad .. \quad (46)$$

If we put $A_1 = 2 \sin \alpha (\cos \beta - \cos \alpha + \beta \sin \beta) - \alpha(\sin^2 \alpha + \sin^2 \beta)$,
$A_2 = \alpha(\sin^2 \alpha + \sin^2 \beta) - 2\beta \sin \alpha \sin \beta$,
$B_1 = 2\alpha(\alpha + \sin \alpha \cos \alpha) - 4 \sin^2 \alpha$,
$B_2 = 2\alpha(\alpha + \sin \alpha \cos \alpha)$,

and let $A = \dfrac{A_2}{A_1}$, $B = \dfrac{B_2}{B_1}$, and $H_0 = \dfrac{A_1}{B_1} P$, we have

$$H = H_0 \frac{1 - A\kappa}{1 + B\kappa},$$

where H_0 is the value of H from the above formula, when terms containing κ are disregarded.

Transforming by series, we have

$$H = 1\tfrac{1}{8} P \frac{(\alpha^2 - \beta^2)^2}{\alpha^3} \left[1 - \tfrac{10}{210} \alpha^2 - \tfrac{11}{30} \beta^2 + \ldots \right]$$

$$A = \frac{12}{\alpha^2 - \beta^2} \left[1 - \tfrac{1}{10} \alpha^2 + \tfrac{11}{30} \beta^2 + \ldots \right]$$

$$B = \frac{45}{\alpha^4} \left[1 - \tfrac{4}{21} \alpha^2 + \ldots \right]$$

From the exact formula (46) above, we have the following tables:

TABLE FOR H_0.

β	$\alpha = 0$	$\alpha = 10°$	$\alpha = 20°$	$\alpha = 30°$	$\alpha = 40°$	$\alpha = 50°$	$\alpha = 60°$	$\alpha = 90°$
0	0.4688	0.4687	0.4683	0.4678	0.4671	0.4661	0.4610	0.4592
0.2	0.4320	0.4317	0.4309	0.4291	0.4272	0.4243	0.4173	0.4017
0.4	0.3308	0.3301	0.3281	0.3244	0.3196	0.3128	0.3012	0.2601
0.6	0.1920	0.1912	0.1887	0.1845	0.1784	0.1703	0.1578	0.1087
0.8	0.0608	0.0603	0.0590	0.0566	0.0534	0.0490	0.0421	0.0181
1	0	0	0	0	0	0	0	0
.a				$.P\dfrac{a}{h}$				

SUPPLEMENT TO CHAP. XIV.

For the values of the quantities **A** and **B** we have the following table:

VALUES OF **A** AND **B**.

	β	α = 0	α = 10°	α = 20°	α = 30°	α = 40°	α = 50°	α = 60°	α = 90°	
A	0	3.00	3.01	3.02	3.06	3.10	3.16	3.25	3.66	$\dfrac{a^2}{h^2}$
	0.2	3.12	3.15	3.16	3.19	3.26	3.34	3.45	4.07	
	0.4	3.57	3.58	3.63	3.70	3.81	3.79	4.20	5.05	
	0.6	4.09	4.70	4.81	4.96	5.20	5.57	6.14	10.57	
	0.8	8.33	8.42	8.67	9.10	9.78	10.87	12.93	35.62	
	1	∞	∞	∞	∞	∞	∞	∞		
B	.α	2.813	2.825	2.861	2.931	3.039	3.198	3.417	5.279	$\dfrac{a^4}{h^4}$

From the above tables it is easy to find the thrust for any given position of load, and any given span and rise. The preceding table gives the reaction; it only remains to determine the moment M_0 at crown.

(c) *Moment at crown.*

From VII. we have

$$M_0 = \tfrac{1}{2} H r \left(2 - \cos\alpha - \frac{\alpha}{\sin\alpha}\right) - \tfrac{1}{2} P r \frac{(\sin\alpha - \sin\beta)^2}{\sin\alpha}.$$

Substituting the value of **H**, already given. eq. (46), we obtain

$$2 M_0 \left[(1 + \kappa)\alpha(\alpha + \sin\alpha\cos\alpha) - 2\sin^2\alpha\right]$$

$$= P r \left[\sin\alpha - \sin\alpha\cos(\alpha-\beta) + 2\sin\alpha(\cos\beta - \cos\alpha) - \sin\alpha(\sin\alpha - \sin\beta)\sin\beta\right.$$

$$- \alpha(\cos\beta - \cos\alpha) - (1 + \kappa)\{\alpha(\sin^2\alpha + \sin^2\beta) - 2\beta\sin\alpha\sin\beta\}$$

$$\left. + (1 + \kappa)(\alpha - \beta)(\alpha + \sin\alpha\cos\alpha)\sin\beta\right].$$

In similar manner, as before, for **H** we have

$$M_0 = M_{00} \frac{1 + C\kappa}{1 + B\kappa},$$

where M_{00} is the value of M_0 when terms containing k are disregarded, and **B** has the same value as above. By series we have

$$M_{00} = -\tfrac{1}{24} P r \frac{(\alpha - \beta)^2}{\alpha^2} \left[3\alpha^2 - 10\alpha\beta - 3\beta^2\right.$$

$$\left. + \tfrac{1}{14}(3\alpha^4 + 6\alpha^3\beta + 45\alpha^2\beta^2 + 308\alpha\beta^3 + 154\beta^4)\right],$$

and

$$C = \frac{360}{\alpha^2(3\alpha^2 - 10\alpha\beta - 5\beta^2)} + \cdots$$

From the exact formula above we have the following tables.

VALUE OF M_{aa}.

β	a = 0	a = 10°	a = 20°	a = 30°	a = 40°	a = 50°	a = 60°	a = 90°
0	−0.09375	−0.09426	−0.09582	−0.09849	−0.10241	−0.10777	−0.11613	−0.15108
0.2	−0.01606	−0.01615	−0.01658	−0.01742	−0.01846	−0.02063	−0.02280	−0.03083
0.4	+0.02025	+0.02021	+0.02008	+0.01973	+0.01942	+0.01888	+0.01719	+0.01441
0.6	0.02400	0.02393	0.02369	0.02326	0.02267	0.02181	0.02001	0.01454
0.8	0.01025	0.01019	0.00999	0.00962	0.00917	0.00850	0.00716	0.00332
1	0	0	0	0	0	0	0	0
.a	.Pa							

VALUES OF B AND C.

	β	a = 0	a = 10°	a = 20°	a = 30°	a = 40°	a = 50°	a = 60°	a = 90°	
	0	+7.50	+7.52	+7.57	+7.78	+7.94	+8.52	+8.61	+11.12	
	0.2	28.13	28.09	27.99	28.14	28.43	27.99	27.76	31.81	
C	0.4	−12.50	−12.61	−13.05	−14.08	−15.40	−17.35	−21.11	−37.47	$\frac{a^4}{h^4}$
	0.6	4.69	4.77	5.00	5.51	6.24	7.31	9.10	20.27	
	0.8	2.74	2.85	3.20	3.89	4.94	6.61	9.77	42.53	
	1	1.88	∞	∞	∞	∞	∞	∞	∞	
B	.a	+2.81	+2.83	+2.86	+2.93	+3.04	+3.20	+3.42	+5.28	$\cdot \frac{a^4}{h^4}$

Here, as always, a negative moment denotes tension in lower or inner flange. We see at once from the table that the maximum compression in this flange at crown does *not* occur for full load, but for load extending from both ends towards the crown as far as about ₄ths of the span or ⅘ths of the half span. Within the middle half of the arch, then, a load anywhere causes *tension* in lower flange at crown—outside of this middle half a load anywhere causes compression in the lower flange at crown. For large central angles, κ may be disregarded, and we have simply $M = M_{aa}$.

19. Intersection Curve.—From Art. 13 we have

$$y = \frac{V r \sin \beta - M_0}{H}.$$

Hence, by substitution of the values of **V**, **H** and M_0,

$$y = \frac{r a^2}{10}\left[1 - \frac{120 (a^2 - 2 a z - z^2)}{a^4 (a + \beta)^2}\kappa\right]$$

or,

$$y = \tfrac{1}{2} h \left[1 - 80 \frac{a^2 - 2 a z - z^2}{(a + z)^2} \frac{I}{A h^2}\right]$$

. . (47)

which is the equation given in the text, for which a table is there given.

20. Direction Segments.—It will in the present case be found most convenient to determine the directions of the resultants by c_1 and c_2 equation (34).

Thus, $c_1 = -\dfrac{M_1}{H}$, $c_2 = -\dfrac{M_2}{H}$.

But $M_1 = M_0 - H h + (P - V) a - P z$, $M_2 = M_0 - H h + V a$.

We have, by series, then the approximate formulæ,

$$c_1 = -\frac{2 h}{15 (a+z)}\left[a - 5 z - \frac{45\, I\, a}{A h^2}\right] \quad c_2 = -\frac{2 h}{15 (a+z)}\left[a + 5 z - \frac{45\, I\, a}{A h^2}\right],$$

where positive values of c_1 and c_2 are laid off upward above, negative values downward below, the centres of gravity of the end cross-sections.

From the preceding tables we can calculate easily in any case **H** and **V** and M_0, and thus check the results obtained by the method of Chap. XIV. The formulæ above for c_1 and c_2 do not admit of tables, nor, in fact, are such needed. They are sufficiently simple for ready insertion.

Thus, by the aid of our tables, having computed **V** and **H**, and, if necessary, M_0 and e_0, we can by the method of moments, as explained in Chap. XIV., Art. 162, readily calculate the strains in the braced arch, whether continuous at crown and fixed or hinged at the ends, or hinged at both ends and crown.

20 (b). Transformation Series.—We have in the preceding repeatedly made use of series in the transformation of angular functions, such as sin, cos, etc., into functions of the arc itself. We group here, for convenience of reference, the series thus used:

$\sin x = x (1 - \tfrac{1}{6}x^2 + \tfrac{1}{120}x^4 - \tfrac{1}{5040}x^6 + \tfrac{1}{362880}x^8 - \tfrac{1}{39916800}x^{10} + \ldots - \ldots)$

$\cos x = 1 - \tfrac{1}{2}x^2 + \tfrac{1}{24}x^4 - \tfrac{1}{720}x^6 + \tfrac{1}{40320}x^8 - \tfrac{1}{3628800}x^{10} + \ldots - \ldots$

$\sin^2 x = x^2 (1 - \tfrac{1}{3}x^2 + \tfrac{2}{45}x^4 - \tfrac{1}{315}x^6 + \tfrac{2}{14175}x^8 - \tfrac{2}{155925}x^{10} + \ldots)$

$\cos^2 x = 1 - x^2 + \tfrac{1}{3}x^4 - \tfrac{2}{45}x^6 + \tfrac{1}{315}x^8 - \tfrac{2}{14175}x^{10} + \ldots$

$\sin x \cos x = x (1 - \tfrac{2}{3}x^2 + \tfrac{2}{15}x^4 - \tfrac{4}{315}x^6 + \tfrac{2}{2835}x^8 - \tfrac{4}{155925}x^{10} + \ldots)$

$\sin^3 x = x^3 (1 - \tfrac{1}{2}x^2 + \tfrac{13}{120}x^4 - \tfrac{41}{3024}x^6 + \tfrac{671}{604800}x^8 - \tfrac{73}{1995840}x^{10} + \ldots)$

$$\cos^2 x = 1 - \tfrac{1}{2}x^2 + \tfrac{1}{3}x^4 - \tfrac{17}{540}x^6 + \tfrac{31}{2241}x^8 - \tfrac{691}{17550}x^{10} + \ldots$$

$$\tan x = x\left(1 + \tfrac{1}{3}x^2 + \tfrac{2}{15}x^4 + \tfrac{17}{315}x^6 + \tfrac{62}{2835}x^8 + \tfrac{1382}{155925}x^{10} + \ldots\right)$$

$$\cot x = \frac{1}{x} - \frac{x}{3}\left(1 + \tfrac{1}{15}x^2 + \tfrac{1}{315}x^4 + \tfrac{1}{1575}x^6 + \tfrac{1}{31185}x^8 + \ldots\right)$$

$$\sin x \sin y = xy\left[1 - \tfrac{1}{6}(x^2+y^2) + \tfrac{1}{360}(3x^4+10x^2y^2+3y^4) - \tfrac{1}{5040}(x^6+7x^4y +7x^2y^4+y^6) + \tfrac{1}{1814400}(5x^8+60x^6y^2+126x^4y^4+60x^2y^6+5y^8)\ldots\right]$$

$$\cos x \cos y = 1 - \tfrac{1}{2}(x^2+y^2) + \tfrac{1}{24}(x^4+6x^2y^2+y^4) - \tfrac{1}{720}(x^6+15x^4y^2+15x^2y^4+y^6) + \tfrac{1}{40320}(x^8+28x^6y^2+70x^4y^4+28x^2y^6+y^8) - \ldots +\ldots$$

$$\sin x \cos y = x\left[1 - \tfrac{1}{6}(x^2+3y^2) + \tfrac{1}{120}(x^4+10x^2y^2+5y^4) - \tfrac{1}{5040}(x^6+21x^4y^2+35x^2y^4+7y^6) + \tfrac{1}{362880}(x^8+36x^6y^2+126x^4y^4+84x^2y^6+9y^8) - \ldots +\ldots\right]$$

CHAPTER V.

INFLUENCE OF TEMPERATURE.

21. General considerations.—When the temperature of a perfectly free body, which possesses in every direction the same coefficient of elasticity and expansion, changes equally at all points, there can be no strains in the body. For were there such strains, then, as there are no outer forces, there could be no equilibrium.

If, however, the change of temperature is not the same at all points; or if the body is not free, so that it is possible for outer forces to act, there are strains.

In the following we assume the change of temperature to be everywhere the same, but that the body is not free.

We assume that at a certain temperature t_0 no strain exists in the body, and call this the *mean* temperature. The deviation above or below the mean temperature we call $+ t$ or $- t$, and denote the coefficient of expansion for one degree by ϵ.

The determination of the strain in a straight beam held at both ends, is very simple. If the length is l, its relative change of length is ϵt. Since, however, it cannot expand, the strain S per unit of area is precisely as great as the force which would be required to produce this relative elongation, or from eq. (4)
$$s = + E \epsilon t \quad \ldots \ldots \ldots (48)$$
If the area of cross-section is A, then the strain at each end is
$$S = s A = E A \epsilon t.$$
In equation (48) it is assumed that a compressive strain, due to $+ t$, is positive, a tensile strain, due to $- t$, is then negative.

22. Influence of Temperature on the Arch.—Since by a change of temperature the length of the arch varies, while the span remains always the same, the shape or curvature must change, which naturally must give rise to strains and outer forces. In the following we have only to determine these outer forces, since, as shown in Chap. XIV., these are all we need to determine the strains themselves.

The relative change of length is from eq. (8), Art. I., $\dfrac{1}{EA}\left(G + \dfrac{M}{r}\right)$.

This change is caused by the outer forces. The relative change of length due to temperature alone is ϵt. Hence the total relative change of length is
$$\frac{\Delta ds}{ds} = \frac{Gr + M}{EAr} - \epsilon t \quad \ldots \ldots (49)$$

Hence the change of length of the axis is

$$\Delta s = \frac{1}{Er} \int_0^s \frac{Gr + M}{A} ds - \epsilon t s \quad \ldots \quad (50)$$

The change of the angle between two infinitely near cross-sections, and the actual turning of a cross-section is from (9) and (12), Art. 1, given by

$$\frac{\Delta d\phi}{d\phi} = \frac{M}{EI} + \frac{Gr + M}{EAr} \quad \ldots \ldots (51)$$

$$-\phi = \frac{1}{E} \int \frac{M}{I} ds + \frac{1}{Er} \int \frac{Gr + M}{A} ds \quad \ldots \quad (52)$$

Finally, from (14) we have

$$\Delta x = -\int \Delta \phi \, dy + \int \frac{\Delta ds}{ds} dx.$$

$$\Delta y = +\int \Delta \phi \, dx + \int \frac{\Delta ds}{ds} dy.$$

Substitute in these last two equations for $\Delta \phi$ and $\frac{\Delta ds}{ds}$ their values from (49) and (52). The double integral thus arising can be resolved by partial integration.

Thus
$$\int \int f(x) \, dx = x f(x) - \int x \, df(x).$$

Applying this, we obtain

$$\left. \begin{array}{l} \Delta x = -y \Delta \phi + \int y \, d\Delta + \int \frac{\Delta ds}{ds} dx \\[6pt] \Delta z = +x \Delta \phi - \int x \, d\Delta \phi + \int \frac{\Delta ds}{ds} dy \end{array} \right\} \quad \ldots \quad (53)$$

We shall assume in the following the axis always circular.

23. Fundamental Equations—General.

Upon this assumption of a circular axis we have generally

$$\left. \begin{array}{l} G = H \cos \phi, \quad N = H \sin \phi \\ M = M_0 + Hr(1 - \cos \phi) \\ Gr + M = M_0 + Hr \end{array} \right\} \quad \ldots \ldots (54)$$

Hence, from the preceding Art.,

$$\Delta s = \frac{Hr + M_0}{EA} \int_0^\phi d\phi - r\epsilon t\phi \quad \ldots \ldots (55)$$

$$\Delta \phi = \frac{M_0 r}{EI} \int d\phi + \frac{Hr^2}{EI} \int (1 - \cos \phi) d\phi + \frac{Hr + M_0}{AEr} \int d\phi + A' \ldots (56)$$

$$\Delta x = -r \Delta \phi (1 - \cos \phi) + \frac{r^2}{EI} \int M (1 - \cos \phi) d\phi$$

$$+ \frac{Hr + M_0}{EA} \int d\phi - r\epsilon t \sin \phi + B \ldots \ldots (57)$$

24. Arch with three Hinges.—If there are three hinges, the moment M_0 at the crown must be zero, and therefore $M = H r (1 - \cos \phi)$. But for $\phi = a$, M must also be zero, hence $H r (1 - \cos a) = 0$, and therefore H is zero. Then for any point $G = 0$, and $M = 0$, and $N = 0$. That is, *for the arch with three hinges there are for a change of temperature no outer forces, and hence no strains.*

25. Arch hinged at Ends.—Here, since for $\phi = a$, $M = 0$, we have from (54)

$$M_0 = - H r (1 - \cos a)$$
$$M = - H r (\cos \phi - \cos a) \quad \} \quad \ldots \ldots (58)$$
$$G r + M = + H r \cos a$$

From (56), since for $\phi = 0$, $\Delta \phi = 0$,

$$\Delta \phi = - \frac{H r^2}{E I} \int_0^\phi (\cos \phi - \cos a) \, d\phi + \frac{H \cos a}{E A} \int_0^\phi d\phi \ldots (59)$$

From (57), since for $\phi = 0$, $\Delta z = 0$,

$$\Delta z = \frac{H r^3}{E I} \left[(1 - \cos \phi) \int_0^\phi (\cos \phi - \cos a) \, d\phi - \int_0^\phi (1 - \cos \phi)(\cos \phi - \cos a) \, d\phi \right]$$

$$+ \frac{H r \cos a \cos \phi}{E A} \int_0^\phi d\phi - r \epsilon t \sin \phi.$$

For $\phi = a$, this becomes zero, and we have for the horizontal thrust

$$H = \frac{E \epsilon t \sin a}{\dfrac{r^2}{I} \int_0^a \cos \phi (\cos \phi - \cos a) \, d\phi - \dfrac{r^2 \cos a}{I} \int_0^a (\cos \phi - \cos a) d\phi + \dfrac{\cos^2 a}{A} \int_0^a d\phi}$$

$$\ldots \ldots (60)$$

Performing the integrations indicated (Art. 7), and putting, for brevity,

$$\kappa = \frac{I}{A r^2}; \text{ we have}$$

$$H = \frac{2 E I \epsilon t \sin a}{r^2 (a - 3 \sin a \cos a + 2 a \cos^2 a) + 2 \kappa r^2 a \cos^2 a} \ldots (61)$$

By series (Art. 20), we obtain the approximate formula

$$H = \frac{15 E I \epsilon t}{r^2 (2 a^4 + 15 \kappa)} = \frac{15 E I A \epsilon t}{8 A h^2 + 15 I} \quad \ldots \ldots (62)$$

The above are the expressions given in Art. 165 without proof. The less h, the greater for equal dimensions is H. For $h = 0$, we have $H = E A \epsilon t$, as we should have for a straight beam.

26. Arch without Hinges.—In this case we have the general equations (54) and (55), which apply directly without change.

From (56), since for $\phi = 0$, $\Delta \phi = 0$, we have

$$\Delta \phi = \frac{M_0 r}{E I} \int_0^\phi d\phi + \frac{H r^2}{E I} \int_0^\phi (1 - \cos \phi) \, d\phi + \frac{H r + M_0}{E A r} \int_0^\phi d\phi.$$

From (57), since for $\phi = 0$, $\Delta x = 0$, we have, inserting the value of $\Delta \phi$, above,

$$\Delta x = \frac{M_0 r^2}{EI}\left[\cos\phi \int_0^\phi d\phi - \int_0^\phi \cos\phi\, d\phi\right]$$

$$-\frac{Hr^3}{EI}\left[(1-\cos\phi)\int_0^\phi (1-\cos\phi)\, d\phi - \int_0^\phi (1-\cos\phi)^2\, d\phi\right]$$

$$+\frac{Hr + M_0}{EA}\cos\phi \int_0^\phi d\phi - r\epsilon t \sin\phi.$$

For $\phi = a$, $\Delta\phi = 0$, hence from the first of these expressions

$$M_0 = -Hr\, \frac{\dfrac{r^2}{I}\int_0^a (1-\cos\phi)\, d\phi + \dfrac{1}{A}\int_0^a d\phi}{\dfrac{r^2}{I}\int_0^a d\phi + \dfrac{1}{A}\int_0^a d\phi} \quad \ldots \quad (63)$$

If the distance at which the horizontal thrust H acts from the crown is e_0, we have $M_0 = -He_0$, whence we see at once that e_0 is the fraction in (63) multiplied by r. For $\phi = 0$, Δx must also be zero, and we thus obtain another relation between M_0 and H which does not contain A. If we multiply the expression thus obtained by $r \cos a$, and then subtract the result from that previously obtained for $\phi = 0$, $\Delta\phi = 0$, we have

$$\frac{M_0 r}{EI}\int_0^a (\cos\phi)\, d\phi + \frac{Hr^2}{EI}\left[\int_0^a (1-\cos\phi)\, d\phi - \int_0^a (1-\cos\phi)^2\, d\phi\right] = -\epsilon t \sin a$$

$$\quad \ldots \quad (64)$$

Performing the integrations indicated in (63), we have (Art. 7)

$$M_0 = -Hr\, \frac{(1+\kappa)a - \sin a}{(1+\kappa)a} \quad \ldots \quad (65)$$

where, as before, $\kappa = \dfrac{1}{Ar^2}$.

From (64) we obtain

$$M_0 r \sin a - \tfrac{1}{2} Hr^2 (a - 2\sin a + \sin a \cos a) = -EI \epsilon t \sin a.$$

Inserting the value of M_0 above, we have

$$H = \frac{2EI\epsilon t (1+\kappa) a \sin a}{r^2[(1+\kappa)(a^2 + a \sin a \cos a) - 2\sin^2 a]} \quad \ldots \quad (66)$$

and hence

$$M_0 = \frac{-2EI\epsilon t \sin a\,[(1+\kappa)a - \sin a]}{r[(1+\kappa)(a^2 + a \sin a \cos a) - 2\sin^2 a]} \quad \ldots \quad (67)$$

From these two we obtain for the point of application of H

$$e_0 = -\frac{M_0}{H} = +\frac{(1+\kappa)a - \sin a}{(1+\kappa)a}\, r \quad \ldots \quad (68)$$

By series, we have (Art. 20)

$$e_0 = \frac{a^2 + 6\kappa}{6} r = \frac{(Aa^2 + 6I)h}{3Aa^2} \quad \ldots \quad (69)$$

without reference to κ, $e_0 = \tfrac{1}{3} h$.

For small central angles, then, for which κ may be disregarded, the thrust given above by (66) acts at $\tfrac{1}{3} h$ *below* the crown for a *rise* of temperature of t degrees above the mean. For a decrease of temperature below the mean it acts above, M_0 is negative, and the strain in the lower flange tensile.

Further, we have, by series, the approximate formulæ

$$\left. \begin{aligned} H &= \frac{45\,EI\epsilon t}{r^2(a^4 + 45\kappa)} = \frac{45\,EIA\epsilon t}{4Ah^2 + 45I} \\ M_0 &= \frac{15\,EI\epsilon t(a^2 + 6\kappa)}{2r(a^4 + 45\kappa)} = \frac{15\,EI\epsilon t(Aa^2 + 6I)h}{(4Ah^2 + 45I)a^2} \end{aligned} \right\} \quad \ldots \quad (70)$$

These are the expressions given in Art. 165 without proof.

CHAPTER VI.

PARTIAL UNIFORM LOADING.

27. Notation.—In the preceding discussion of the arch we have considered the influence of a single concentrated load only, and this, as we have repeatedly seen in the case of the simple and continuous girder, etc., is sufficient for full and accurate solution. When once we are able to find and tabulate the strains in every piece due to a single load in *any* position, the thorough solution becomes simply a question of time.

It may often happen, however, that we may wish to determine the strains for a full load only, or for a uniformly distributed load extending from one end to some given point. In such case it would be unnecessarily tedious to obtain our result by the successive determination and addition of all the intermediate apex loads. We may easily deduce from the preceding the general formulæ for partial loading also.

As before, we shall let $a=$ the half span, $h=$ the rise, I the moment of inertia, and A the area of the cross-section. But we shall represent by p the load per unit of length of horizontal projection, and by z the distance of the end of the load extending from the *left*, from the crown. This distance z, from the crown to the end of load, is then positive towards the *left*. In the circular arch the angle subtended by this distance z we call β. The angle β is then positive to the *left* of the vertical. The angle subtended by the half span is, as before, a. For $\beta = a$, then, or for $z = a$, there is no load upon the span. For $\beta = 0$, or $z = 0$, the load extends from the left to the centre. For $\beta = -a$, or $z = -a$, the load covers the whole span. Pls. 23 and 24, Figs. 91 and 92, still hold good, therefore, for our notation. We have only to conceive, instead of the concentrated load **P**, a uniformly distributed load, per *horizontal* unit, extending from left end as far as the position of **P**. This much being premised as to notation, we shall treat, as before, the three cases of arch hinged at crown and ends, hinged at ends only, and without hinges.

A. ARCH HINGED AT CROWN AND ENDS.

28. Reaction.—This case is too simple to demand any extended notice, in view of what has already been said. We have from eq. (16), Art. 3, for the reaction at the left or loaded end, for concentrated load,

$$V = P \frac{a+z}{2a}.$$

If now we put $P = p\,dz$, and integrate, we have

$$V = \int p\,dz\, \frac{a+z}{2a} = p\,\frac{2az+z^2}{4a} + C,$$

where C is the constant of integration. By taking the proper limits, we can eliminate this constant, and thus obtain the reaction for load covering any desired portion of the span. As we shall in every case suppose the load to extend from the left end up to any point, we shall take the limits of $z = a$ and z, and therefore obtain

$$V = \frac{3a^2 - 2az - z^2}{4a} \quad \ldots \ldots \quad (71)$$

For $z = a$, this is zero, as it should be, since then the load has not come on. For $z = -a$, the load extends over the whole span, and $V = pa$, or half the whole load, as it should. We might have obtained this result at once by moments. Thus,

$$V \times 2a = p(a-z)(a+z+\frac{a-z}{2}) = \frac{3a^2 - 2az - z^2}{2},$$

but have preferred the above method as showing how uniform loading is deduced directly from concentrated by inserting $p\,dz$ for P and integrating.

29. Horizontal Thrust.—In precisely similar manner we have from (21), Art. 4, for the thrust due to concentrated load P, $H = P\dfrac{(a-z)}{2h}$.

Put $P = p\,dz$ and integrate between the limits $z = a$ and z, and we have

$$H = P\frac{a^2 - 2az - z^2}{4h} \quad \ldots \ldots \quad (72)$$

For $z = a$, this is zero, as should be. For $z = -a$, or for full load over whole span, $H = \dfrac{pa^2}{2h}$. We may also deduce the above equation directly by moments.

The above formulæ (71) and (72) are all that we need either for calculation or diagram. They apply evidently equally well, whether the arch be circular or parabolic, or, in general, whatever its shape may be. The form has no influence upon either the thrust or the reaction.

For the moment at any point whatever, whose distance horizontally from crown is x and vertically below crown y, we have at once

$$M = H(h-y) - V(a-x) + \frac{p}{2}(a-x)^2.$$

If this point is an apex, then the moment divided by depth of arch at this point is the strain in flange opposite that apex. A positive moment throughout this work *always* indicates compression in the inner or lower flange.

B. ARCH HINGED AT ENDS ONLY.

30. Reaction.—The vertical reaction at the end is precisely the same as before for three hinges, and is given by equation (71). This reaction is evidently independent of the shape of the arch, and the above formulæ holds good generally.

31. Horizontal Thrust—Parabolic Arch.—We must here distinguish the shape of the arch, and treat first the parabola. We have already from eq. (27), Chapter III., Art. 8, for a single load,

$$H = \frac{5}{64} P \frac{5a^4 - 6a^2 z^2 + z^4}{a^3 h}.$$

We have, as before, simply to make $P = p \, ds$, and then integrate between the limits $s = a$ and s indeterminate.

We thus find at once

$$H = \frac{p}{64 \, a^3 \, h} \left[16 \, a^5 - 25 \, a^4 \, s + 10 \, a^2 \, s^3 - s^5 \right] \quad . \quad . \quad (73)$$

For $s = a$, this reduces to zero, as it should. For $s = -a$, the load covers the whole span, and we have $H = \frac{p \, a^2}{2 \, h}$. For $s = o$, the load reaches from the left as far as the crown, and $H = \frac{p \, a^2}{4 \, h}$. The formulæ is simple, and requires no table. Numerical values may be easily inserted.

32. Horizontal Thrust—Circular Arch.—
As already noticed, the vertical end reaction for this case has been given in eq. (71). It remains to determine the thrust. We have, as before, simply to insert $p \, dx = p \, r \cos \beta \, d\beta$ in place of P in the expression for the thrust for concentrated load of Art. 10, and then integrate between the limits $\beta = a$ and β indeterminate.

We have thus similarly to that Art.

$$H = H_0 \frac{1 - A \kappa}{1 + B \kappa} \quad . \quad . \quad . \quad . \quad . \quad (74)$$

where H_0 is the value of H when terms containing κ are neglected, or

$$H_0 = \frac{p \, r}{12} \frac{A_1}{B_1}; \text{ and } A = \frac{A_2}{A_1}, \; B = \frac{B_2}{B_1}$$

The quantities A_1, B_1, A_2 and B_2 are as follows:

$A_1 = 7 \sin^3 a + 3 \, a \cos a - 3 \sin a - 6 \, a \cos a \sin^2 a - 6 \sin^2 a \sin \beta$
$\quad + 2 \sin^3 \beta - 3 \beta \cos a - 9 \cos a \sin \beta \cos \beta + 12 \cos^2 a \sin \beta$
$\quad + 12 \, a \cos a \sin a \sin \beta - 6 \beta \cos a \sin^2 \beta.$

$A_2 = 3 \, [2 \, a \cos a \sin^2 a + a \cos a - \sin a \cos^2 a - 4 \, a \cos a \sin a \sin \beta$
$\quad + 2 \beta \cos a \sin^2 \beta - \beta \cos a + \cos a \sin \beta \cos \beta].$

$B_1 = a - 3 \sin a \cos a + 2 \, a \cos^2 a. \quad B_2 = 2 \, a \cos^2 a.$

These expressions can be tabulated as in Art. 10, or developed into series as in that Art., and the formula thus made practically available.

For $\beta = a$, we have H·zero, as should be. For $\beta = -a$, we have the load *covering the entire span*.

For this case we have

$$H = \frac{1}{6} p \, r \, \frac{\sin^3 a - 3 \, (1 - 2 \sin^2 a) \, (\sin a - a \cos a) - 3 \kappa \cos a \, (a + 2 \, a \sin^2 a - \sin a \cos a)}{a + 2 \, a \cos^2 a - 3 \sin a \cos a + 2 \kappa \, a \cos^2 a}$$

For the SEMI-CIRCLE, this reduces simply to

$$H = \frac{4}{3 \, \pi} p \, r = 0.424 \, p \, r.$$

In any case where exact results are desired, eq. (74) must be used, and a table calculated for the central angle a. We have approximately by series also, more especially for small central angles, or for a large in respect to h, *for total load over whole span*:

$$H = \frac{p \, a^2}{2 \, h} \frac{8 \, A \, h^2}{15 \, I + 8 \, A \, h^2} = \frac{p \, a^2}{2 \, h} \frac{8 \, h^2}{15 \, g^2 + 8 \, h^2} \quad . \quad . \quad (75)$$

where **A** is the area and **I** the moment of inertia of cross-section, and y the *radius of gyration*. In framed arches this may be taken as approximately equal to the half depth from centre to centre of flanges.

C. ARCH WITHOUT HINGES—FIXED AT ENDS, CONTINUOUS AT CROWN.

33. Parabolic Arch—Formulæ for V, H and M.—In this case the reactions no longer follow the law of the lever, and eq. (71), therefore, no longer holds good.

(a) *Vertical reaction at unloaded end.*

We have from eq. (28), Art. 14, for the reaction at the *right end* for a single load,

$$V = \frac{1}{4} P \frac{(a-z)^2 (2a+z)}{a^3}.$$

Making $P = p\,dz$, and integrating between the limits $z = a$ and z, we find the reaction for a load coming on from *left*,

$$V = \frac{p}{16\,a^3} \left[3\,a^4 - 8\,a^3 z + 6\,a^2 z^2 - z^4 \right] \quad \ldots \quad (76)$$

for a full load $z = -a$ and $V = pa$, as should be.

(b) *Horizontal thrust.*

In like manner we have for the horizontal thrust at end from (36), Art. 14,

$$H = \frac{15}{32} P \frac{a^4 - 2\,a^2 z^2 + z^4}{a^3 h}.$$

Replacing **P** by $p\,dz$, and integrating as before, we obtain directly

$$H = \frac{p}{32\,a^3 h} \left[8\,a^4 - 15\,a^4 z + 10\,a^3 z^3 - 3\,z^5 \right] \quad \ldots \quad (77)$$

for a full load $z = -a$, and $H = \frac{p\,a^2}{2\,h}$.

(c) *Moment at unloaded end.*

In precisely similar manner we have from (37), Art. 14,

$$M_s = -\frac{1}{32} P \frac{(a-z)^2 (3\,a^2 - 10\,a z - 5\,z^2)}{a^3}.$$

Putting $P = p\,dz$, and integrating, we have for the moment, always at the right end, or for load not extending past the centre, at crown

$$M = \frac{p}{32\,a^3} \left[3\,a^4 z - 8\,a^3 z^2 + 6\,a^2 z^3 - z^5 \right] \quad \ldots \quad (78)$$

For $z = a$, this is zero, as should be. For $z = 0$, or for load extending as far as crown, it is also zero. For $z = -a$, the moment *at the end* is $-\frac{p\,a^2}{2}$.

A negative moment, as always, denotes tension in lower flange.

Just as for concentrated load, as shown in Art. 14, as the load comes on, the moment at crown is positive, and increases with increasing load up to a certain point, beyond which any load causes a negative moment, and beyond which the moment at crown, therefore, decreases, until, when the load reaches the crown, it becomes zero. This point, which gives M_s, the moment at crown, a positive maximum, is at a distance $z = -a + a\sqrt{\frac{5}{3}} =$ 0.264911 a, or nearly $\frac{1}{4} a$ from the crown.

The values of V_1, H and M_1 (M_2 and V_2 being always the moment and reaction at *unloaded end*), for various values of z, are given in the following table:

z	V_1	H	M_1	z	V_2	H	M_2
1	0	0	+	−0.1	0.2349	0.29656	−0.01206
0.9	0.0002437	0.000579	0.0001096	−0.2	0.3024	0.34128	−0.03024
0.8	0.0014	0.00421	0.00076	−0.3	0.3707	0.38242	−0.055611
0.7	0.00624	0.01643	0.002185	−0.4	0.4459	0.41846	−0.08918
0.6	0.0144	0.02889	0.00432	−0.5	0.5273	0.44824	−0.131836
0.5	0.0273	0.051757	0.006836	−0.6	0.6144	0.47104	−0.18482
0.4	0.0459	0.08346	0.00918	−0.7	0.7062	0.48667	−0.24718
0.3	0.07074	0.11777	0.010611	−0.8	0.8014	0.49884	−0.32076
0.2	0.1024	0.15872	0.01024	−0.9	0.90024	0.49942	−0.405109
0.1	0.1412	0.20315	0.007062	−1	1.0000	0.5000	−0.5000
0	0.1875	0.25	0				
a	pa	$\frac{pa^2}{h}$	pa^2	a	pa	$\frac{pa^2}{h}$	pa^2

It will be seen that the moment at the unloaded end, which, as long as the load is left of crown, is the moment at crown also; increases as the load passes on, is positive and increases up to about $z = .25 a$. Then it diminishes, becomes zero when the load reaches the crown, changes to negative as the load passes the crown, and this negative value increases up to full load when it is $-\frac{1}{2} p a^2$. For full load, then, the lower end flanges are in tension. At the crown the moment is zero, and the compression there in both flanges is due to H only.

34. Circular Arch—Formulæ for V, H and M.

(a) Vertical Reaction.

Here we have $r \sin \beta = z$, $r \cos \beta\, d\beta = dz$, and $P = p\, dz = p\, r \cos \beta\, d\beta$. Inserting this in place of P in eq. (43), Art. 18, and integrating between the limits $\beta = a$ and β indeterminate, we have for the reaction at *unloaded* end, or for reaction at crown when load does not extend past the crown,

$$V = \frac{pr}{2(a - \sin a \cos a)} \left[\frac{\cos^3 a}{3} - \cos a - a \sin \beta + \cos \beta + \beta \sin \beta \right.$$
$$\left. + \sin a \cos a \sin \beta - \frac{\cos^3 \beta}{3} - \cos a \sin^2 \beta \right] \ldots (79)$$

For $\beta = a$, this is zero, as should be, since then the load is not upon the span. For $\beta = -a$, $V = pr \sin a$, as should be, for full load over whole span. For the *semi-circle*, $a = 90° = \frac{\pi}{2}$, $\sin a = 1$, $\cos a = 0$, and

$$V = pr \frac{-\frac{\pi}{2} \sin \beta + \cos \beta + \beta \sin \beta - \frac{\cos^2 \beta}{3}}{\pi}.$$

If the semi-circle is uniformly loaded over whole span, $\beta = -a = -90° = -\frac{\pi}{2}$, $\sin \beta = -1$, $\cos \beta = 0$, and $V = pr$, as should be. The formula (79) above is precisely the same as that given by Capt. Eads in his *Report to the Illinois and St. Louis Bridge Co.*, May, 1868.

(*b*) *Horizontal Thrust.*

In similar manner, from eq. (46), Art. 18, by inserting $p\,dx = pr \cos \beta\, d\beta$ in place of **P**, and integrating between $\beta = a$ and β. we have similarly to Art. 32

$$H = H_0 \frac{1 - A\kappa}{1 + B\kappa} \quad \ldots \ldots \quad (80)$$

where $\quad H_0 = \frac{pr \sin a}{12} \frac{A_1}{B_1}, \quad A = \frac{A_2}{A_1}, \quad B = \frac{B_2}{B_1}, \quad$ and

$A_1 = 3a - 3 \sin a \cos a - 2a \sin^2 a - 3\beta - 9 \sin \beta \cos \beta + 12 \cos a \sin \beta$

$\quad - 6 \beta \sin^2 \beta + 6 a \sin a \sin \beta + 2a \frac{\sin^3 \beta}{\sin a},$

$A_2 = 3a - 3 \sin a \cos a + 2a \sin^2 a + 6\beta \sin^2 \beta - 3\beta + 3 \sin \beta \cos \beta$

$\quad - 6a \sin a \sin \beta - 2a \frac{\sin^3 \beta}{\sin a},$

$B_1 = a(a + \sin a \cos a) - 2 \sin^2 a, \quad B_2 = a(a + \sin a \cos a).$

Formula (80) agrees exactly with that given by Capt. Eads in the Report above quoted, if terms containing κ are neglected. Since $\kappa = \frac{I}{Ar^2}$, where I is the moment of inertia and **A** is the area of cross-section; r being the radius; for small central angle r is very large in proportion to $\frac{I}{A}$, or the square of the half depth. In such case, then, κ may be neglected. For $\beta = a$, we have $H = 0$, as should be. For $\beta = -a$, we have load over entire span, and

$$H = \tfrac{1}{6} pr \sin a \frac{3a - 2a \sin^2 a - 3 \sin a \cos a - \kappa (3a + 2a \sin^2 a - 3 \sin a \cos a)}{(1 + \kappa) a (a + \sin a \cos a) - 2 \sin^2 a}$$

Approximately we have, by series, for *full load*, from (80):

$$H = \frac{pa^2}{2h} \frac{4Ah^2}{45 I + 4Ah^2} = \frac{pa^2}{2h} \frac{4h^2}{45 g^2 + 4h^2} \quad \ldots \quad (81)$$

For $a = 90°$, or for *semi-circle*, we have from (80)

$$H = \frac{pr}{24\left[(1+\kappa)\frac{\pi^2}{4} - 2\right]}\left[3\pi - 6\beta + 18\sin\beta\cos\beta - 12\beta\sin^2\beta + 6\pi\sin\beta\right.$$
$$\left. + \kappa(-3\pi - 12\beta\sin^2\beta + 6\beta - 6\sin\beta\cos\beta + 6\pi\sin\beta + 2\pi\sin^2\beta)\right].$$

For $\beta = -90°$, or for *full load upon semi-circle*,

$$H = \frac{pr}{24\left[(1+\kappa)\frac{\pi^2}{4} - 2\right]}\left[6\pi - 8\pi\kappa\right].$$

(c) *Moment at unloaded end.*

From Art. 18 (c) we have, for concentrated load,

$$M = \tfrac{1}{2}Hr\left(2 - \cos a - \frac{a}{\sin a}\right) - \tfrac{1}{4}Pr\frac{(\sin a - \sin\beta)^2}{\sin a}.$$

The value of H we have already given in (80).

Inserting in the second term $p\,dx = pr\cos\beta\,d\beta$ for P, and then integrating, we have

$$M = AH - B \quad \ldots \ldots \quad (82)$$

where $\quad A = \tfrac{1}{2}r\left(2 - \cos a - \dfrac{a}{\sin a}\right) = \dfrac{a}{2\sin a}\left(2 - \cos a - \dfrac{a}{\sin a}\right),$

and $\quad B = \dfrac{pa^2}{12\sin^3 a}\left(\sin a - \sin\beta\right)^3.$

For a uniform load over whole span, $\beta = -a$, and

$$M = \tfrac{1}{2}Hr\left(2 - \cos a - \frac{a}{\sin a}\right) - \frac{2pr^2\sin^2 a}{3} \quad \ldots \quad (83)$$

We have from (83), by series, the approximate formula for moment at *crown*

$$M_o = -\tfrac{1}{75}ph^2\frac{4Ah^4 + 175Ia^2}{4Ah^4 + 45Ih^2} = -\tfrac{1}{75}ph^2\frac{4h^4 + 175g^2a^2}{4h^4 + 45g^2h^2}\ldots\,(84)$$

where A is the area and I the moment of inertia of cross-section, g the radius of gyration, or, approximately, the half depth for framed arch—as always a negative moment indicates tension in lower or inner flange.

Equations (79), (80) and (82) may, if desired, be tabulated as in Art. 18. For *small* central angles, or for h small with respect to a, κ may be disregarded, and the results already given for parabola (Art. 33) may be taken as sufficiently exact.

CHAPTER XV.

THE STONE ARCH.

167. Definitions, etc.—In the stone arch we have a system of bodies in contact with each other, and so supported between certain fixed points, that they are not only in equilibrium among themselves, but also with the exterior forces. The surfaces of contact we call the *bed-joints*; the fixed points are the *abutments*; the central or highest arch stone is called the *key-stone*, and those resting upon the abutments, the *imposts*. The inner and outer limiting surfaces of the arch, generally curved, are designated as the *intrados* and *extrados*, and the arch stones generally are called *voussoirs*.

168. Line of Pressures in Arch.—We have already indicated (Art. 28, Fig. 16) the manner in which a number of successive forces are resisted by an arch. We see from the force polygon in that Fig. that the horizontal pressure is the same at every point, and that the vertical pressure is equal to the sum of the weights between the crown and any point. The pressure line is then an equilibrium polygon formed by laying off the weights of the arch stones, choosing a pole, and drawing lines from this pole, etc., as described in our second chapter.

If the weights are very small, and their number very great, the equilibrium polygon becomes a curve. This curve for equilibrium should never pass outside the limits of the arch.

169. Sliding of the Arch Joints.—The arch is properly, then, nothing but a curved wall. Upon a vertical wall, which may also support loads, but which has no horizontal thrust, only vertical forces act, and the resultant is known in position and direction. We may, then, investigate the stability of an ordinary wall, and apply the results directly to the arch.

We assume the wall divided by plane bed-joints extending through its entire breadth, whose distances apart depend upon the dimensions of the stones. These joints are the weak places of the wall, since separation here is not resisted by the greatest strength of the stone. Neglecting the influence of the mortar, we assume that any section along a bed-joint resists only a perpendicular pressure due to the parts above, and a force parallel to the joint which must not exceed the resistance to sliding due to friction. If this parallel force is greater than the resistance of friction, the upper part will slide upon the joint.

If we represent the greatest angle of repose by ϕ, then the resultant of the vertical forces, acting upon the joint in question, must make an angle with the normal to the joint *less* than the angle ϕ. Thus at the joint **A** (Pl. 24, Fig. 95), this angle is greater than ϕ, and the upper part will slide along this joint. At **B** this angle is less than ϕ, and no sliding can take place.

The *ratio* of the force of friction due to the component of **P** normal to the joint, to the component of **P** parallel to the joint, we call the *coefficient of safety* against sliding. It is evidently equal to $\dfrac{\tan \phi}{\tan \mathbf{PN}}$, or to the distance **GN** divided by **PN**.

Since we can dispose the bed-joints at pleasure, we may always make them perpendicular to the direction of the pressure, for instance in Fig. 95 horizontal; or at least so place them that their normals vary from the direction of the resultant of the outer forces, at most by an allowable angle **PN**.

The sliding of the joints can then always be prevented by the position of the bed-joints.

170. Forces acting upon a Cross-section—Neutral Axis. —Let us consider what happens when the resultant of the outer forces acting upon a joint, instead of acting at the centre of gravity, approaches the edge of a joint, under the assumption that sliding cannot take place, or that the direction of this resultant is perpendicular to the joint. There is no reason for assuming the distribution of pressure upon the joint surface any different from the case of a beam. The stone, as well as the mortar, is elastic, though in a less degree than wood or iron, and accordingly the pressure at any portion of the joint is proportional to the approach of the limiting surfaces of the upper and lower portions of the wall. If, then, we assume that these surfaces are plane before and after loading, if the resultant pressure does not act at the centre of gravity, but near to one edge, the pressure at different points will vary, and there will be a neutral axis, or line of no pressure, either within or wholly without the joint surface.

Every cross-section is therefore acted upon by a system of parallel forces whose intensities are directly as their distances from a certain axis.

Now, neglecting the influence of the mortar, the wall can resist compression only. No tension can exist at any point of the joint surface.

Clearly, then, the neutral axis should lie wholly without the cross-section, or at most only touch it. It should never be found *within* the cross-section, as in that case all the material on the other side is useless, and might be removed entirely without affecting the pressure upon the actual bearing surface.

The neutral axis, then, should always lie without the cross-section of the joint.

171. System of Parallel Forces whose Intensities are proportional to their Distances from a certain Axis—The Kernel of a Cross-section.—If in a system of equal and parallel forces we find the *moment* of each of these forces with reference to a certain axis, and then *consider these moments as themselves forces*, we shall have a system of the kind referred to, since each moment force will be directly proportional to its distance from a given axis.

Now, as we have seen in Art. 60, Chapter VI., the centre of action of such a system of moment forces does not coincide with the centre of gravity of the original simple forces, but for any given axis is found from the *central curve* of the cross-section. In Pl. 11, Fig. 35, we have already given the construction for finding this centre of action, the semi-diameter of the central curve being known, for any given axis.

Suppose now this axis to envelop in all its different positions the outline of the given cross-section, and find the corresponding positions of the centre of action of the moment forces. These different points lie in a closed figure which we may call the *kernel* of the cross-section. Then, in order that we may always have compression in every part of the joint surface of our wall, the resultant of the forces acting upon it should always act *within the kernel.*

In Plates 11 and 12, Figs. 36, 37, 38 and 40, we have constructed the *kernels* of the various cross-sections represented.

Thus in Fig. 36, according to the construction of Art. 62, for an axis at **A**, we describe upon **O C** a semi-circle. Then with **O** as a centre and radius equal to semi-diameter of the central ellipse on **A C**, describe an arc intersecting the semi-circle in *a*.

From a drop a perpendicular upon **A C**, and we obtain the centre of action for axis at **A**. A similar construction for other axes, as **A B**, **B C**, etc., give us other points, and we thus find the small central parallelogram, which is the *kernel* or locus of the centres of action of the moment forces for all positions of the axis enveloping the parallelogram **A B**, **C D**. A similar construction gives us the kernel for the other figures.

We have from Art. 60

$$m = \frac{a^2}{i},$$

where m = the distance of the resultant **P** of the forces acting upon the cross-section from the centre of gravity, and a = the semi-diameter of the central curve, and i = the distance of the neutral axis from the parallel diameter of the central curve.

If we call c the distance of an outer fibre from this diameter measured on the side of **P**, its distance from the neutral axis is $i + c$. If the strain in this fibre is **S**, we have

$$S : i + c :: \frac{P}{A} : i,$$

where **A** is the area of the cross-section. Hence

$$S = \frac{P}{A}\left(1 + \frac{c}{i}\right).$$

If **P** acts at the centre of gravity of the cross-section, $i = \infty$ (Art. 60), the neutral axis is infinitely distant, and $S = \frac{P}{A}$. If **P** moves away from the centre of gravity, the neutral axis approaches, and is always parallel to the conjugate diameter in the central ellipse. When **P** reaches the perimeter of the kernel, the neutral axis touches the perimeter of the cross-section, and at least, then, in one point of this perimeter, the pressure is zero. If **P** passes beyond the kernel, the neutral axis enters the cross-section, and tensile strains enter on one side to balance the compressive strains on the other. The *kernel then forms a limit beyond which the resultant* **P** *must not act.*

172. Position of Kernel for different Cross-sections.—If the cross-section is symmetrical with reference to the centre of gravity, we have $\frac{c}{i} = 1$, and therefore $S = 2\frac{P}{A}$; that is, when the neutral axis touches the cross-section, or **P** acts in the kernel, the strain **S** is *twice* as great as when **P** passes through the centre of gravity of the joint surface and is uniformly distributed.

As **P** passes beyond the kernel, the neutral axis, as we have seen, enters the joint area, and on the side away from **P** occasions, or would occasion in a beam, tensile strains. But as the assumption is that the joint (neglecting mortar) cannot resist tensile strains, we may remove all that portion on the opposite side of the neutral axis without increasing the pressure on the other side.

In this case, then, the central ellipse is not that for the whole joint area, but only for that portion on the same side as **P**, and **P** is upon the *kernel for that portion*.

This portion can be determined directly for a certain position of **P** only in a few individual cases; generally, it must be found by trial. We must first find for the central ellipse of the entire joint area the neutral axis corresponding to given position of **P**, and then draw a parallel cutting off somewhat more of the area. Then determine the central ellipse of the cut-off portion, and see if the pole lies symmetrically to the pole of the cutting line.

The *parallelogram* is one of the areas in which we can determine directly the amount cut off when **P** acts at a point upon the line joining the centres of two opposite sides. For if we cut off by a parallel to these sides a portion so that **P** is at $\frac{1}{3}$d of the *line* joining the centres of the opposite sides of the new parallelogram, then **P** *lies upon the kernel for this new area*. The proof is easy. The moment of inertia of the parallelogram is $\frac{1}{12} b h^2$, with reference to the diameter b. The square of the radius of gyration a^2 is then $\frac{1}{12} h^2$. The distance of the point of application of **P** from one of the sides is

$$i + m = i + \frac{a^2}{i}.$$ Hence

$$i + \frac{a^2}{i} = \frac{1}{2}h + \frac{\frac{1}{12}h^2}{\frac{1}{2}h} = \frac{1}{2}h + \frac{1}{6}h.$$

The half height of the kernel is, then, $\frac{1}{6}$th the height of the parallelogram, or the kernel occupies the inner third. (See Fig. 36; also *Woodbury: Theory of the Arch*, p. 328, Art. 3.)

For any given position of **P**, then, three times its distance from the nearest side on a line parallel to the other two, gives the position of the fourth side of the parallelogram for which **P** is upon the kernel.

173. The resultant pressure should therefore act within the middle third of the joint area.—As this principle is most important, and the demonstrations of Chapter VI., upon which the above result is based, may appear to some too purely mathematical, we give here the demonstration of the same principle as given by *Woodbury*, in the work above cited.

"Suppose the pressure to be nothing at the intrados a, and to increase uniformly from that point to the extrados b (Pl. 24, Fig. 96). It is plain that the pressure at any point along ab will be represented by the ordinate of a certain triangle. The whole pressure will be represented by the surface of that triangle; and the point of application of the resultant of all the pressures will be at c opposite the centre of gravity of that triangle. We have then $cb = \frac{1}{3}ab$. *Vice versa*, if the point of application be at c, $cb = \frac{1}{3}ab$, we know that the pressure is nothing at a.

"If the point of application be at c, cb being *less* than $\frac{1}{3}ab$, c being still opposite the centre of gravity of the triangle whose ordinates represent the pressure, we know that the vertex of that triangle and point of no pressure are at e, $be = 3 \times bc$.

"In this case, the joint ab will open at a as far as e; the adjacent joints will also open until we come to one where the curve of pressure passes within the prescribed limit.

"This reasoning is, of course, applicable to all the joints; and we readily conclude that the curves of pressure should lie entirely between two other curves which divide the joint into three equal parts."

Thus, in Pl. 24, Fig. 97, suppose the resultant **P** of the upper part of the wall to have the position as represented, so that it intersects the joint **B D** in **C** outside of the middle third of the cross-section. The entire pressure is distributed over $3\, \mathbf{CB} = \mathbf{AB}$, and the area **D A** does not act at all. Moreover, the pressure at **B** is *twice* as great as when **P** passes through the centre of gravity and is uniformly distributed over **AB**, or is $\frac{4}{3}$ds of the uniformly distributed pressure of **P** upon **CB**.

Beyond **A** the pressure is zero, and the conditions of load and equilibrium would not be changed if the stone beyond **A** were removed.

If **C** approaches still nearer **B**, so that the pressure is distributed upon an ever-decreasing area, the resistance of the mortar will be finally overcome; it will be forced out, and stone will come in contact with stone, and there will be rotation about the edge at **B**. This rotation can never occur if the pressure **P** is distributed over the whole joint area. If, then, we consider rotation to commence at the moment when **P** is no longer distributed over the entire area—when, therefore, the neutral axis just enters the joint—then, *in order that no rotation may occur,* **P** *must pierce the joint area inside the kernel.*

174. Line of Pressures in the Arch.—When the dimensions and form of a wall are given, we can determine directly the resultant **P** of the outer forces acting upon a joint, and then by the two preceding Arts. can determine the condition of stability of the wall. In the arch, however, we cannot determine **P** directly for a given cross-section, but must first make certain assumptions.

In the first place, it is clear that an arch is stable when it is possible in two joints to take two reactions P_1 and P_2, (Pl. 24, Fig. 98) such that, with the weight of the intervening portion of the arch and its load, the resulting line of pressure shall lie so far in the interior of the arch that rotation about a joint edge cannot take place. If the arch is so feeble and the resistance of the material so slight that only *one* such assumption of P_1 and P_2 can be made, and *only one* such pressure line drawn, this is plainly the true pressure line for stability, and by it P_1 and P_2, as also the pressure at every joint, are determined.

If, however, the arch is so deep and the resistance of the material so great that by variation of P_1 and P_2 several such pressure lines may be drawn, none of which causes rotation about a joint edge, which of all these possible pressure lines is the true pressure line of the arch?

We assert: *That is the true pressure line which gives the least thrust consistent with stability, or which causes the pressure in the most compressed joint to be a minimum.*

If we assume the material so soft that the pressure line approaches the axis so near that only one assumption of P_1 and P_2 is possible, then this would evidently be the true pressure line. If now the material hardens without altering any of its other properties, such as its specific weight or modulus of elasticity, then the position of the pressure line is not altered. As there is no reason for supposing the pressure line different in an arch built of hard material from that in one originally soft which has afterwards gradually hardened, it follows that the pressure line in all arches of same form and loading has the same position which it would have had if the arch had been originally of the softest material; that is, that position which gives the least thrust consistent with stability, or makes the pressure in the most compressed joint edge a minimum.

In order to draw the pressure line in an arch, we may then seek by means of the formula

$$S = \frac{P}{A}\left(1 + \frac{c}{i}\right)$$

this pressure in the joint, where the pressure line approaches nearest the edge, and ascertain whether it can be still further diminished by change of position of the pressure line. This is, however, not necessary. We have only to ascertain whether it is possible to draw a pressure line whose sides cut the corresponding joint area, *within the kernel*, for then, since we know that there can be a still more favorable position, there is no danger of rotation.

175. The Line of Support.—The curve formed by joining the intersections of the sides of the pressure line with the joint areas we call the *support line*, or line of support.

If the joints of an arch answer to the condition of Art. 169, so that sliding of the joints cannot occur, we see at once from the position of the support line on what side and where rotation will take place. If at any point this line passes beyond the *kernel*, we have theoretical beginning of rotation; if it passes outside of the arch, there is actual rotation, and if it lies within the *kernel*, there is no rotation.

The manner of determining from the position of the support line all the possible motions of an arch is illustrated in the following Figs.

In Pl. 24, Fig. 99, we have a possible support line touching the extrados at crown and springing, and the intrados between these points. We have accordingly rotation at crown, and at the points between crown and springing, so that the joints at these points open on the sides of the arch opposite the support line. The crown will sink, and as at the crown and flanks the support line is approximately parallel to the extrados and intrados, there will be several joints in the same condition, and several will open, as indicated in the Fig.

In Pl. 25, Fig. 100, we have the condition of stability of a pointed arch, not loaded at the crown. The support line is horizontal at crown, and there is no angle there, as in the arch itself. The rotation at various points is indicated in the Fig. We shall soon see that the support line deviates but very little from the pressure line. From the direction of the tangent to the support line at any point, therefore, we may conclude as to the conditions of sliding.

From Fig. 101 we may conclude that the arch will slide outwards upon the right abutment. The rotation at various points is given by the Fig. It is sufficient, as we see, to make the abutment surface more nearly perpendicular to the support line, as shown in the left abutment, to prevent this sliding, and at the same time a more favorable support line can be drawn. Since, as we have seen in Art. 100, sliding can and must in this manner be always prevented, we shall give no more examples of arches unstable in this particular.

The arches of Figs. 99 and 100 can be made stable by sufficiently increasing their thickness, or conforming their shape more nearly to that of the support line.

176. Deviation of the Support from the Pressure Line.
This deviation is not great. In order to make it apparent, we must draw a pressure line for slight pressure in the lower part of an arch with very long and inclined voussoirs [Pl. 25, Fig. 102]. Thus, if we combine the weights of the voussoirs 1, 2, 3, 4, etc., acting at their centres of gravity, with the pressure **Q** in the first joint, we have the pressure line shown by the broken line 1, 2, 3, 4, 5, 6, 7, 8, whose sides 1 2, 2 3, 3 4, etc., give the direction of the pressure in the corresponding joints between the voussoirs 1 and 2, 2 and 3, etc. Thus 5 6 is the *direction* of the pressure upon the joint between voussoirs 5 and 6. This direction cuts the joint at 5', which is therefore the point of application of the pressure, or a point upon the line of support. Thus we find 3', 4', 5', 6', and the line joining these points is the *support line*. In general, then, the support and pressure lines coincide when the vertical through the centre of gravity of any very small element coincides with the joint, and they deviate when this vertical does not coincide with the joint.

In the ordinary form of joint, as shown in Fig. 103, the support line varies from the pressure line, since the vertical through the centre of gravity **S** does not coincide with the joint under **S**. If, however, we should conceive the arch divided into vertical laminæ, then the support and pressure lines fall together. This is precisely the assumption always made in the analytical discussion of the theory of the arch.

Thus we take the area $\mathbf{A} = \int y \, dx$, and this expression supposes the arch divided into vertical laminæ.

The first to make clearly this distinction between the lines of pressure and support, was *Mosely* (*Civil Eng.*). Other authors have after him adopted this distinction, and then proved that the two lines always coincide, without remarking that this coincidence is *only* because of the adoption of the above integral. The same assumption simplifies greatly the graphical construction also (the analytical treatment is without it well-nigh impossible). We shall therefore assume vertical laminæ where it is at all permissible. This is always permissible at the crown of arches with horizontal tangent, because there the joints are vertical, and over all, when the pressure line lies *below* the axis of the arch; for the support line lies always *above* the pressure line, and therefore, in this case, the conditions of stability are more favorable for it than for the pressure line itself, when considered as the line of support.

Moreover, it is easy at any point of the pressure line constructed with vertical laminæ to pass to that line for another form of joint, and to the corresponding support line. Thus, if for the point **A** (Pl. 25, Fig. 104) we have found the pressure **Q**, and if now we wish to pass to the joint **A B C**, we prolong **Q** till it meets **P**, the weight of the voussoir **A B C D**, and resolve **P** and **Q** at this point into **Q'**. Then **Q'** is a side of the new pressure line, and it cuts **A B** in a point of the *support line*.

In this way we can easily determine whether the error committed when we substitute the pressure line for vertical laminæ for that for the actual joints, which is given by the segment of the joint **A B** between $\mathbf{Q}_1$ and the pressure line, can be disregarded.

177. Dimensions of the Arch.—The object of the construction of the pressure or support line in the arch is to determine the stability and the joints of the abutments. When the live load of the arch can be neglected with respect to its own weight, and when the material of the arch possesses the usual strength, and the pressure line lies within the inner third, then the lower point of rupture lies so low that the back masonry reaching from this point beyond the pressure line completely encloses it.

There is, therefore, nothing arbitrary, when the form of the arch is given, except the depth. Since in an arch of less depth than is allowable in practice a support line can still be inscribed, the graphical method is unable to determine the proper depth. We must then leave to theory the development of formulæ by which this can be determined, and assume that not only the form of the arch is given, but also its proper depth and the lower joint of rupture. It is required to determine the stability of the abutments.

The stability of the abutments can be regarded from two points of view. We may consider it as a continuation of the arch, as in many English and French bridges, in which the arch is continued as such, clear to the foundation; or we may regard it as a wall whose moment about the joint of rupture resists the rotation about this joint due to the thrust. Both views are identical, as the entire theory of the support line rests upon the investigation of the rotation. They differ only in the method of expressing the safety of the abutment.

If the arch is continued to the foundation, and the space between it and the road line filled up by spandrels; or if the thickness of the abutment increases from above as the support line requires; or, as is often the case in England, the abutment consists of walls parallel to the crown, separated by hollow spaces; still, in every case the abutment is not to be distinguished from the arch proper—it is stable when the support line lies in the interior. If the prolonged arch is separated entirely from the adjacent masonry, there is no reason for not giving the axis of the prolongation the form of the support line itself.

If, on the other hand, there is no separation of the arch and abutment, as in the English hollow abutments, it is sufficient that the support line lie in the inner third, and the abutment will be certainly stable.

The supposition that the resistance of the mortar is sufficiently great to unite the whole abutment as a single block which turns about its under edge, gives too small dimensions. To ensure safety it is assumed that equilibrium exists with reference to rotation about the lower edge, when the thrust of the arch is 1.5 greater than the actual. Investigations of French engineers have shown that this coefficient of safety for very light arches is not less than 1.4. The old tables of *Petit* give 1.9. We assume it, therefore, = 2.

If, therefore, the *double* thrust of the arch at the lower point of rupture is united with the weight of the abutment, the resultant should *still* fall within the base. Since it is indifferent in what order the elements of the abutment are resolved, it is best to divide it into vertical laminæ, and unite these with the double thrust. The equilibrium polygon thus obtained should cut the foundation base within the edge of the abutment.

When the thickness of the abutment is thus determined, we must construct the actual pressure line for the simple thrust in order to determine the joints. In drawing this second pressure line, we should properly take the divisions of the arch by the joints themselves. If, however, we take the division in vertical laminæ, the deviation, as we have seen, is insignificant. The normals to the actual joints must, then, not deviate from the sides of this pressure line by more than the angle of repose.

178. Construction of the Pressure Line.—In Pl. 25, Fig. 105, we give the method of construction of the proper width of abutment for an arch. We first divide the arch into vertical laminæ, and determine their weight. If the surcharge has vacant spaces, or is generally of different specific weight from the material of the arch itself, it must first be reduced. Thus, if the surcharge (spandrel filling, etc.) weighs, for instance, only $\frac{2}{3}$ds as much as an equal area of masonry in the arch, we have simply to diminish the vertical height above the arch by $\frac{1}{3}$d. We thus obtain the dotted line given in the Fig., which forms the *limit* of the reduced laminæ, and we can treat the areas bounded by this line—the vertical lines of division and the intrados—as homogeneous. We have then only to determine the centres of gravity of the various laminæ according to the construction for finding the centre of gravity of a trapezoid (Art. 33), and suppose at these points the weights, which are proportional to the reduced areas of the trapezoids to act.

Laying off these weights in their order, we have the force line (Fig. to left). The weights of the abutment laminæ 9, 10 and 11 are laid off to same scale *one-half* of their proper intensities. The reason will soon appear.

1st. To determine the thrust **H**, and also the joint of rupture.

We first inscribe a pressure line by *eye*, and assume the point of the intrados to which this line most nearly approaches as the edge of the joint of rupture. Draw next from the corresponding point of the force line a line parallel to the assumed pressure line *at this point*. This line will cut off from the horizontal through the beginning of the force line our first approximate value of **H**.

Thus, suppose we have inscribed by eye the pressure line 1, 2, 3, 4, etc., which gives us the point *a* for the position of the edge of the joint of rupture. Then a line drawn from 5 on the force line, parallel to the side 4 5 of the pressure line, gives us our first value for **H**.

Now assuming this value of **H**, we *erase* the first assumed pressure line, and proceed to construct the pressure line corresponding to this value of **H**, and the force line divisions 1, 2, 3, 4, etc. If *this* pressure line lies always within the middle third of the arch, it may be taken as the proper pressure line, and **H** as the true thrust. In general, however, this will not be the case. The pressure line thus determined may even pass without the arch entirely. We then determine the new point of rupture, as given by the point of exit of this pressure line, and *produce the side at this point back to intersection with* **H** *prolonged*. From this point of intersection draw a line which *does* lie within the middle third of the arch at the lamina of rupture, and *then* in the force polygon from the corresponding point of the force line draw a parallel to this line, thus cutting off a new value for **H**. Erasing now the preceding pressure line, we construct a third with this new value of **H**, producing it right and left from the new line just drawn, which *does* lie within the middle third, which will in general give us a pressure line lying everywhere within the middle third of the arch. If not, another approximation may be made. . We thus find by successive approximation the true joint of rupture and the corresponding thrust.*

---

* Instead of the "middle third" we may allow the pressure line to approach

2*d*. *Width of abutment.*—Since we have laid off the arch weights to scale in their true value, the pressure line thus obtained is the true pressure line for the arch. But we have laid off the abutment laminæ 9, 10, and 11, *one-half* their true value, and the pressure line thus obtained with the same thrust and pole O is the same as if we had taken their true value and *twice* H. Its intersection with the foundation gives us, then, the proper width of the abutment for stability, according to our assumption of 2 for the coefficient of stability (Art. 177).

179. Thus we can easily determine for any given case of arch and surcharge the horizontal thrust and the proper width of abutment, and then from the pressure line can easily so dispose the joints as to prevent sliding. If the dimensions of the arch as given are not such as to be stable, it will be found impossible to inscribe, as above, a pressure line which shall lie within the middle third, and the curve of extrados or intrados will have to be altered so that this shall be the case. The pressure line thus obtained, it is true, does not exactly correspond with the true one, as it is still possible to inscribe another which shall deviate still less from the true line. We have also taken the double thrust for the abutment laminæ alone, instead of for all laminæ from the joint of rupture of the arch. Both deviations are made on account of the far greater ease and rapidity of construction. It would be found very tedious to take first the force polygon up to somewhere *about* the section of rupture, then by long trial find the innermost support line, and finally, after the section of rupture is by this line determined, to lay off the remainder of the force polygon, and prolong the pressure line through the abutments.

It is far simpler to proceed, as above, by assuming the point of application of the horizontal thrust, as also temporarily the section of rupture. We obtain thus a somewhat smaller value for the width of abutment, but, on the other hand, we have taken the coefficient of stability at 2 instead of 1.9, as assumed in *Petit's* tables.

Moreover, the widths of abutment thus obtained are greater

---

the edge *as near as the strength of the material will allow*. In many arches the pressure line does pass outside of the middle third. The condition is not essential for stability, except in the absence of data determining the proper limits. If the pressure line will not lie within the required limits, whatever they are, the dimensions of arch must be changed, so that it will.

than those obtained by these tables, as it is assumed in them that the point of application of the horizontal thrust is at the upper edge of the abutment. Thus in every respect the construction gives results reliable and even more accurate than the tables, as we take the arch as it *really is* in any given case, while in the tables suppositions are made with reference to spandrel filling, etc., which do not hold good for every case.

180. Proper Thickness of Arch at Crown.—The proper depth of the arch at the key depends not only upon the rise and span, but also upon the load. The pressure at the extrados at the key, which is, in general, the most exposed part of joint, should not, according to the best authorities, exceed $\frac{1}{10}$th the ultimate resisting power of the material. If **P** is the pressure per unit of surface, **H** the thrust, and d the depth of key stone joint, then
$$\mathbf{P} = \frac{2\,\mathbf{H}}{d},$$
since, on the assumption that the curve of pressure does not pass outside the kernel, the maximum pressure is twice the mean pressure $\frac{\mathbf{H}}{d}$. This mean pressure, then, should not exceed $\frac{1}{20}$th the ultimate resistance of the material. In the best works of *Rennie* and *Stevenson* the thickness at key varies from $\frac{1}{30}$th to $\frac{1}{33}$d the span, and from $\frac{1}{26}$th to $\frac{1}{30}$th the radius of the intrados. The augmentation of thickness at the springing line is made by the Stevenson's from 20 to 40 per cent., by the Rennie's at about 100 per cent.

Perronet gives for the depth at crown the empirical formula
$$d = 0.0694\,r + 0.325 \text{ meters},$$
in which r is the greatest length in meters of the radius of curvature of the intrados.

For arches with radius exceeding 15 meters, this gives too great a thickness. According to *Rankine*,
$$d = 0.346\,\sqrt{r}$$
for circular arches, and
$$d = 0.412\,\sqrt{r},$$
where r is the radius of curvature of the intrados at the crown in feet.

"The London Bridge is in its plan and workmanship perhaps the most perfect work of its kind. The intrados is an ellipse, the span 152 ft., the rise $\frac{1}{4}$th as much, the depth of key $\frac{1}{50}$th the span. The crown settled only two inches upon removal of centres."—[*Woodbury: Theory of the Arch.*]

In general, we must first assume the depth at key in view of the strength of the material, the character of the workmanship, the load, etc. Then the thrust being found, we find the mean pressure per unit of area as above. If this mean pressure exceeds $\frac{1}{10}$th the ultimate resisting power of the material, make a new supposition, increase the thickness, find the thrust and pressure anew, and so on, till the results are satisfactory.

The ultimate resisting power of granite may be taken at 6,000 lbs., brick 1,200, sandstone 4,000, limestone 5,000 lbs. per square foot. These values are, of course, very general, and subject to considerable variations, according to the kind and quality of the stone. The strength of the material to be used must, for any particular case, be determined by actual experiment.

The weight of a cubic foot of stone may, in general, be assumed at 160 lbs., brick masonry at 125 lbs.

181. Increase of thickness due to change of form.—Having obtained a thickness which satisfies all the conditions, we must, if the arch be very light, make some further provision for the change of form which is sure to take place after the removal of the centres. By this change of form the pressure line is altered, and the thickness may need to be increased. In general, we need only to increase the depth from the key to the springing. This increase need not exceed fifty per cent. at the joint of rupture and weakest intermediate joint. [*Woodbury: Theory of the Arch.*]

182. Thus, by a simple and rapid construction, we can determine, for any particular case, the thrust, joint of rupture, and proper thickness of the abutments, without the use of tables or the intricate formulæ usually employed. There is no difficulty in laying down on paper and verifying all the elements of the most complex case. The method is entirely independent of all particular assumptions, and is therefore especially valuable when irregularities of outline or construction place the arch almost beyond the reach of calculation. It is general, and may be applied with equal ease to loaded and unloaded, full circle, segmental, or elliptical arches with any form of surcharge.

CHAPTER XVI.

THE INVERTED ARCH—SUSPENSION SYSTEM.

183. The inverted arch forms the supporting member of chain or cable suspension bridges. Whether the cable be composed of chains, links, or wires, we suppose them so flexible that they can perfectly assume the curve of equilibrium. As, therefore, disregarding the dead weight, any partial load would cause a change of shape, the cables must be *stiffened* in order to prevent the motion which would otherwise take place.

We may stiffen the chains, as shown in Pl. 26, Fig. 106, by triangular bracing, thus making a rigid system; or we may have two parallel chains and brace them to each other, as shown by Fig. 90 inverted; or we may introduce an *auxiliary truss*, the office of which is not to add in any degree to the supporting power of the combination, but simply to *distribute* a partial load over the whole span, so as to cause it to take effect as a distributed load, and thus prevent change of shape.

As in the first and last cases the structure is commonly hinged at the centre in order to eliminate the effects of temperature, the method of resolution of forces explained in Arts. 8-13 will, in general, be applicable for the determination of the strains.

In the second case, we can apply the principles of Arts. 158-161.

The rear chains, anchorages, and stiffening truss deserve, however, special notice.

184. Rear Chains and Anchorages.—The greatest tension in the *main* chains occurs, of course, for full load. To find the tension at top of tower, as also the horizontal pull, we have simply to lay off half the whole load vertically from o to d [Pl. 26, Fig. 106], and then draw **O** o horizontal and **O** d parallel to the last side at tower. Then **O** d is the tension in that side, and o **O** the horizontal pull. This pull is neutralized by the opposite and equal pull of the rear chain leading to the anchorage; provided, as should always be the case, it makes an equal angle with the vertical. We have thus acting upon the tower simply the half load; and the tension in the rear chain is equal to that in the last link, Od.

If from O we draw parallels to the other links, we have at once the strains in these links, Oc, Ob, Oa, etc.

Now, if the anchorage is a solid block of masonry, its condition of stability is, of course, very easily determined. The moment of the tension in the rear cable, with reference to the edge of rotation, must be more than balanced by the moment of the weight of the block acting at its centre of gravity, with reference to this edge. The case is too simple to need further notice.

It is, however, more economical to make the anchorage hollow—that is, in the form of an arch. The preceding method for determining the stability of the arch has then here direct application.

Thus, laying off along the vertical through the centre of the tower the weights of segments of the arch, we form with these segment weights and the *double* tension in the chain an equilibrium polygon. For this we have the pole O_1, AO_1 being double the tension Od already found. We then draw O_11, O_12, O_13, etc., and then from A parallels to these to the segment verticals 1, 2, 3, etc. We thus have the polygon A 1, 2, 3, 4, 5. [*Note.*—We take the *double* tension, as before, for the arch, we took 2 H instead of H, in order to ensure stability.]

The last line of this polygon 4 5 prolonged must, for stability, pass within the pier abutment, and its resultant, when it is combined with the weight of the pier and pier abutment, must pass within the abutment foundation. Through its intersection with the vertical line through the axis of the tower the curve of pressure for the arch must pass.

Drawing now $O_2 4$ parallel to the rear chain, and making it also equal to the double tension, or twice Od, we find the pole O_2, and from it draw $O_2 1$, $O_2 2$, $O_2 3$, etc., and then construct the pressure line for the arch. It must, for stability, lie within the middle third.

To ensure stability when the tension in the rear chain diminishes, or when the bridge is unloaded, the arch must be stable *by itself.* We must, therefore, construct the curve of pressure for the arch alone, neglecting the tension of the rear chains, as explained in the preceding chapter.

If this also passes within the middle third of the arch, as represented by the dotted line, the arch is, under all circumstances, stable, and can fully resist the tension of the rear chains.

We can now, finally, so dispose the joints as to prevent sliding.

185. Stiffened Suspension System.—We have already referred to the methods of stiffening the cable or chain so as to prevent the changes of shape due to partial loading. Of these methods, it only remains to notice particularly the last, viz., by means of an auxiliary truss. The office of this truss is to distribute a partial load over the whole length. We have now to investigate the forces which act upon the truss.

In Pl. 27, Fig. 107, let the chain be acted upon by the truss represented by **A B**, which is called into action only by a partial load, and not at all by a total uniform load. We can neglect then the weight of the truss itself, as this is borne by the cable. At the apices 3, 4, 5, 6, 7 let us suppose partial loads indicated by the small arrows pointing down. Then, at every point of connection with the chain, we have the reactions 1′, 2′, 3′, etc., acting upwards. Now the truss must prevent deformation, and hence these forces are dependent upon the form of the cable itself. Indeed, if we take any point, as **O**, as a pole, and draw lines parallel to the respective sides of the cable, these lines will cut off upon a vertical **P′** these forces. The absolute value of these forces will, it is true, vary according to the position of the pole assumed, but their *relative proportions* remain always the same. The resultant **P′** of all these forces passes then through the intersection of the two outer sides of the catenary.

Since the truss distributes its load **P** upon the cable, the reaction **B** at the right support is here zero. The reaction, however, at **A** cannot be zero unless **P** and **P′** coincide, as is the case for total uniform load. These, then, are all the forces which are kept in equilibrium by the truss. If **P** is given, **P′** and the reaction at **A** can be easily found, and if we then divide **P′**, according to the form of the chain, into the portions 1′, 2′, 3′, etc., we have the forces at each apex.

Thus we lay off to scale the given forces 3, 4, 5, 6, 7 = **P**, and with a pole distance any convenient multiple of the height of truss draw lines to these points of division, and then construct the corresponding equilibrium polygon **A** 3, 4, 5, 6, 7, **B**. Prolong then the outer side **B** 7 to intersection with **P'**, and draw the closing line **A P'**. A parallel through **O** to this line cuts off from the force line **P** the reaction at **A** and the cable reaction **P'**.

Now **P'** being thus found and the form of cable given, we have only to lay off **P'** vertically, draw from its extremities lines parallel to the two outer sides of the *given cable arc*, and from the pole thus determined, lines parallel to the other sides will give us the forces 1', 2', 3', etc. These when thus found we lay off on our force line for the pole **O**, as shown in the Fig., and then construct the corresponding equilibrium polygon **A** 1', 2' . . . . . 9', 10', **B**.

Thus the vertical ordinates between **A P'**, **P' B** and this polygon give us the moment at any point for a truss acted upon by the forces 1', 2', 3', etc., alone. The ordinates between **A P'**, **P' B** and the polygon **A** 3, 4, 5, 6, 7, give, in like manner, the moments for a truss acted upon by the forces 3, 4, 5, 6, 7, whose reactions are **A** and **P'**. The ordinates, then, included between *both* polygons give us the moment at any point of the stiffening truss. Thus the ordinate y, multiplied by the pole distance, gives us the moment in the truss at the point o. If we had taken the pole distance **O** equal to the height of the truss, then these ordinates would give us at once the strain in the flanges. We can thus easily find the strains in the stiffening truss for any weight or system of weights in any position.

186. Most unfavorable method of Loading.—Let us investigate the action of a single weight **P** at any point. In Pl. 27, Fig. 109, we have a single weight **P** acting between **A** and **P'**.

The Fig. is nothing more than a repetition of Fig. 108, only we have a single load **P** instead of a system of four loads, and therefore the polygon for **P** consists only of two straight lines instead of having as many angles 3, 4, 5, etc., as there are apex loads in the first case. All lines have the same position as in Fig. 108, and hence the construction needs no further explanation.

We see at once from the Fig. that any load between **A** and **P'** increases the moment at every point of the span **A B**, and therefore at the point of rupture or of maximum moment also. So also for the shearing force. When, therefore, the moment

CHAP. XVI.] THE INVERTED ARCH. 335

at any point, and the sum of the forces between that point and **A**, is a maximum, at least the entire distance from **A** to **P'** must be covered with the load.

In Fig. 110 we have the weight **P** on the *other* side of the but centre **P'**. The construction is identical with Figs. 109 and 108, the position and the direction of action of the forces is now different. Since the resultants **A** and **P'** now lie on the same side of **P**, **A** and **P'** act in opposite directions, and since **P'** must still act upwards, **A** must act downwards. In the neighborhood of 5' the moment is zero. Between this point and **B** the moments have the same signification as in Fig. 109 ; on the other side the moments have then a different sign. In order, then, that the moment at 5' shall be a maximum, the load must cover the length from **A** to **P**, this last point being the point at which a load causes no moment in 5'; for if any point between **A** and **P** were not loaded, as we have seen, a load at that point would increase the moment at 5'. A load beyond **P**, however, would diminish the moment at 5'.

The above holds good for every point between **A** and **P'**, and therefore for the point of rupture or of maximum moment itself. In order that this maximum moment can be no more increased, the load must extend from **A** beyond the centre to that point at which a load being placed causes no moment at the cross-section of rupture.

As for the shearing force, at the end **A** it will evidently be greatest for load from **A** to **P'**, or over the half span, since every load on the other side of **P'** diminishes this reaction. Hence we have the following principles established:

The moment at any cross-section of the stiffening truss is a maximum, when the load reaches from the nearest end beyond the centre to a point for which the moment at this cross-section is zero.

The above condition holds good, therefore, for the maximum of all the maximum moments, or for the cross-section of rupture itself.

The maximum shearing force is at one end of the truss when the adjacent half span is loaded.

If the arc is unsymmetrical, we must understand by "*half span*" the distance between the end and vertical through the intersection of the outer arc ends produced.

187. Example.—As an illustration of the above principles, let us take the structure represented in Pl. 28, Fig. 111. Span, 60 ft.; depth of truss and panel length, 5 ft. Scale, 10 ft. to an inch. We suppose the live load to be 2 tons per ft., giving thus 10 tons for each lower apex, and take the scale of force 50 tons per inch.

On the left we have laid off the force lines for the loads 2, 3, 4, 5, 6 and 7 to 11, and have taken the poles for each, so that the first lines are all parallel to each other and to the first link of the cable; the common pole distance being $2\frac{1}{2}$ times the height of truss, or 1.5 inches. The *moment scale* is then $1.5 \times 10 \times 50 = 750$ ft. tons per inch. Since the full load is entirely supported by the cable, we have only to investigate the effect of the live load upon the truss.

Precisely as in Fig. 108, we construct the polygons for forces 2–11, 3–11, 4–11, etc., and draw the closing lines as indicated by the broken lines radiating from the centre O. Parallels to these from the poles cut off from the force lines the end and chain reactions. The upper portions are the chain reactions, the lower the reactions at the right end for the loads 2–11, 3–11, etc.

Now we have to divide these chain reactions into as many parts as there are load apices by lines parallel to the sides of the chain. This we have done by drawing two lines parallel to the two chain ends, inserting the chain reactions between these lines, and then drawing parallels to the chain sides. If, as in this case, the curve of the chain is a parabola, these reactions are divided into 11 equal parts. If the chain has any other form, the parallels to the chain sides determine the relative lengths of these portions.

It will only be found necessary to construct the moment polygons for 4, 5 and 6–11; the other polygons already drawn are necessary for the determination of the shearing forces only.

Thus, on the force line for loads 4 to 11 we can now lay off the 11 equal parts just found, into which the chain reaction is divided. So for 5–11 and 6–11. These portions we have indicated by *Roman* numerals. We can now draw the corresponding polygons precisely as in Fig. 108, which are indicated also by Roman numerals.

It is then easy with the dividers to pick out the maximum moment at any apex. These moments, laid off as below, give the curve of moments for the truss, which being scaled off and divided by the depth of truss, give at once the strains in the flanges. Since the moment scale is 750 tons per inch and the depth of truss 5 ft., the moment ordinates scaled off at 150 tons per inch will give at once the strains in the flanges, without division.

For the shearing forces, we know from the preceding that the maximum reaction at right end is for loads 6–11. This reaction we have already found in the corresponding force line by means of the closing line already drawn. We lay it off then right and left, half-way between the ends and first apex, that being the effective length of load, the two half-end panels resting directly upon the abutments.

The maximum shear at any point is evidently when the load reaches from right support to that point, and is equal to the sum of the chain reactions at the unloaded apices. Thus, maximum shear at 3 is equal to the interval I II for the line 3–11; at 4, I III for line 4–11; at 5, I IV for line 5–11; and at 6, I V for line 6–11. Laying off the shear at 6, we can draw the line 6–11, as indicated in the diagram, and thus determine the shear at 2. This we cannot find, as above, for 3, 4, etc., as for the load 2–11; owing to the shape of the chain as represented, there is *no* upward reaction at 1, as there is no angle of the chain at 6.

The shear diagram is, of course, symmetrical on each side of the centre. We can therefore construct it as represented, and then the determination of the strain in the diagonals is easy. We have only to multiply the shear at any apex by the secant of the angle which the diagonals make with the vertical. This we may do by properly changing the scale at once, and thus scale off the strains directly.

188. Analytical Determination of the Forces acting upon the stiffening Truss.—Assuming that the truss distributes the partial loading uniformly over the whole arc, we may deduce very simple formulæ for the forces acting upon the truss. As we have already seen, for a maximum moment at any point, the load must always extend out from one end.

Let us represent, then, the ratio of the loaded part from left to the whole span by k.

Let the entire span be $2l$, then the loaded portion is $2kl$. Let m be the load per unit of length; then the whole load $\mathbf{P} = 2klm$ [Fig. 108].

The distance of **P** from the left is then half the loaded portion, or kl. Its distance from **P'**, which acts at the centre of the span, is $l(1-k)$.

Hence we have for the left reaction **A**

$$\mathbf{A} \times l = \mathbf{P}\, l\, (1-k) \quad \text{or} \quad \mathbf{A} = \mathbf{P}\,(1-k) = 2\,klm\,(1-k).$$

Also $\quad\mathbf{P}'\, l = \mathbf{P} \times kl \quad \text{or} \quad \mathbf{P}' = 2\,k^2\, lm.$

The chain reaction per unit of length is then

$$\frac{\mathbf{P}'}{2\,l} = k^2\, m.$$

Now let x be the distance to the point of maximum moment.

Now since at this point the shear must be zero, the weight of the portion x must be equal to **A** (Art. 38).

We have then

$$\mathbf{A}\, x - \frac{\mathbf{A}\, x}{2} = \mathbf{P}\,(kl - x) = 2\,klm\,(kl - x),$$

whence, by substituting the value of **A**,

$$x = \frac{2\,kl}{1+k}.$$

But the maximum moment is $\mathbf{A}\, x - \dfrac{\mathbf{A}\, x}{2} = \dfrac{\mathbf{A}\, x}{2}$, and therefore, substituting the value of x above,

$$\mathbf{M}\text{ max.} = \frac{k^2\,(1-k)}{1+k} \cdot 2\,l^2\, m.$$

This becomes a maximum for $1 - k - k^2 = 0$, or for

$$k = \tfrac{1}{2}\sqrt{5} - \tfrac{1}{2} = 0.618034.$$

Therefore, *the greatest moment occurs when 0.62 of the span is covered with the load.*

We have then the

Length of the loaded portion, $= 2\,kl = 0.61803 \times 2\,.\,l.$

Reaction, $\mathbf{A} = 2\,klm\,(1-k) = (\sqrt{5} - 2)\,2\,lm = 0.23607 \cdot 2\,lm$

Chain reaction, $\mathbf{P}' = 2\,k^2\,lm = (\tfrac{3}{2} - \tfrac{1}{2}\sqrt{5})\,2\,lm = 0.38196 \cdot 2\,lm.$

Load per unit in loaded portion, or the difference between the load m and the chain reaction $m\,k^2$ per unit of length $= m\,(1-k^2) = \tfrac{1}{2}(\sqrt{5} - 1)\,m = 0.61803\,m.$

The distance of the point of maximum moment is

$$x = \frac{2\,kl}{1+k} = (\tfrac{3}{2} - \tfrac{1}{2}\sqrt{5})\,2\,l = 0.38196 \cdot 2\,l = \frac{\mathbf{A}}{m\,(1-k^2)}.$$

THE INVERTED ARCH. 339

The maximum moment itself is

$$\frac{A^2}{2m(1-k^2)} = \frac{k^2(1-k)}{1+k} 2l^2 m = (5\sqrt{5}-11)l^2 m = 0.18034\, l^2 m.$$

For a simple girder uniformly loaded, the maximum moment is $\frac{1}{8}p\, l^2$. The maximum moment is then reduced from $\frac{1}{8}$ to 0.18, or to about $\frac{1}{4}$d, *or is $\frac{9}{25}$ths the maximum moment for a simple girder of same span and load.**

If we represent the dead load by p, then, since the stiffening truss sustains only the moving load, we have

$$\frac{0.18\, m\, l^2}{\frac{1}{8}(p+m)\, l^2} = 0.36\, \frac{m}{p+m} = \frac{\frac{1}{8}m\,(0.6\, l)^2}{\frac{1}{8}(p+m)\, l^2}.$$

That is, *the maximum moment in the stiffening truss is the same as for a simple girder of $\frac{6}{10}$ths the span, loaded only with the moving load.*

189. Summary.—The reaction at the end abutment and the chain reaction at each apex having been found, as above, for any given load, we might have found the strains in every piece by the method of Arts. 8–13. This would, however, in this case have proved long and tedious. The construction of the curve of maximum moments and shear is preferable.

We can therefore readily determine the strains in such a combination as that represented in Fig. 111. We have already, Arts. 90–94, given practical and simple methods for the determination of the strains in braced arches of the usual forms of construction.

It will be observed that it is by no means necessary that the arrangement of bracing and flanges should be the same as that shown in Figs. 90 and 94.

Thus we may treat the arch represented in Fig. 5 (c) according to Art. 158, as hinged at both abutments and crown, or, making the lower flange continuous at the crown, we may find the resultant pressures at the abutments by Art. 159, and then follow these pressures through precisely as shown in the Fig.

The combination of Fig. 111 being of considerable importance, as the more usual form of construction of suspension bridges, and not falling under our classification of "braced arches," we have considered it desirable to discuss it somewhat

---

* *Rankine* gives $\frac{1}{16}$ths for a girder whose *ends are fixed*, the greatest moment occurring for a load over $\frac{2}{3}$ds the span.

fully. A better form of construction is that shown in Fig. 106, which is perfectly rigid, and the strains in which are easily found by Art. 158 or 159, according as we hinge it in the centre or not.

Reviewing now the preceding, we see that the graphical method, as here developed, furnishes us with a simple, accurate and practical solution of nearly every class of structure occurring in the practice of the engineer or builder. In our first chapter we have a method by the resolution of forces applicable to any framed structure, however irregular or unsymmetrical, provided only there are no moments at the ends to be determined.

In Art. 125 we have explained fully the application of the method for this case also, when these moments are known, and in Chaps. VIII. to XIV. inclusive we have given practical constructions for the determination of these moments for all the important classes of structures in which this condition occurs, such as the continuous girder, braced arches, etc.

When the structure is *not* framed, or composed of pieces the strains in which can be definitely determined, we have the *method of moments* of Chap. V., which, as we have seen, may be extended so as to completely solve the difficult case of the continuous girder, and which may, of course, be applied to framed structures also, as illustrated in Fig. 111 (Art. 187) in the case just discussed. Thus we have *two* distinct graphical methods by which our results may be checked. The first method includes a great variety of the most important and usual structures, such as bridge girders, roof trusses, cranes, etc., and in view of its ease and accuracy will undoubtedly be found of great service by the engineer and architect. The second method has important mechanical applications, as noticed in Art. 41; and aside from these, and its application to structures having end moments, such as the continuous girder, etc., furnishes us with ready determinations of the centre of gravity of areas (Chap. III.), the moment of inertia of areas (Chap. VI.), and also gives us a very complete solution of the *stone arch* (Chap. XV.).

We have also the analogous methods of *calculation*, viz.,

both by resolution of forces and by moments (Arts. 9 and 16 of Appendix). The latter being so general and simple in its application, we have not felt justified in leaving it entirely out of sight, and in those cases where it seemed of especial service, or assisted the graphical solution, we have illustrated it more or less fully (Chap. XII.). In this latter chap. we have also given constructions as well as formulæ, and developed principles which, it is believed, render possible, for the first time, the complete and accurate solution of the important case of the "*draw span.*" (Arts. 118-121.)

The formulæ of Chap. XIII. in connection with the method of calculation by moments, render the calculation of the continuous girder generally as simple, and but little more tedious than for the simple girder itself. Whatever may be thought of the advantages or disadvantages of this class of structures by engineers generally, it is at least time that such structures as *draws* or *pivot spans* should be calculated under suppositions which approach somewhat more nearly the actual case than is at present the practice. As to the relative economy of continuous girders, we have endeavored to enforce the fact that the saving over the simple girder is from 15 to 20 and even 50 per cent. We give in the Appendix a tabular comparison of a few cases sufficient to show the point beyond dispute, and any one may easily add to the list, or verify the calculations.

The "graphical arithmetic," as it might be called, such as graphical addition, subtraction, multiplication, division, extraction of roots, determination and transformation of areas, etc., we have entirely omitted in the present work, judging it of but little practical value, except in rare cases, when we have explained the necessary constructions as they occur, and unnecessary for the development of the graphical method proper. [See Chap. IV. of Introduction.]

APPENDIX.

GRAPHICAL STATICS.

A. JAY DUBOIS

APPENDIX.

NOTE TO CHAPTER VIII. OF THE INTRODUCTION—UPON THE MODERN GEOMETRY.*

It is to be regretted that, notwithstanding its beauty of form, simplicity, and many happy applications in the technical and natural sciences, the *Modern Geometry* is yet hardly known, scarcely by name even, in our schools and colleges.

The work of Gillespie upon Land Surveying, already cited in the Introduction, and a treatise on Elementary Geometry by William Chauvenet (Phil., 1871), are the only ones which occur to us in this connection.

It has already been stated that the modern or *pure* geometry of space differs essentially from the ancient, and from analytical geometry, in that it makes no use of the idea of *measure* or of metrical relations. We find in it no mention of the bisection of lines, of right angles and perpendiculars, of areas, etc., any more than of trigonometrical quantities, or of the analytical equations of lines. We have nothing to do with right-angled, equilateral, or equiangular triangles, with the rectangle, regular polygon, or circle, except in a supplementary manner. So also for the centre, axes, and foci of the so-called curves of the second order, or the conic sections.

On the contrary, we obtain much more general and comprehensive properties of these curves than those to which most text-books upon analytical geometry are limited.

A new path is thus opened to the conic sections, without the aid of the circular cone, after the manner of the ancients, or of the equations of analytical geometry.

As a direct consequence, the principles and problems of the modern geometry are of great generality and comprehensive-

---

* The following remarks and illustrations are taken from the *Geometrie der Lage*, by *Reye*. Hannover, 1866.

ness. Thus the most important of those properties of the conic sections which are proved in text-books of analytical geometry are but special cases of its principles. A few particular examples taken from the *Geometrie der Lage*, by *Reye*, which could not well have been inserted in the Introduction to this work, will best explain and illustrate our general remarks—the more so as these examples are of special interest and value to the engineer.

It is a problem of frequent occurrence in surveying to pass a line through the inaccessible and invisible point of intersection of two given lines. The Geometry of Measure, or ancient geometry, gives us any required number of points upon this line by the aid of the principle, that the distances cut off from parallel lines by any three lines meeting in a common point are proportional. The Geometry of Position furnishes us with a simpler solution.

FIG. 1.

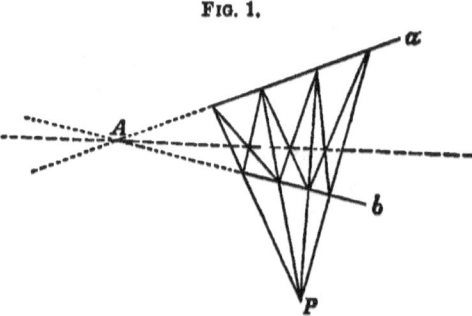

Thus the two lines a, b being given [Fig. 1.], we have simply to choose any point we please, as P. From this point draw any number of lines desired, in any direction intersecting the given lines. Now, in any quadrilateral which any two of these lines form with the two given lines a and b, we have simply to draw the diagonals. The intersections of all these diagonals lie in the same straight line passing through the intersection A of the two given lines, and therefore determine the line required. Observe that the construction is entirely independent of all metrical relations, and depends solely upon the relative position of the two given lines.

Again: If we take upon any straight line three points, A, B and C [Fig. 2.], and construct any quadrilateral, two opposite sides of which pass through A, one diagonal through B, and the other two opposite sides through C, then will the other diagonal intersect the line in a point D, which for the same three points,

FIG. 2.

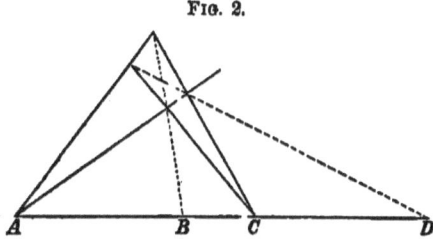

A, B and C, is always the same for every possible construction. Moreover, these four points, A, B, C and D, are always *harmonic* points, so that D is harmonically separated from B by the points A and C. Thus, $AB : BC :: AD : CD$. This construction may also be applied in surveying, as in passing around an obstruction, as a wood, etc., into the same line again.

Again: We may notice the following principle concerning the triangle [Fig. 3]:

FIG. 3.

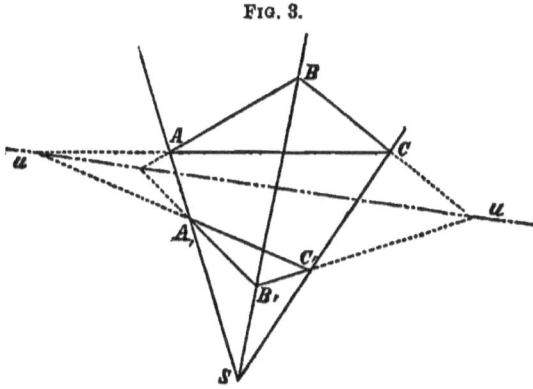

If two triangles, ABC and $A_1 B_1 C_1$, are so situated that the lines joining corresponding angles, as $A A_1$, $B B_1$, $C C_1$, meet in a common point S, then will the intersections of corresponding sides, as AC and $A_1 C_1$, AB and $A_1 B_1$, BC and $B_1 C_1$, meet in a common line, as $u\,u$. The inverse also, of course, holds good: that if the sides intersect on a line, the lines through the angles intersect in a point.

Another series of principles are connected with the curves of the second order, or conic sections. From analytical geometry, as is well known, a curve of the second order is completely determined by five points or five tangents. But the length of the calculation or construction of a curve thus determined is also well known. The geometry of position, however, proves two very important principles, which render it easy to construct to the five given points or tangents any number of new points or tangents, and thus quickly draw the curve itself. The reader already acquainted with these principles will also probably remember how much auxiliary demonstration their proof in the analytical geometry requires. The first of these, due to *Pascal*, is, that the three pairs of opposite sides of a hexagon inscribed within a conic section intersect upon a straight line. The second, due to *Brianchon*, is, that the three principal diagonals of the circumscribing hexagon, which unite every pair of opposite angles, intersect in one and the same point. Both principles are easily deduced from the circle. It will be observed that they are independent of the relative dimensions, centre, axes, and foci of the curves. For this very reason they are of the greatest generality and significance, so that an entire theory of the conic sections can be based upon them. Thus Pascal's principle solves the important problem of tangent construction from a given point, even when the curve is given by five points only, without completely constructing it.

This problem of tangent construction to curves of the second order can in many cases be solved by the aid of a principle which expresses one of the most important properties of the conic sections, but which, nevertheless, is seldom found in textbooks upon analytical geometry, because its analytical proof is somewhat complicated, and little suited to set forth the property in its proper light.

For example: If through a point **A** [Fig. 4] in the plane of but not lying upon a curve of the second order, we draw secants, every two secants determine four points, as **K, L, M, N**, upon the curve. Any two lines joining these four points, as $\overline{LM}$ and $\overline{KN}$ or $\overline{KM}$ and $\overline{LN}$, intersect in a point of a straight line aa, which is the *polar* of the given point **A**; that is, which intersects the curve in the two points of tangency **G G**. Thus the lines through **A** and the intersections of aa with the curve are the tangents to the curve through **A**. If the point **A** were within the curve, this line aa would not intersect it. This construction can be used in order to draw through a given point tangents to a conic section by the simple application of straight lines. Upon every secant through

FIG. 4.

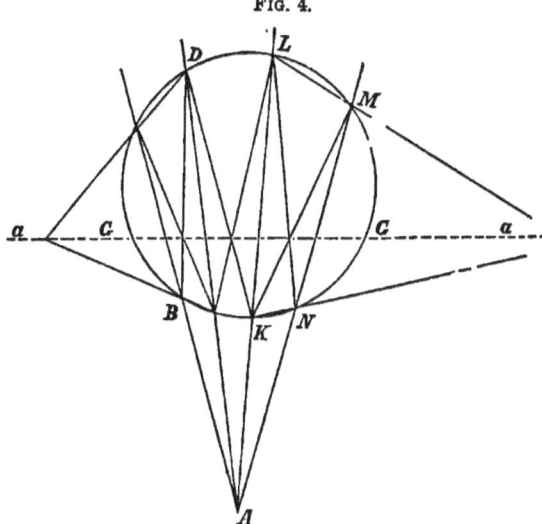

A, moreover, there are four remarkable points, viz.: the point A itself, the first intersection B with the curve, the intersection with the polar, and, finally, the second intersection D with the curve. These four points are *harmonic* points, and the polar *a a* contains, then, every point which is harmonically separated from A by the two curve points. The important principles relating to the centre and conjugate diameter of conic sections are merely special cases of the above important principles. These last can be easily extended to surfaces of the second order, as the intersection of these by a plane is, in general, a curve of the second order.

From these few examples, which might be indefinitely multiplied, it may easily be seen how very different, but not less important than those of analytical geometry, are the theorems of the geometry of position. Thus the latter are generally proved by aid of the angle which the tangents make with the line through the focus, or by the distances cut off from the axes—that is, by *metrical relations*. We refer, of course, to the elements of analytical geometry as contained in most text-books, and not to those most fruitful and later methods whose existence are chiefly due to the sagacity of *Plücker* (Introduction, VIII.).

NOTE TO CHAPTER I.

1. The method by the resolution of forces developed in Chapter I. is so simple and easy of application, and its principles are so few and self-evident, that we have not considered it advisable to tax the patience of the reader by any great variety of practical applications. A large number of such applications are to be found in a most excellent little treatise by *Robert H. Bow*, entitled *The Economics of Construction in Relation to Framed Structures*. There are, however, a few important practical points of detail, and a few general considerations, which we think it well to notice here, and to which, in illustration of the remarks in Chap. I., the reader will do well to attend.

2. In Pl. 1, Fig. I. (Appendix), we have represented the "*Bent Crane*" given by *Stoney* in his *Theory of Strains*, p. 121, Art. 200.

We assume the following method of notation. Let all that space above the Fig. be indicated by **X**, and all that space below by **Y**, and the triangular spaces enclosed by the flanges and diagonals by the numbers 1, 2, 3, 4, etc. The first upper flange is then denoted by **X** 2, the second by **X** 4, and so on. So also the first lower flange is **Y** 1, the next **Y** 3, etc. The first diagonal is then **X** 1, the next 1 2, the next 2 3, etc.*

The flanges are equidistant, forming quadrants of two circles whose radii are respectively 20 and 24 feet. The inner flange is divided into four equal bays, on which stand isosceles triangles, and a weight of 10 tons is suspended from the peak. The scale for this and all the Figs. of Pl. I. is 20 tons to an inch and 10 feet to an inch. Laying off, then, the weight $XY = 10$ tons, we form, according to the method of Chapter I., the strain diagram. It will be seen at once that all the lower flanges, **Y** 1, **Y** 3, etc., radiate from **Y**, all the upper flanges, **X** 2, **X** 4, etc., from **X**, and everywhere the letters in the one diagram indicate the corresponding pieces in the other.

---

* For this very elegant method of notation, we are indebted to the work of R. H. Bow, above alluded to.

We can now at once take off the strains to scale in the various pieces.

A comparison of our method with that given by *Stoney* for the same case will be instructive, as illustrating the comparative merits of the two.

3. Character of the Strains in the Pieces.—One of the most important points of our method is the ease and certainty with which the *character* of the strains in the pieces may be determined. We have only, as detailed at length in Chapter 1., to follow round any closed polygon in the direction of the forces, and then refer back to that apex of the frame where for the moment we may happen to be.

Thus for the peak, since we know that the weight acts down, we follow down from **X** to **Y**, and then from **Y** to 1, and 1 back to **X**. Referring back now to the frame, and remembering that a force acting away from the apex means tension, and towards, compression, we have at once **Y** 1 compression and **X** 1 tension.

Now for apex a, since **X** 1 is tension, with respect to *this* new apex, it must act away. We go round then from **X** to 1, 1 to 2, and 2 back to **X**, and then, referring these directions to the corresponding pieces meeting at a, we have 1 2 compression and **X** 2 tension.

We find thus all the outer flanges in tension, as evidently should by simple inspection be the case. Also all the inner flanges compression. As for the diagonals, they alternate, the first being tension, the next compression, until we arrive at 4 5, which we find to be *also* compression.

A glance at the strain diagram shows how this comes about. The line **X** 4 *crosses* **Y** 5, and thus gives us a *reverse* direction for 4 5.

In such a simple structure as the present, the character of the strains would present no especial difficulty in any case; but in more complicated ones, the aid of such a simple and sure criterion as the above is indispensable, and we have been thus even prolix upon this point, the more so as it is not so much as alluded to, as far as we are aware, in those few works which notice the above method at all.

4. There are other points which we may here illustrate by our Fig.

According to our first principle (Art. 3, Chapter I.), when

any number of forces are in equilibrium, the force polygon is closed. Inversely, then, a closed force polygon indicates forces which, if applied at a common point, would hold each other in equilibrium.

Thus **Y** 3, 3 4, **X** 4, and the weight are, or would be, if all applied at a common point, in equilibrium. This we see directly from the Fig. Thus we know that when any number of forces are in equilibrium, the algebraic sums of their vertical and horizontal components must be zero, otherwise there must, of course, be motion. Now the vertical component of **Y** 3 plus that of 3 4 minus that of **X** 4 is exactly equal and opposed to the weight, while the horizontal component of **Y** 3 plus that of 3 4 is equal and opposed to that of **X** 4, and there is then equilibrium.

Again, according to the principle of Art. 5, Chap. I., any line, as the one joining 2 and 6 (broken line in Fig.), is the resultant of **X** 2 and **X** 6, as also of 2 3, 3 4, 4 5 and 5 6.

The Fig. also well illustrates the points to be *avoided* in making a strain diagram, already alluded to in Art. 13, Chap. I. The scale to which the frame is taken is here altogether out of proportion to the scale of force. The first should be increased or the second diminished, or both. The present length of the diagonals and flanges is inadequate to give with sufficient accuracy the directions of strain lines of such length.

Nevertheless we have experienced no difficulty in checking to tenths of a ton the results given by *Stoney* for this structure.

5. In Pl. 1, Fig. II., we have represented a roof truss, span 30 ft., rise 8 ft., camber 1 ft.; and the strain diagram illustrates in its two symmetrical halves (one full, the other dotted) the remarks of Art. 13, Chap. I., upon the *check* which in such cases our method furnishes of its accuracy.

We lay off the weights 1, 2, 3, 4, 5, and then the reactions at **A** and **B**, which should bring us back to the point of beginning, and thus complete the force polygon. The strains are then easily found, and the two halves should be perfectly symmetrical, and give the same results.

In Fig. III. we have given another form of truss with strain diagram, the other half of which the reader can complete and letter for himself.

6. In Fig. IV. we have a form called the French roof truss

and two strain diagrams—the larger for vertical reactions, the smaller for *inclined* reactions.

This last brings out the force polygon in perhaps a clearer shape than before. The weights 1 to 7 being laid off downwards, the two reactions must always bring us back to the starting-point, and thus close the polygon—in this case a triangle, in the preceding case a straight line, and in the case of Fig 6, Art. 10., Chap. I., a true polygon. Both strain diagrams illustrate the check we have upon the accuracy of the work. The second half should be perfectly symmetrical with the first, and the lines $Y k$ and points k in each should *coincide*.

We have here also to notice a point which in roof trusses is of frequent occurrence, and may, if not noticed, cause difficulty.

We have already observed in Art. 9, Chap. I., that we can always find the strains in the pieces which meet at an apex, provided only *two* are unknown. Now in the strain diagram to Fig. IV., we readily determine the strains in $X a$, $Y a$, $X b$, $a b$, $Y c$ and $b c$ successively, and arrive finally at apex 2, where we have the two known strains in $X b$ and $b c$, and wish to find the strains in *three* pieces, viz., $X d$, $d h$ and $c h$. At first sight this seems impossible. If, however, we assume that the pieces of the frame can take only strains of a certain kind, as, for instance, $h d$ only tension, and *not* compression, the problem is perfectly determinate. This assumption is easily realized in practice. Thus if $h d$ is a *rod* of small diameter, it cannot act as a compression member at all. Moreover, the strain of tension in $h d$ must evidently be *precisely equal to that in $b c$, already found*. We have then to form a closed polygon with the weight at 2 and the known strains in $X b$ and $b c$, whose other three sides shall be parallel to $X d$, $h d$ and $c h$ respectively, and in which, moreover, the strain in $h d$ shall be *equal* to that in $b c$, and where *both* these strains must be, when the polygon is followed round according to rule, *tensile*. We have evidently, then, in accordance with these conditions, only the polygon $2 X d h c b X$, thus finding the point d, from which we can now proceed to find e, etc. The points a, b, d and e are evidently in the same straight line parallel to $c h$. This point is one of importance, and the reader should carefully follow the above remarks with the aid of the Fig.

The strain diagram thus constructed shows us many facts

about the system not otherwise apparent. Thus bc, ch and hd are in equilibrium with the load at 2. Again, ab and bc are in equilibrium with $\mathbf{Y}a$ minus $\mathbf{Y}c$, as also are hd and de with ke minus kh. Also kh, hc, $\mathbf{Y}c$ and $\mathbf{Y}k$ are in equilibrium, and $\mathbf{Y}c$, cb and $\mathbf{X}b$ are in equilibrium with the reaction minus the weight at 1, or with the *shear* to the right of 1. This last principle is general. *When a section can be made entirely through a structure, the strains in the pieces cut are in equilibrium with the shear at the section.* If only three pieces are cut, then, by taking as a centre of moments the point of intersection of any two, we can easily find, knowing the moment of the shear, the strain in the third.

Thus we have the general and easy method of calculation given in Art. 14, Chap. I. The moment of the *shear* is, of course, the sum of the moments of all the exterior forces between the section and one end.

We have then two methods, one graphic and one by calculation, by which we can find the strains in every kind of simple truss which can ever occur in practice. By "simple" we mean merely resting at the supports, or not acted upon at the ends by a couple or *moment*, as is the case, for instance, in the continuous girder.

When the structure is unsymmetrical, or complex, the determination of the different lever arms is often very tedious, involving a good deal of trigonometrical computation. On the other hand, the frame can always from its known proportions be easily and accurately drawn to scale, and then the exterior forces, whatever their relative intensity or directions, can be laid off, and the strains at once determined. Here we see, then, one of the great advantages of our graphical method. An unsymmetrical frame and different directions of the forces requires no more time or labor than a more simple case.

7. Application to Bridges—Bow-string Girder.—In Art. 12, Chap. I., we have alluded to this application, and shown how by two strain diagrams only we can completely calculate a bridge of any length. As this application is so important, and as the method is stated by several authors to be *inapplicable* to bridges,[*] or, at best, to be unsatisfactory, we will here call more

---

[*] *Iron Bridges and Roofs*—Unwin—p. 143. *Economics of Construction*—Bow- p. 61.

special attention to the points to be observed in the tabulation of the strains. There is, indeed, no more satisfactory, complete and rapid method for the solution of bridge girders generally than that afforded by the graphic method.

As an example, let us take the Bow-string Girder given by *Stoney*, p. 131. Span, 80 ft., divided into 8 panels; rise of bow, 10 ft. Load, 10 tons at each lower apex.

We construct the two strain diagrams * given in Fig. V., Pl. 2, viz., one for the load P_7 at the first apex, and one for the load at the last apex, P_1. Referring, if necessary, to Art. 12, Chap. I., the reader can easily follow out these diagrams. We then scale off the strains, and obtain, for the strains in the diagonals—

	ab	bc	cd	de	ef	fg	gh
P_7	− 2.7	− 11.4	+ 4.8	− 4.3	+ 2.4	− 2.3	+ 1.4
P_1	− 0.4	+ 0.23	− 0.56	+ 0.51	− 0.9	+ 0.88	− 1.4

Now from the strains thus obtained for these two weights we can easily obtain all the others.

Thus, as the end reactions are inversely as the distances of the weight from the ends, the reaction at the left end due to P_2 will be *twice* that due to P_1. For P_3 three times that due to P_1. The strains will therefore be twice and three times those due to P_1, *until we arrive* at the weights P_1 and P_2 respectively. So also for P_6 the reaction at the *right* is twice that due to P_7, and the strains are therefore double *up to* the weight P_6. To the right, then, of P_6 the strains are twice those due to P_7, and to the *left* of P_6 they are six times those due to P_1. Take, for instance, P_5. The right reaction is ⅜ths of the apex load, and the right reaction of P_7 is ⅛th of that load. For P_5, then, the strains in all pieces to the right of that weight are 3 times those due to P_7. Again, the left reaction is for P_5 ⅝ths the apex load. But the left reaction for P_1 is ⅛th the same load. The strains then in all the pieces to the left of P_5 are 5 times those due to P_1. So for any other load. We can therefore form at once the following table:

---

* Strain diagrams in Fig. V., and also in Fig. VI., are, for obvious reasons, drawn to different scales.

356　　　　　　　　　NOTE TO CHAP. I.　　　　　　　[APPENDIX.

		P_1	P_2	P_3	P_4	P_5	P_6	P_7	Uniform Load.	Max. Comp. +	Max. Tens. −	Total Strains.
Bracing.	ab	−0.4	−0.8	−1.2	−1.6	−2.0	−2.3	−2.7	−8.25		−11.0	−19.25
	bc	+0.23	+0.5	+0.7	+0.9	+1.1	+1.4	−11.4	−4.9	+4.8	−11.4	−16.3
	cd	−0.56	−1.1	−1.7	−2.2	−2.8	−3.4	+4.8	−5.2	+4.8	−11.8	−17.0
	de	+0.51	+1.0	+1.5	+2.0	+2.6	−8.6	−4.3	−3.97	+7.6	−12.9	−16.9
	ef	−0.9	−1.8	−2.7	−3.6	−4.5	+4.7	+2.4	−4.8	+7.1	−13.5	−18.3
	fg	+0.88	+1.8	+2.6	+3.5	−6.9	−4.6	−2.3	−6.75	+8.8	−13.8	−20.5
	gh	−1.4	−2.8	−4.2	−5.6	+4.2	+2.8	+1.4	−4.2	+8.4	−14.0	−18.2

In the columns for P_1 and P_7 we put the strains already found by diagram. The strains for P_2 on the entire left half will be double those for P_1; for P_3 three times, and for P_4 four times those for P_1. We have therefore at once the columns for P_1, P_2, P_3 and P_4. Now for P_5 we see from the Fig. that the strains in diagonals ef and fg must both be tension. From the left end, then, as far as ef, the strains are 4 times those due to P_1, and from the right, as far as fg, 3 times those due to P_7. We thus obtain the column for P_5. In the same way for P_6, all above or to left of cd are 6 times P_1, all below or to right of de twice P_7. Thus we fill out the whole table. Adding now all the tensions and compressions in each piece, we obtain the maximum strains of each kind due to the live load, as given in the last two columns but one. Suppose now the dead load or weight of the girder itself to be ¾ths of the rolling or live load. We have only to take, then, ¾ths the sum of these last two columns and we have the strains due to uniform or dead load, as given in the fourth column from the right.

We can now easily obtain the total strains. Thus the tension in ab due to the live load only is 11 tons. The tension due to the dead load is 8.25 tons. Total greatest strain which can ever come upon ab, then, is 19.25 tons tension. No compression can ever come on this piece; it does not need, therefore, to be counterbraced. On the other hand, all the other diagonals, except perhaps cd, must be counterbraced, as the maximum compression due to the live load overbalances the constant tension of the dead. Had the dead load been taken much greater than the live, the diagonals might always have been in tension. Hence the appropriateness of this class of girder for long spans.

We see also from the table just what weights, and where placed, give the greatest strain of each kind in any piece.

8. Strains in the Flanges.—The method is precisely similar for the flanges. Thus we scale off from our diagrams—

	X a	X b	X d	X f	X h
P₁	+ 2.82	+ 3.08	+ 3.47	+ 4.11	+ 5.11
P₇	+ 19.7	+ 21.6	+ 10.4	+6.8	+ 5.1

	Y a	Y c	Y e	Y g
P₁	− 2.52	− 3.01	− 3.62	− 4.46
P₇	− 17.6	− 13.1	− 7.9	− 5.7

Tabulating these, we obtain the following table:

		P₁	P₂	P₃	P₄	P₅	P₆	P₇	Uniform live load.	Uniform dead load.	Total Strains.
Flanges	X a	+2.82	+ 5.6	+ 8.5	+11.3	+ 14.1	+ 16.9	+19.7	+ 78.9	+ 59.1	+ 138
	X b	+3.08	+ 6.2	+ 9.2	+12.3	+ 15.4	+18.5	+21.6	+ 86.3	+ 64.7	+ 151
	X d	+3.47	+ 6.9	+10.4	+13.9	+ 17.3	+20.8	+10.4	+ 83.2	+ 62.4	+ 145.6
	X f	+4.11	+ 8.2	+12.3	+16.4	+20.5	+ 13.7	+6.8	+ 82.0	+ 61.5	+ 143.5
	X h	+5.11	+10.2	+15.3	+20.4	+15.3	+ 10.2	+5.1	+ 81.6	+ 61.2	+ 142.8
	Y a	−2.52	−5.0	− 7.6	−10.1	− 12.6	− 15.1	−17.6	− 70.5	− 52.8	− 123.3
	Y c	−3.01	−6.0	− 9.0	−12.0	− 15.0	−18.1	−13.1	− 76.2	− 57.1	− 133.3
	Y e	−3.62	−7.2	−10.9	−14.5	−18.1	−15.9	−7.9	− 78.1	− 58.5	− 136.6
	Y g	−4.46	−8.9	−13.4	−17.8	−17.1	− 11.4	−5.7	− 78.8	− 59.1	− 137.9

This table is obtained precisely as before. Thus for P_6 the strains in **X** a and **X** b are multiples of P_1, while those in the other flanges are multiples of P_7. So also for **Y** c and **Y** e. We see at once that the greatest strains are for full load, since for all loads the upper flanges are always compressed and the lower extended.* The above is sufficient to illustrate fully the application of our method to bridges. It is evidently appli-

---

* A more convenient form of tabulation is to put the weights in the left vertical column and the pieces in the top horizontal line. The numbers can then be more easily added.

cable to any structure where the reactions are inversely as the distances from the end. The strains due to the first and last weights are all that we need in order to thoroughly solve any case of the kind. It is advisable, however, to construct a third diagram for an intermediate weight, in order to serve as a check upon the others.*

9. Method of Calculation by Moments.—We may illustrate here the method of calculation by moments referred to in Art. 14, Chap. I., a little more fully. Thus in the example above, Fig. V., suppose we wish the strains due to P_1. Reaction at left end is evidently $\frac{1}{8}$th of 10 tons = 1.25 tons. Conceive the lower flange $Y\,a$ cut. Rotation would evidently take place about apex a, and we have, therefore, strain in $Y\,a$ × its lever arm from apex $a = 1.25 \times 5$. The depth of truss, or lever arm of $Y a$, from apex a, is 2.58 feet. Hence we have

$$\frac{1.25 \times 5}{2.58} = \text{strain in } Y\,a = 2.42 \text{ tons.}$$

This strain is evidently, by reason of the direction in which the two portions of the truss would rotate about a, tension. In like manner, for upper flange $X\,b$, if we know the lever arm of this flange from the opposite apex, we can easily find the strain; for the diagram shows that $X\,b$, $b\,c$ and $Y\,c$ are in equilibrium with the reaction, and hence, if we take the point of moments at the intersection of the two pieces $b\,c$ and $Y\,c$, the moments of these pieces are zero, and we have remaining only the moment of the strain in $X\,b$ balanced by the moment of the reaction.

Again, if $X\,b$ and $Y\,c$ are thus found, and if these two strains, together with the reaction, are in equilibrium with the diagonal $b\,c$, we can find the strain in this diagonal by taking the apex d as a centre of moments. The moment of $Y\,c$ then is

---

* It may also be well to notice here that the practice of deducing in the tabulation the dead load from the live load strains is not strictly accurate, as the live load acts at the lower apices only [or at the upper apices only, if the bridge is under grade], while the dead load is distributed along both flanges, and acts at both upper and lower apices.

In every case, however, the greater portion of the dead load, say, for instance, $\frac{4}{5}$ds of the whole, owing to the track, platform, cross-girders, etc., acts at the same apices as the live load itself; and the error is in any case very slight, and practically of no account.

zero, and we have the moment of the strain in bc balanced by the moment of the reaction and the moment of $\mathbf{X} b$; the first causing compression, the second tension, and the difference then giving the resultant moment strain, which, divided by the lever arm of bc, gives the strain itself.

The method is easy of application, but, as we have already remarked, the determination of the lever arms for each case is frequently tedious. These may, however, be scaled off from the frame diagram with sufficient accuracy in practice.

As before, we need only the strains due to the first and last weight, and can then form our tabulation as above. This tabulation we can also check by finding the strains due to uniform load independently, and seeing whether it agrees with the sum of the separate apex weight strains.

Thus, for *all* the weights acting, suppose we wish the strain in $\mathbf{Y} g$. The lever arm of $\mathbf{Y} g$ is 9.85 feet. We have then reaction $= 35$ tons, multiplied by 35 feet $= 1225$. This must be diminished by $\mathbf{P}_7 \times 25 = 250$, $\mathbf{P}_6 \times 15 = 150$, and $\mathbf{P}_5 \times 5 = 50$. We have then $1225 - 450 = 775$, which, divided by 9.85, gives 78.7 tons tension in $\mathbf{Y} g$, agreeing with our tabulation above.

For the method of calculation by resolution of forces, see Art. 16 of this Appendix.

10. Girder with Straight Flanges.—In such a case, as the lever arms are at once known and are constant, the above method is of very easy application. In this case the strains in the diagonals are best found by multiplying the shear by the secant of the inclination of the diagonal with the vertical.* Thus, if this angle is 45°, we have simply to multiply the shear at any point by 1.4142, and we have at once the strain in the

---

* This is but a particular result of the general method of moments. Thus, for any diagonal, as ab (Fig. VII.), according to our rule, we take the centre of moments at the intersection of the two other sides cut by a section through the truss, viz., the flanges. But these two sides are here *parallel*, hence their intersection is at an infinite distance. The lever arm of ab is then $\infty \times \cos \phi$, ϕ being the angle with the vertical. If the weight $\mathbf{P}_1$ acts, we have then, calling the reaction $\mathbf{R}$, $\mathbf{R} \times \infty - \mathbf{P} \infty = \mathbf{S} \cos \phi \times \infty$, where $\mathbf{S}$ is the strain in ab. This can be put $(\mathbf{R} - \mathbf{P}) \infty = \mathbf{S} \cos \phi \times \infty$, hence $\mathbf{S} = \dfrac{(\mathbf{R} - \mathbf{P}) \infty}{\cos \phi \times \infty}$. But $\mathbf{R} - \mathbf{P}$ is the *shear* at b, $\dfrac{\infty}{\cos \phi \times \infty} = \dfrac{1}{\cos \phi} = \sec \phi$; hence we have only to multiply the shear by the secant.

diagonal at this point. The *shear* is always in such cases the reaction at the end minus the weights between that end and the apex in question.

The flanges are easily obtained by moments, as above.

The following points need attention, however. First, if there are two or more systems of diagonals, as represented in Pl. 2, Fig. VII., by the full and dotted diagonals (omitting the upright lines), we must find the strains for each system separately, and then add them together. Thus, if the strains found in ac and ce, etc., for one system, are 50 and 60 tons, and those in df and fg, for the other, are 40 and 70 tons, we have, when the two systems are combined, $dc = ac + df = 50 + 40 = 90$, $cf = df + ce = 40 + 60 = 100$, $fe = ce + fg = 60 + 70 = 130$, and so on. This holds true, of course, whether the strains are obtained by calculation or diagram. Thus, for a lattice girder, we calculate or diagram *each system by itself*, and then the strain in any flange, when the two are combined, is equal to the sum of the strains on that flange due to each system of triangulation which includes it.

There is another point to be observed in connection with the system known as the *Howe* or *Pratt Truss*. Inserting the dotted verticals into our Fig., we have this system of square panelling. Let us suppose that the *diagonals take tension only*, and the verticals *compression* only.

Now for a weight at apex 9 of 10 tons, we have a right reaction of 1 ton, which, running through the system, causes strain in the diagonal of $f P_4$. For the flange D, then, our point of moments is at f, and if the height of truss is equal to panel length, viz., 10 feet, we have the strain in $D = \dfrac{1 \times 50}{10} = 5$ tons, for P_9. In the same way for P_8, we have for D 10 tons; for P_7, 15 tons; for P_6, 20 tons; for P_5, 25 tons. For P_4, on the other hand, we have a left reaction of 4 tons, which causes strain in diagonal ek, and for this weight and all succeeding weights our point of moments for D is then at e. We have then P_4 $\dfrac{4 \times 60}{10} = 24$ tons; for P_3, 18 tons; for P_2, 12 tons; and for P_1, 6 tons.

For all these weights, then, acting together, we have 135 tons strain in D.

But for all the weights acting together, it is evident that only all the braces sloping each way from the centre are strained. Hence ek is *not* strained, and our point of moments is for **D** *always* at f. Thus for *total load* we have strain in

$$D = \frac{45 \times 50 - 10 \times 40 - 10 \times 30 - 10 \times 20 - 10 \times 10}{10} = 125 \text{ tons},$$

whereas we found by addition of the several weights 135 tons.

There is thus an ambiguity in this class of bracing as to the way in which the strains may go. Two symmetrical weights, as 9 and 1, may either go left and right directly to the abutments or a portion of each go towards the centre. The intermediate diagonals may be either all strained or not strained at all. The strains may go partly in one way or partly in the other. We should then not rely on our summation of the separate weights, but always check them by calculation or diagram for the total load also, and take the greatest strain. Practically, for long spans, it is very rare that the difference is of any importance.

In *diagraming* by our method such a system of bracing as the above, we should consider but one series of braces, viz., *those strained by the uniform load* alone. Thus, for our Fig. and loads on the lower apices, we should take only the diagonals parallel to fh on the left of centre, and $f\mathbf{P}_4$ on the right. If, on the other hand, the verticals are ties and the diagonals struts, we should retain only those parallel to ck on the left, and those parallel to ke on the right of centre. The others are to be omitted. Then, the tabulation being formed, if in any diagonal a strain may occur of reverse character to that which it is intended to resist, a *counterbrace* must be inserted in this panel to take this reverse strain.

As in our examples we have taken always a triangular system of bracing, it is important that the reader clearly understand the method to be pursued in other forms. For the rectangular system of bracing generally, *the point where for uniform load the shear is zero* is the point from which the braces must slope both ways. The other diagonals, or the *counterbraces*, are then omitted in both calculation and diagram, and replaced from the tabulation when necessary to replace a strain of the reverse character to that which the *braces* are intended to sustain.

Attention to the above points will enable us to both calculate and diagram with ease and accuracy any form of truss which occurs in engineering practice.

11. In the bow-string girder represented in Fig V. it is evident that the bottom flange serves merely to resist the thrust of the bow and keep it from spreading. It *adds nothing* to the supporting power of the combination. We might remove it entirely and replace it by abutments which would equally well sustain this thrust, and if we then introduced a horizontal flange at crown, and inserted diagonals between for stiffness, we should have the form of braced arch given in Chap. I., Fig. 5 (c). If, however, we should resist the thrust of the bow by an *inverted* arc, it would answer the same purpose as the bottom flange, and we should, in addition, *double the supporting power*.

We have illustrated this in Fig. VI.

The span is the same as before. The lower apices only are supposed to be loaded, for comparison. [Properly, we should have distributed the load over both upper and lower apices.] The rise of each arc is *one-half* as great as before, or 5 ft. only, thus making the total depth the same as in the preceding case.

By means of two strain diagrams, we find the strains due to P_1 and P_7. Thus:

	Xa	Xb	Xd	Xf	Xh	Ya	Yc	Ye	Yg
P_1	+2.3	+2.6	+3.2	+3.7	+5.0	−2.25	−2.87	−3.48	−4.32
P_7	+16.1	+17.8	+9.1	+5.7	+4.5	−15.75	−12.35	−7.3	−4.8

Then, precisely as in the preceding Art., we can fill out our table of strains. This the reader can now easily do for himself. We thus find, for a uniform dead load ⅜ths the live load, the total maximum strains below.

Xa	Xb	Xd	Xf	Xh	Ya	Yc	Ye	Yg
+112.7	+126.7	+133.5	+127.	+134.7	−110.1	−126.7	−129.3	−125.6

Comparing these with the corresponding strains for the bow-string, we find that they are very much less in every piece. In fact, there is a total gain of over 10 per cent., and that, too, notwithstanding that the rise of each arc is *only half* that in the first case. Had we taken a double depth, the saving would have been very great, and as in this case also, for a long span and relatively large dead load, the diagonals would always be in tension, the increased length of these last would be no disadvantage.

12. The above construction is worthy of the careful consideration of the bridge builder. It peculiarly recommends itself for long spans, and has several important advantages possessed by no other form of truss. For long spans the strains in the flanges are *nearly uniform*. The diagonals are less strained than in any other system, and are *always* in *tension*. Every member acts to *support*, as well as to strengthen. The height is everywhere proportional to the maximum moment of the exterior forces. *The load is distributed along the neutral axis*, thus securing the maximum of rigidity ; while the *neutral axis itself passes through the points of support.*

This construction is known in Germany, from the name of its inventor, as *Pauli's Truss*. Upon this system are the double track bridge over the *Isar* at *Grossheselohe*, 2 spans of 170.6 ft. ; a large number of smaller bridges, such as one over the *Rodach*, 109 ft. span ; over the *Main* in *Schweinfurt*, 116.4 ft. span ; and especially one over the *Rhine* at *Mayence*, of 32 *spans*, 4 of 345 ft., 6 of 116 ft., 20 of 50 ft., and 2 of 82 ft. ; all upon the same system.

In England, we might notice the famous bridge over the *Tamar* at *Saltash*, near *Plymouth*, whose two principal spans are 455 ft., which is also constructed upon this system. Finally, we may mention the bridge over the *Elbe*, near *Hamburg*, the three principal spans of which are 325 ft. each.

In this latter structure both the upper and lower members are *braced* or *ribbed arches*, of a constant depth of about 10 ft., a combination which, for long spans, seems most excellent. A single arch alone, similar, for example, to the steel arch over the *Mississippi*, by *Capt. Eads*, would have required heavy abutments.

The same arch inverted would have required equally heavy anchorages. The combination does away with both. The thrust of the upright arch is opposed by the pull of the inverted one, all the advantages of Pauli's system are obtained, and *there are no temperature strains* such as occur in the single arch, while the bracing is reduced to a minimum. At the same time all the rigidity due to the arch is obtained.*

13. In the construction of the diagrams, care should be exercised in the selection of the scales, that the frame diagram may be large enough to secure the desired accuracy. Lines should be drawn very fine with a hard, sharp-pointed pencil, so as to be scarcely discernable, and their intersections accurately marked by needle point.

With an accurate scale and good instruments, strains can be taken off in nearly every practical case to hundredths of a ton

---

* Compare *Long and Short Span Railway Bridges*, by *John A. Roebling*, C.E. —In this work, Mr. Roebling proposes a system in principle essentially the same as the above, to which he gives the name of "*Parabolic Truss.*" He, however, constructs the arch of channel irons bolted to the sides of a straight truss, the sole office of which is to give rigidity to the system. Also, claiming that iron in the shape of wire will safely sustain three times as much as in the shape of bars or rods, he introduces a wire cable in place of the inverted braced arch.

It will thus be seen that for rigidity the system is wholly dependent upon extraneous members, such as the auxiliary truss and the *tower stays*, which are liberally introduced. By dividing the material composing the upright arch into *two* portions, bracing between them, and thus forming a *braced arch* similar to Capt. Eads, the stays and stiffening truss might be entirely dispensed with, the construction greatly simplified in the number of its members, and the bracing reduced to a minimum. If, also, as claimed by *Capt. Eads*, the conditions for *cast steel* are just the reverse of iron, and it is most advantageous to use it in compression, then it seems that such a modification of Mr. Roebling's design with wire cable and a *cast-steel braced arch* would better sustain the thesis with which his work, above quoted, opens, viz. : that "*the greatest economy in bridging is only to be obtained by a judicious application of the Parabolic Truss.*"

Such a combination of the suspension and upright arch would seem to avoid the principal objections urged against each separately. The anchorages and abutments are dispensed with, the greatest rigidity is secured with the minimum of bracing, and the material is used in the most advantageous way. In addition to the advantages of *Pauli's system* being secured, we have the ease of erection of the suspension system combined with the rigidity of the arch. The system is self-balancing, and practically *unaffected by changes of temperature.*

For the practical details of construction of such a system, the reader can with profit consult Mr. Roebling's work, above quoted. They will be found to be neither expensive nor difficult of execution.

with perfect accuracy. The use of parallel rulers is not to be recommended. The **T** square, triangle and drawing-board are far preferable. It should be remembered, finally, that careful habits of manipulation, while they give constantly increased skill and more accurate results, affect in no degree the rapidity and ease with which those results are obtained.

NOTE TO CHAPTER II.

14. The reader will observe that in Chapter I. we had given forces acting at certain points of a given frame, and we found by simple resolution of forces the strains in the pieces of that frame. In Chapter II. we have given forces acting in certain directions, and having *assumed* the strains, we find the equilibrium polygon or *frame*, which, having its angles on these force directions, and having these strains, will hold the given forces in equilibrium. Thus in Figs. 12 (*b*) and (*c*), Pl. III., of the text, by choosing a pole and drawing lines to the forces in the force polygon (*a*), we virtually assume the strains which are to act upon our frame. Then lines parallel to these strains in (*b*), forming a polygon whose angles are upon the forces, must give us the frame which holds these forces in equilibrium, provided we close the polygon by a line and apply at the ends forces which balance each other horizontally, and whose components parallel to the resultant of the forces balance the forces.

Thus the polygon *m a b c d e n m* is a *frame* along whose sides the forces S_0 S_1, etc., act, and whose reactions at the supports *m* and *n* must then be *a o* and *5 a*, as given in (*a*).

This *frame*—keeping the same pole, that is, the same strains—we may put anywhere in the plane, its angles being always on the forces, and its sides always respectively parallel, though varying in length according to the position assumed.

We might also have assumed different strains, that is, taken a different pole, and constructed a different frame; but evidently the end reactions will not be altered, and will be always equal to *a* 0 and 5 *a*, as given in (*a*).

The peculiarities of the frame thus obtained are, as we see further on, that its end sides always intersect upon the resultant of the forces; its depth is always proportional (for parallel forces) to the moment at any point; its area to the moment of inertia of the forces; while, finally, in a loaded beam the *deflection curve* itself is but a polygon or frame of this character, when the curve of loading follows the law of the moments in the beam.

It is upon this polygon and its properties that the entire system of Graphical Statics is based.

NOTE TO CHAPTER V., ART. 51.

15. In Fig. VIII. (Appendix) we have given the construction referred to in Art. 51 of the text for a system of loads of given intensities. The span $s_0\,s_0$ is supposed to shift to $s_1\,s_1$, $s_2\,s_2$, etc., and a certain cross-section k_0 to shift with it to k_1, k_2, etc. The intersections of the respective closing lines with verticals through k_0, k_1, k_2, etc., gives us a curve between which and the polygon the greatest ordinate gives the maximum moment for the assumed cross-section. The place of this ordinate is the position of the cross-section from which we determine the ends of the span, and thus have its position with reference to the loading when the moment in k is the greatest possible.

Thus if this greatest ordinate is at the angle VIII. in the Fig., the weight P_8 must rest upon the cross-section. The distance then from P_8 to the left end of span s, is the distance from s_0 to k_0 on left, and to right end of span s, is the distance from k_0 to s_0 on right.

The ends s and s being thus found, perpendiculars through them determine the closing line **L**, and the parallel to this in the force polygon gives the end reactions **L** 0 and 20 **L** for the position of span which makes moment at k a maximum.

NOTE TO CHAPTER XII., ART. 124.

16. In Arts. 120 and 121 we have given the formulæ and principles necessary for the complete solution of the pivot span. We propose here to illustrate more fully their application by a simple example.

Fig. IX. represents such a structure. The two outer spans **A B** = **C D** = 40 ft. The central or turn-table span, **B C** = 20 ft. Centre height at **B** and **C** = 10 ft. End height = 6 ft. Panel length, 10 ft.; each apex live load, 10 tons, or 1 ton per foot. Dead load, *half* as much. Two systems of triangulation, as shown in the Fig.

Our proportions are taken for the sake of illustration merely, and not as an example of actual practice. All the points to be observed are, however, illustrated as well as by a much longer span, and more usual proportions.

It is to be observed that the end verticals are compression members *only*, and cannot take tension. This is necessary to prevent ambiguity as to the way in which the strains go. A negative reaction might otherwise cause tension in 1 2, and compression in **F**, or tension in 1 5, compression in 5 6, and tension in **A**. If 1 5 cannot take tension, we have but one course for the strains, and the problem is determinate.

We also, for similar reasons, construct the centre span so that the diagonals take *tension only*, and the verticals compression only. These points as to construction being settled, let us proceed, first, to determine the reactions.

1st. REACTIONS.

We shall consider the case of the "*Tipper*," or secondary central span only [Art. 120], as this case most nearly approaches the true state of things. The method of procedure for four *fixed* supports is precisely similar, only taking the formulæ for that case from Art. 122.

The less the span **B C**, the nearer the case approaches to three fixed supports; and when the distance **B C** is zero, n is zero, and our formulæ are the same as for beam over three supports.

For a load in the left span distant a from **A**, these formulæ are as follows [Art. 120]:

$$R_A = \frac{P}{2H}\left[2H - (10 + 15n + 3n^2)k + (2+n)k^3\right],$$

$$R_B = R_C = \frac{P}{2H}\left[(6 + 9n + 3n^2)k - (2+n)k^3\right],$$

$$R_D = \frac{P}{2H}\left[(2+n)k^3 - (2 + 3n + 3n^2)k\right],$$

in which $k = \frac{a}{l}$, $l = AB = CD$, $nl = BC$ and $H = 4 + 8n + 3n^2$.

We have first to put these formulæ into the most convenient shape for use in the particular case under consideration. Thus in this case $l = 40$, $nl = 20$; hence $n = \frac{1}{2}$ and $H = \frac{35}{4}$, and $k = \frac{a}{40}$, where a has the successive values of 10, 20, 30, 40 for P_1, P_2, P_3, P_4. k is therefore successively $\frac{1}{4}, \frac{2}{4}, \frac{3}{4}$ and $\frac{4}{4}$.

Our equations for reactions are then, after reducing,

$$R_A = \frac{5}{35}\left[70 - 73k + 10k^3\right],$$

$$R_B = R_C = \frac{5}{35}\left[45k - 10k^3\right],$$

$$R_D = \frac{5}{35}\left[10k^3 - 17k\right].$$

Now, as we may notice, the denominator of k is always 4, of k^3 always 64; the numerator only changing according to the position of the weight. These equations can then be written

$$R_A = \frac{1}{224}\left[2240 - 584a + 5a^3\right],$$

$$R_B = R_C = \frac{5}{224}\left[72a - a^3\right],$$

$$R_D = \frac{1}{224}\left[5a^3 - 136a\right],$$

where a has the values 1, 2, 3 for P_1, P_2, P_3, etc.

These, then, are the practical formulæ for this case, and from them we can easily find the reactions for the apex loads of 10 tons each.

Thus, for P_1 make $a = 1$, and we have

$R_A = 7.415$, $R_B = R_C = 1.58$, $R_D = -0.585$.

For P_2 make $a = 2$, and

$R_A = 4.964$, $R_B = R_C = 3.035$, $R_D = -1.035$.

For P_3 make $a = 3$, and

$R_A = 2.78$, $R_B = R_C = 4.22$, $R_D = -1.22$.

For P_4 make $a = 4$, and

$R_A = 1$, $R_B = R_C = 5$, $R_D = -1$.

Loads upon the centre of the span BC acting, that is, at apex 10, give no reactions, but are supported directly by the turn-table. Hence, for P_5; R_A, R_B, R_C and R_D are zero. For the first load, P_7 to the right of C, the reactions at A and B are the same as for P_3 at D and C, already found. For the next load, P_8, the reactions at A and B are the same as for P_2 at D and C, already found. For P_9, the same as for P_1. For P_6, as for P_4, etc.

We thus have the reactions at A and B due to every individual apex load, and can now proceed to find the strains.

Our formulæ, it will be observed, thus become very simple and easy of application for any particular case.

2d. FLANGES—BRIDGE SHUT.

Let us first find the strains in the flanges. We have only to apply the method of moments, and the work is so simple that an example or two will suffice.

We repeat again the rule. Conceive a section cutting only three strained pieces. Take the intersection of two of these as the centre of moments for finding the strain in the third. The moment of the strain in this last about this point must be equal to the algebraic sum of the moments of *all* the forces acting between the section and one end. Take P_1 for example. Its upward reaction at A is 7.415. [A negative reaction acts *down*. Thus, for P_7 above, the reaction at A is, from our formulæ, -1.22. The minus sign indicates that the reaction is down, and that, neglecting the dead load, the girder must be *held down* to the support A. If the reader will draw roughly the curve of deflection, he will see that this is so.]

Conceive a section through the girder at, say, the centre of flange **A**. It cuts 4 pieces, but, since the weight P_1 acts only *through its own system* of diagonals, only three are strained. The point of moment for **A** is then at 6, the intersection of the other two strained pieces. The strain, then, in **A** × by its lever arm $= 7.415 \times 10$. The lever arm of **A** is 6.965; hence

$$\mathbf{A} \times 6.965 = 7.415 \times 10,$$

or $\quad\quad \mathbf{A} = +10.64$ tons *compression*,

because the upward reaction acting with 6 as a centre of rotation tends to compress **A**.

This strain evidently acts through both **A** and **B**, since both these flanges are included by the two diagonals of the system for P_1; hence also, $\mathbf{B} = +10.64$ tons.

For flanges **C** and **D**, since 7 8 is the strained diagonal, 8 is the centre of moments. The same reaction acts now with the lever arm 30 to cause compression, and P_1 acts with the lever arm 20 to cause tension. We have then

$$\mathbf{C} \times 8.955 = +7.415 \times 30 - 10 \times 20,$$

or $\quad\quad \mathbf{C} = \mathbf{D} = 2.5$ tons compression.

Now we come to the centre span, and must carefully observe the following points. Since **D** has been found to be compression for P_1, we see at once that the whole upper flange for the span **A B** is for this weight in compression. Diagonal 8 9 is therefore in tension. Were there no vertical strut at **B**, this would cause compression in 9 10. But brace 9 10 *cannot* by construction take compression. The strained pieces cut by a section through **E** are then **E**, **B** 11 and **K**, which give us the centre of moments at **B** for strain in **E**. Observe, that were it not for the vertical, we should have had 10 for the centre of moments; or, *with* the vertical, had **D** been found *tension*, 8 9 would have been compression; there would then have been no strain in the vertical, that being incapable of tension, and diagonal 9 10 would have been strained, thus giving us also 10 for the centre of moments. Attention to the above is necessary in order to properly pass from the span **A B** into the middle span.

We have then for strain in **E**

$$\mathbf{E} \times 10 = 7.415 \times 40 - 10 \times 30, \text{ or } \mathbf{E} = -0.34,$$

or in tension, as indicated by the sign, since the moment of P_1 overbalances that of the reaction.

[*Note.*—The different *lever arms* are easily obtained from the known dimensions of the truss. We have considered it unnecessary to detail how they are to be found. They may either be measured to scale from the frame or computed trigonometrically.]

The lower flanges are found in similar manner.

Thus, strain in **F** is zero, since it passes through the point of moments.

For **G** and **H**, we have

$$\mathbf{G} \times 8 = -7.415 \times 20 + 10 \times 10, \text{ or } \mathbf{G} = -6.04 \text{ tension.}$$

In like manner, for **I**,

$$\mathbf{I} \times 10 = -7.415 \times 40 + 10 \times 30 = +0.34.$$

For **K**, for similar reasons as above for **E**, we have centre at 11, and therefore the reaction at **B** *also* enters into the equation of moments, and

$$\mathbf{K} \times 10 = -7.415 \times 50 + 10 \times 40 - 1.58 \times 10, \text{ or } \mathbf{K} = +1.34.$$

We have then, finally, for the strains in the flanges due to P_1

	A	B	C	D	E	F	G	H	I	K
P_1	+10.04	+10.04	+2.5	+2.5	−0.34	0	−6.04	−6.04	+0.34	+1.34

In a precisely similar manner we find the strains due to P_2, P_3 and P_4.

We have only to observe that for P_7, the first weight to the right of **C** in the other span, the reaction at **A** is negative and equal to the reaction of P_3 at **D**, already found, or −1.22.

Now as we suppose the end **A** *bolted down*, this reaction acts as a weight of 1.22 tons suspended from the end. So for the reactions of P_8 and P_9, viz., −1.035 and −0.584. These reactions, moreover, must all take effect through diagonal 12 and flange **F**, as the end vertical cannot take tension.

Finding then the strains due to each of the other weights, we can, finally, tabulate our results as on next page:

STRAINS IN FLANGES—LIVE LOAD—BRIDGE SHUT.

	A	B	C	D	E	F	G	H	I	K
P_1	+10.64	+10.64	+2.5	+2.5	−0.34	0	−6.04	−6.04	+0.34	−1.34
P_2	0	+12.47	+12.47	−0.14	−0.14	−7.1	−7.1	−5.32	−5.52	+2.14
P_3	+3.9	+3.9	+9.31	+9.31	+1.12	0	−6.95	−6.95	−1.12	+1.88
P_4	0	+2.5	+2.5	+4.0	+4.0	−1.42	−1.42	−3.33	−3.33	0
P_5	0	0	0	0	0	0	0	0	0	0
P_6	0	−2.5	−2.5	−4.0	−4.0	+1.42	+1.42	+3.33	+3.33	0
P_7	0	−3.1	−3.1	−4.88	−4.88	+1.74	+1.74	+4.06	+4.06	+1.88
P_8	0	−2.6	−2.6	−4.4	−4.14	+1.46	+1.46	+3.45	+3.45	+2.14
P_9	0	−1.4	−1.4	−2.34	−2.34	+0.83	+0.83	+1.95	+1.95	+1.34
Total Strains	+14.54	+29.51	+26.78	+15.81	+5.12	+5.45	+5.45	+12.79	+13.13	+9.38
		−9.6	−9.6	−15.76	−15.84	−8.52	−21.51	−21.64	−9.77	−1.34

In the two horizontal lines at bottom, we have the total strains of each kind caused by the live load.

3*d*. FLANGES—BRIDGE OPEN—DEAD LOAD.

We have next to find the strains due to the dead load when the span is open.

We have then 5 tons at each apex, except the ends, where we have $P_0 = 2.5$ tons.

These strains are easily found by moments as above, and we have then the following table:

	A	B	C	D	E	F	G	H	I	K
P_0	0	−6.2	−6.2	−10.1	−10.1	+3.57	+3.57	+8.3	+8.3	+10
P_1	0	0	−11.1	−11.1	−15.0	0	+6.1	+6.1	+15.0	+15.0
P_2	0	0	0	−10.1	−10.0	0	0	+5.5	+5.5	+10.0
P_3	0	0	0	0	−5.0	0	0	0	+5.0	+5.0
Total Strains	0	−6.2	−17.4	−31.2	−40.0	+3.6	+9.7	+20.1	+33.8	+40.0

If now, as should be the case, we suppose the centre supports *raised* above the level of the ends, so that the ends just bear, then these strains above act *even when the bridge is shut*. As we have already seen in Art. 131, our formulæ for the reactions are not affected by this state of things. The strains due to live load will then be increased by those above, and we thus have for the total maximum strains which can ever occur,

A	B	C	D	E	F	G	H	I	K
+14.54	+29.51	+26.78	+15.81	+5.12	+9.05	+15.15	+32.89	+46.93	+49.38
	−15.8	−27.0	−46.96	−55.84	−8.52		−21.51	−21.64	−9.77	−1.34

Of course, for this condition of things the ends must *always* be bolted down.

It is sometimes customary to raise the ends by an apparatus for that purpose, after closing the draw, until the proper proportion of the dead load takes effect also as a positive reaction.

We can easily find the strains in this case also by adding the numbers in the last horizontal line of our table for bridge shut, with their proper signs, and taking half the results for a new line for dead load strains. The resulting strains can then be found precisely as in the table of Art. 8 (Appendix). We must also find the strains for bridge open as above, and then take the greatest strains of each kind from these two tables.

In this case the strains would be differently distributed. Flange E will be always in tension, A and K always in compression; the compression in B C and D will be somewhat greater than above, and the tension in the same flanges less. The reader can easily deduce the strains for this case from the two preceding tables.

If the truss *may* act as a girder over four *fixed* supports, we should, in order to be certain of the maximum strains, make the calculation for this case also, using the formulæ of Art. 122. This is unnecessary, however, if the supports B and C can never sink far enough to strike the turn-table, or be impeded in their motion.

4th. STRAINS IN THE DIAGONALS.

We may find the strains in the diagonals also for each weight separately, both for bridge open and shut; and a precisely similar method of tabulation will give the strains.

It will here be found preferable to make a series of *diagrams*, as illustrated in Fig. 86, Art. 124, for each weight and its own system of triangulation. We obtain thus the diagonal strains, and at the same time check the results obtained for the flanges above.

If we wish to calculate the diagonals, it will be better to find the *resultant* shear acting upon the diagonal, and multiply it by the secant of the angle the diagonal makes with the vertical.

We can also, if we wish, apply the method of moments. Thus, if we determine the point of intersection in the present case of the inclined upper flange with the horizontal lower flange, this point will be a common centre of moments for the diagonals. The lever arms of the diagonals with reference to this point must next be determined, and then we are ready.

This point above for centre of moments is easily found; thus

$$4 : 40 :: 10 : 100.$$

It is therefore 60 ft. to the left of **A**, or 100 ft. left of **B**. Take now any diagonal, as 3 4. Its angle with the horizontal is very nearly 42°, and with the vertical 48°. Its lever arm is then $80 \sin 42° = 53.5$, and sec. of angle with vertical is 1.49.

Now take the weight P_2. Its upward reaction at **A** is 4.964, P_2 being 10. We have then

[str. in 3 4] × 53.5 = 10 × 80 − 4.964 × 60 = + 502.16.

The resultant rotation is then positive, or from left to right. The point P_2 then sinks and 4 rises, and 3 4 is in tension and

$$= \frac{502.16}{53.5} = -9.38 \text{ tons.}$$

This is sufficient to illustrate the method.

For the first method referred to above, viz., that by resultant shear, the following points are to be observed:

When a piece slopes towards the nearest support, we say it is *sloped as a strut*, whatever the real strain in it may be. When it slopes *away* from the nearest support, it is *sloped as a tie*.

The *simple shear* is the reaction at the support minus the weights between any point and that support.

If any three strained pieces are cut by a section through the structure, the strains in these pieces are in equilibrium with the simple shear at this section. Hence the algebraic sum of the vertical components of these pieces must be equal and opposite to the *shear* itself.

In order to add these vertical components with proper signs, we must remember that if a flange is in tension and sloped as a strut, or in compression and sloped as a tie, we *add* the vertical component of the strain in it to the simple shear already obtained. If in compression and sloped as a strut, or tension and sloped as a tie, we *subtract*.

---

* In general, if we take compression as plus and tension as minus, and then measure the angle θ made by the piece with the vertical *always* from that vertical

The *resultant shear* thus obtained then, multiplied by the secant of the angle with vertical, gives the strain in diagonal.

If the sign of the result is negative (−), it shows that the strain on the diagonal is contrary to that indicated by its slope.

To illustrate, let us again take the weight P_2 and consider diagonal 3 4.

The simple shear at apex 4 is $4.964 - 10 = -5.036$. The strain in C for P_2 we have found to be compression, and equal to $+ 12.47$. It is sloped as a strut, and its vertical component is therefore to be *subtracted* from the shear above. Since its angle is nearly 5° 43′ with the horizontal, this vertical component is $12.47 \times \sin 5° 43′ = 1.24$

Since H is in this case horizontal, it has no vertical component. The *resultant* shear is then $-5.036 - 1.24 = -6.276$.

As the secant of the angle of 3 4 with the vertical is 1.49, we have for the strain in 3 4, $-6.276 \times 1.49 = -9.35$.

This result being minus, and 3 4 *being sloped as a strut*, the strain is 9.35 tons *tension*, agreeing closely with the value found above by moments.

The above method is preferable to the method by moments for the diagonals, as we have only to determine the secants for the verticals and the sines for the flanges, which is in most cases easier than to find the lever arms for the diagonals and the points of intersection of the upper and lower flanges in each panel. It is, like the method of moments, of general application to any framed structure whose outer forces are known.

The method of diagram in Art. 124 will be found preferable to both.

It is unnecessary to pursue our example further. With the mutual checks of the two methods of calculation explained above, as well as the diagrams, correct results cannot fail to be obtained. The diagrams should always be made first, as they settle by mere inspection many points which may at first cause trouble—such as whether the shear in a piece is subtractive or not according to our rule, the character of the strains in different pieces, etc. It is well to indicate on the diagrams compressive strains by double or heavy lines.

---

from right to left, thus ↺ ; we shall have for any case *vertical component* = *strain* × cos. θ. The cosine will change its sign according to the quadrant it is in, according to the above rule, and the sign of the vertical component will take care of itself.

NOTE TO ART. 128, CHAPTER XII.

17. We wish here to call more particular attention to the *relative economy* of the continuous as compared with the simple girder. This, we think, is greater than is generally supposed. It may reach from 18 to 25, and *even as high as* 50 *per cent*.

Take the example worked out in Art. 128, Fig. 88. We have obtained the maximum strains in that Art. upon every piece.

We give them below, compared with the strains in the same pieces for a simple girder of same dimensions and load:

```
              Aa      Ac      Ae      Ag      Ak      Bb      Bd      Bf      Bh
Continuous..-203.5   +63.6   +115.3  +63.6  −203.5   +89.3  −115.9  −115.9   +89.3
Simple......  0     +180    +240    +180     0       −90    −210    −210    −90

              ab      bc      cd      de      ef      fg      gh      hk
Continuous..+189.3  −109.9  +109.9  +45.5   +45.5   +109.9  +109.9  +189.3
Simple......+127.3  −127.3  +56.5   +56.5   −56.5   +56.5  −127.3  +127.3
```

It will be seen at once that there is a saving in the flanges—about 11 per cent. in all—but the bracing is heavier, giving little or no saving. The span is too short to properly represent the relative economy of the two systems.

If we take a truss such as represented in Pl. 2, Fig. VII., Appendix, by the full lines only, omitting the dotted verticals and diagonals—height 6 ft., span 50 ft., panel length 10 ft., dead load 5 tons per panel, live load 7 tons per panel—and calculate the strains in the pieces for a simple girder, and then as a continuous girder of two spans and three spans, we have the following results:

	1 span strain in tons.	2 spans	3 spans end.	3 spans middle.
Diagonals........	205.2	231.6	229.8	239.4
Lower Chord....	200.	168.7	152.9	122.6
Upper Chord....	200.	158.7	159.9	135.2
Total........	605.2	559.	542.6	497.2
Per cent. saving..	.	8 per cent.	10 per cent.	18 per cent.

We have then, in the first case, a saving of 8 per cent., in the second, 10, and in the third, 18 per cent. over a simple girder. Quite a notable saving, although the spans are very short, and although, in the first two cases (the spans being end spans), we do not obtain the full advantages of continuity. If, then, instead of three simple girders of the above dimensions, we should construct the girder continuous over the piers, we should save in strain, and hence in material, 10 per cent. in each end span, and 18 per cent., or nearly twice as much, in the centre span.

The advantage of continuity is rendered still more apparent by taking a longer span. Thus for a girder of 200 ft., height 20 ft.—10 panels, and double system of triangulation, similar to Fig. VII.—for a live load of 20 tons per panel, and dead load of 10 tons—we have the following results:

	One span.	Two spans.	Five spans. centre.
Bracing	1398.6	1428.2	1596.2
Lower Chord	2400.	1793.2	1395.7
Upper Chord	2550.	1981.6	1622.6
Total	6348.6	5203.0	4614.5
Per cent. saving		18 per cent.	27 per cent.

That is, we have a saving of 18 per cent. instead of only 8 per cent., as before, for two spans, and of 27 per cent. for the centre span of five spans. For three spans, then, of this length we should save 18 per cent. on the end, and at least 20 per cent. on the centre span.

If we suppose the same girder as above *fastened* or fixed horizontally at the ends, we shall have the case of a middle span in a very great number, and may expect to find the greatest saving possible for this length.

The formulæ of Chapter XIII., as also the simple graphical method for this case, given in Chapter XII., Art. 114 [Fig. 80], enable us to solve this case easily. The reader will find, on making the calculation, the following strains:

	Strain in tons.
Diagonals	1279.2
Upper Chord	940.2
Lower Chord	965.4
Total	3184.8
Per cent. saving	49.8

The above will serve to illustrate the point in question quite as well, perhaps, as an extended theoretical discussion. We see that the saving increases rapidly with the length of the span, and may easily rise as high as 30 or 40 per cent., while in some cases even 50 per cent. may be realized.

THE DISADVANTAGES OF THE CONTINUOUS GIRDER ARE:

1st. The fact that the various pieces, especially the chords, undergo strains of opposite character.

This, in wrought-iron structures, we venture to think of little importance. The extra work and cost of chords and chord connections necessary to secure the flanges against both compressive and tensile strain, can hardly amount to 10, 18, 30, or even 50 per cent. of the cost of girder!

2d. Difficulty of calculation.

We have, we trust, in what precedes, and in Chapter XIII., succeeded in removing this objection.

The opinion is widespread among engineers that the determination of strains in the continuous girder is impracticable and involved in mystery. No opinion could well be more unfounded. The accurate and complete calculation for all possible loading, live or dead, is precisely similar to and offers no more difficulty than the simple girder itself.

The formulæ for moments and shears are, as we have seen, simple and easy of application.

The graphic method here developed offers also a thorough solution. In view of both, and of the extensive literature upon the subject (which seems, by the way, to have been so generally ignored), we can finally pronounce the problem to be *fully* solved.

3d. The changes of strain, unforeseen and often considerable, which a small settling of the piers or change of level of the supports may occasion.

This, be it observed, is *only* of importance when the piers settle *after* the erection of the superstructure. If piers are to be considered as settling indefinitely, or continuously during a succession of seasons, continuous girders are not to be thought of. If, however, as is generally the fact, the piers take their permanent set during the first season, and afterwards are immovable, the above objection has no weight. It is *not* necessary that the piers should be exactly on level or even on line, or even that the differences of level be known.

As shown in Art. 121, these differences produce no effect, provided the girder be built to the profile of the supporting points.

If in any case these differences are required, and it is considered difficult to determine them over water with sufficient accuracy, then the proper reactions at the several piers may be *weighed off*,* and the girder thus left in position under precisely the circumstances for which it has been calculated.

THE PRINCIPAL ADVANTAGES OF THE CONTINUOUS GIRDER ARE:

1st. Ease of erection, where false works are difficult or expensive. The girder may be built on shore, and then pushed out over the piers.

2d. Saving in width of piers, as compared with width required for separate successive spans. The girder may be placed upon *knife edges* at the piers. In fact, such a construction is preferable, as better ensuring the calculated strains. Width of piers is undesirable.

3d. Saving in material—usually from 25 to 30 per cent.

18. Continuous Girder—Supports not on a level. — In Chapter XIII. we have all the formulæ required for the solution of the continuous girder for supports on a level, or all on line, when the deviation from level is small, whatever may be the number or relative length of the spans. If for a continuous girder of *constant cross-section*, the supports are properly lowered after the girder is placed upon them, we may obtain a saving of 23 per cent., or more, in material over the same girder with supports all on level. If, however, the cross-section *varies* according to the strain—in other words, if the girder is of constant strength—no advantage is gained from thus lowering intermediate supports. Such disposition of the supports may even act injuriously.

The formulæ for shear and moments which we have given are, indeed, based upon the hypothesis of constant cross-section, but the strains in every piece of the girder being found for the shears and moments thus obtained, each piece is proportioned

---

* An idea first suggested by Clemens Herschel, C.E.: *Continuous, Revolving Draw Spans.* Little, Brown & Co., Boston, 1875.

to its strain, and the actual girder erected is *not* of constant cross-section, but more nearly one of uniform strength. Formulæ for the case of supports out of level, as well as determinations of the best differences of level, are hence of but little practical importance, and have not been given. If, however, it be required to find the effect due to the sinking of any one pier, the following may be found of service.

Let the nth support be depressed below the level of the others by the distance h. Then the moments at all the supports are changed. The moments at n and at each alternate support from n are diminished, and at the others increased.

Let
$$H = \frac{36\, h_n\, E\, I}{l^2}.$$

Then, *when all the spans are equal*, the following formulæ give the moment at any support:

1st. *All spans equal.* $n =$ number of lowered support, *from left*,

when $m < n$, $\quad M_m = \dfrac{c_m\, c_{s-n+2}}{c_{s+1}} H.$

when $m = n$, $\quad M_m = -\dfrac{H}{12} - \dfrac{c_{n-1}\, c_{s-n+2} + c_n\, c_{s-n+1}}{4\, c_{s+1}} H,$

when $m > n$, $\quad M_m = \dfrac{c_{s-m+2}\, c_n}{c_{s+1}} H,$

where $c_1 = 0$, $c_2 = +1$, $c_3 = -4$, $c_4 = +15$, $c_5 = -56$, $c_6 = +209$, etc.

From the moments at the supports the *shears* can be readily determined from the formula of Art. 148, viz.:

$$S_r = \frac{M_r - M_{r+1}}{l_r} + q, \quad S'_{r+1} = \frac{M_{r+1} - M_r}{l_r} + q',$$

$$S_m = \frac{M_m - M_{m+1}}{l_m}, \quad S'_m = \frac{M_m - M_{m-1}}{l_{m-1}},$$

where $q = P(1-k)$ for concentrated load and $\dfrac{w\, l_r}{2}$; for uniform, $q' = P\, k$ and $\dfrac{w\, l_r}{2}$.

2d. *Spans all unequal.*

when $m < n$, $\quad M_m = \dfrac{c_m \left[\dfrac{d_{s-n+1} - d_{s-n+2}}{l_n} + \dfrac{d_{s-n+3} - d_{s-n+2}}{l_{n-1}} \right] 6\, h_n\, E\, I}{l_2\, d_{s-1} + 2(l_1 + l_2)\, d_s}$

when $m = n$, $M_m = -\dfrac{3 h_n E I}{l_n l_{n-1}} - \dfrac{M_{n-1} l_{n-1} + M_{n+1} l_n}{2 (l_{n-1} + l_n)}$,

when $m > n$, $M_m = \dfrac{d_{s-m+2} \left[\dfrac{c_{n-1} - c_n}{l_{n-1}} + \dfrac{c_{n+1} - c_n}{l_n} \right] 6 h_n E I}{l_{s-1} c_{s-1} + 2 (l_{s-1} + l_s) c_s}$

in which expressions

$c_1 = 0$, $c_2 = 1$, $c_3 = -2 \dfrac{l_1 + l_2}{l_2}$, $c_4 = \dfrac{4 (l_1 + l_2)(l_2 + l_3) - l_2^2}{l_2 l_3}$,

$c_5 = -2 c_4 \dfrac{l_3 + l_4}{l_4} - c_3 \dfrac{l_3}{l_4}$, etc.

$d_1 = 0$, $d_2 = 1$, $d_3 = -2 \dfrac{l_s + l_{s-1}}{l_{s-1}}$,

$d_4 = \dfrac{4 (l_s + l_{s-1})(l_{s-1} + l_{s-2}) - l_{s-1}^2}{l_{s-1} l_{s-2}}$,

$d_5 = -2 d_4 \dfrac{l_{s-2} + l_{s-3}}{l_{s-3}} - d_s \dfrac{l_{s-2}}{l_{s-3}}$, etc.

The reader who has learned the use of the formulæ of Chapter XIII. will have no difficulty in applying the above to any particular case. In the same way as there explained, by making l_1 and l_s zero, we may fix the girder at the ends, etc. The formulæ for shear at any support are, of course, the same as before (Art. 150).

Ex. 1. *Let a beam of two equal spans be uniformly loaded throughout its whole length, and let the centre support be lowered by an amount* $h_2 = \dfrac{w\, l_4}{48\, E\, I}$. *What are the moments and reactions?*

The moments due to the full load alone before the support is lowered are $M_1 = 0$, $M_2 = \dfrac{w\, l^2}{8}$, $M_3 = 0$ (Art. 150). For the moment due to the lowering of the support alone, we have from the above formulæ, since

$$H = \dfrac{3\, w\, l^2}{4}, \quad s = 2, \quad m = n = 2,$$

$$M_1 = 0, \quad M_2 = -\dfrac{w\, l^2}{16}, \quad M_3 = 0.$$

Hence the *total* moment is

$$M_2 = \dfrac{w\, l^2}{8} - \dfrac{w\, l^2}{16} = \dfrac{w\, l^2}{16}.$$

For the shears, then, we have

$$S_1 = -\frac{M_2}{l} + \frac{wl}{2} = \frac{7}{16}wl, \qquad S'_2 = \frac{9}{16}wl,$$

$$S_2 = \frac{9}{16}wl, \qquad\qquad S'_3 = \frac{7}{16}wl.$$

Hence, $\quad R_1 = \dfrac{7}{16}wl, \quad R_2 = \dfrac{18}{16}wl, \quad R_3 = \dfrac{7}{16}wl.$

Ex. 2. *How much must we lower the second support in the above example, in order that the reaction at the centre support may be just zero?*

In this case we have

$$M_2 = \frac{wl^2}{8} - \frac{H}{12} = \frac{wl^2}{8} - \frac{3h_2 EI}{l^2}.$$

$$S'_2 = \frac{2M_2}{l} + wl = \frac{wl}{4} - \frac{6h_2 EI}{l^3} + wl = \frac{5}{4}wl - \frac{6h_2 EI}{l^3}.$$

If this is to be zero, we have

$$h_2 = \frac{5wl^4}{24 EI}, \quad \text{and} \quad M_2 = -\frac{1}{2}wl^2,$$

and $\qquad R_1 = wl, \quad R_2 = 0, \quad R_3 = wl,$

or precisely as for a beam of single span and length $2l$.

Ex. 3. *A beam of four equal spans is unloaded, and the third support is lowered by an amount* $h_3 = \dfrac{wl^4}{24 EI}.$ *What are the reactions?*

Ans. $\quad R_1 = \dfrac{42}{112}wl^2, \quad R_2 = \dfrac{147}{112}wl^2, \quad R_3 = \dfrac{70}{112}wl^2,$

$$R_4 = R_2, \quad R_5 = R_1.$$

Ex. 4. *A beam of five equal spans rests as a continuous girder over six supports. Having given the dimensions of the beam, length of spans, and coefficient of elasticity; to determine the reactions due to a sinking of the third support one-eighth of an inch.*

Let the beam be of wood, 1 foot wide, 1.5 deep,

$l = 20$ feet, $\quad s = 5, \quad r = 3, \quad E = 288{,}000{,}000$ lbs. per sq. ft.,

$\qquad h_3 = \tfrac{1}{8}$ in. $= 0.010417$ ft.

APPENDIX.] SUPPORTS OUT OF LEVEL. 385

Then, $c_2 = 1$, $c_3 = -4$, $c_4 = 15$, $c_5 = -56$, $c_6 = 209$,

and $M_2 = \dfrac{540\,E\,I\,h_3}{209\,l^2}$, $M_3 = -\dfrac{906\,E\,I\,h_3}{209\,l^2}$, $M_4 = \dfrac{576\,E\,I\,h_3}{209\,l^2}$,

$$M_5 = -\dfrac{144\,E\,I\,h_3}{209\,l^2}, \quad M_1 = M_6 = 0,$$

or, inserting the constants above, and $I = \dfrac{1}{12}\,b\,d^3 = \dfrac{3.375}{12}$,

$M_1 = M_6 = 0$, $M_2 = 5448$, $M_3 = -9142$, $M_4 = 5812$,
$M_5 = -1453$ ft. lbs.

If all the spans are unloaded. For the reactions necessary to bend it and keep it down to the supports, $R_1 = -272$ lbs., $R_2 = 1002$, $R_3 = -1477$, $R_4 = 1111$, $R_5 = -436$, $R_6 = 73$.

If the beam weigh 75 lbs. per foot, what deflection of third support will raise the left end from the abutment?

Ans. $R_1 = \dfrac{15}{33}\,w\,l = \dfrac{540\,E\,I\,h_3}{209\,l^3}$, or $h_3 = 0.0226$ ft. $= 0.2712$ in.

It will be observed that a small difference of level has then considerable effect.

Ex. 5. *Two equal spans are uniformly loaded. How high must the centre be raised in order that the ends may just touch?*

This is the case of the pivot span with centre support raised. (See Art. 121.)

The reactions at the ends are zero. At pier, then, $R_2 = 2\,w\,l$, hence moment at pier $M_2 = \tfrac{1}{2}\,w\,l^2$. But the moment when the supports are on level is $M_2 = \tfrac{1}{8}\,w\,l^2$, hence $\tfrac{3}{8}\,w\,l^2$ must be due to the elevation of the support. Then from our formulæ,

$$\tfrac{3}{8}\,w\,l^2 = -\dfrac{3\,E\,I\,h_2}{l^2}, \quad \text{or} \quad h_2 = -\dfrac{w\,l^4}{8\,E\,I},$$

which is precisely the same as the *deflection* of an horizontal beam, fastened at one end, and free at the other (Supplement to Chap. VII., Art. 13).

25

NOTE TO CHAPTER XIV.

THE BRACED ARCH.

19. The subject of braced arches is an important one, and is treated in no work with the fulness and completeness it deserves. The methods and formulæ of Chapter XIV. will, we believe, render the determination of the strains in this class of structure easy, and we propose in the following to illustrate their use, so far as may be necessary to render their application clear.

In Pl. 4, Fig. X., we have represented a braced circular arch with parallel flanges. Span of centre line $= 175$ ft.; radius, 201.4 ft.; rise, 20 ft. In practice, the panels would be taken of equal length; for convenience of calculation, however, we suppose the panel length to vary so that the horizontal projection is constant, and equal to 25 ft. Depth of arch, 10 ft. Hence, span of lower flange $= 170.6$ ft.; rise, 19.5 ft.; radius, 196.4 ft. Span of upper flange, 179.34 ft.; rise, 20.5 ft.; radius, 206.4 ft.

Since the flanges are, in practice, broken lines, and not true curves, the depth or lever arm for upper flanges is 9.43 ft., for lower flanges, 10.4 ft.

The determination of the other dimensions required is then easy, and a simple question of trigonometry.

Thus we have for the half central angle $a = 25° \ 45'$, and for the distances of the apices from the chord of the centre line:

For 1 ... —4.5   3 .... 4.7   5 .... 11.3   7 .... 14.6 ft.
" 2 .... 10.8   4 .... 18.5   6 .... 23.4   8 .... 25 "

We suppose the load at each apex 10 tons, and shall consider

1st. Arch hinged at crown or apex 8, and at the *ends of the lower flange*—the flanges **H** and **A** being removed.

2d. Arch hinged at apex 8, and at the ends of the *centre line*—the flanges **A** and **E** butting against a *skew back* or pivoted plate, and the flange **H** only being removed.

3d. Arch continuous at crown—the flange **H** being retained and hinged at ends of lower flanges.

APPENDIX.] THE BRACED ARCH. 387

4*th*. Arch, as in 3*d*, but pivoted at ends of centre line.

5*th*. Arch without hinges, or continuous at crown and fixed at abutments.

These cases will illustrate all the principles of Chapter XIV., and a comparison of the results obtained in each case may prove instructive.

20. *Arch hinged at apex* 8, *and at the extremities of the lower flange—flanges* **H** *and* **A** *being removed.*

From Art. 158 we can easily find the reaction and horizontal thrust at left end either by construction or formula for every weight. Thus

$$V = \frac{P(a+x)}{2a} \quad \text{and} \quad H = \frac{P(a-x)}{2h}.$$

For the first weight P_1, then,

$$V = \frac{10(85.3 - 75)}{170.6} = 0.603, \quad \text{and}$$

$$H = \frac{10(85.3 + 75.3)}{2 \times 29.5} = 2.73 \text{ tons.}$$

For the weight P_{10},

$$V = \frac{10(85.3 + 37.5)}{170.6} = 7.2, \quad \text{and}$$

$$H = \frac{10(85.3 - 37.5)}{2 \times 29.5} = 8.1 \text{ tons.}$$

In similar manner, we find

$P_1 = 0.603,$ $\quad P_2 = 1.33,$ $\quad P_3 = 2.06,$
$H_1 = 2.74,$ $\quad H_2 = 3.84,$ $\quad H_3 = 5.9,$
$P_4 = 2.8,$ $\quad P_5 = 3.53,$ $\quad P_6 = 4.2,$
$H_4 = 8.1,$ $\quad H_5 = 10.2,$ $\quad H_6 = 12.1,$
$P_7 = 5.0,$ $\quad P_8 = 5.7,$ $\quad P_9 = 6.47,$ $\quad P_{10} = 7.2,$
$H_7 = 14.4,$ $\quad H_8 = 12.1,$ $\quad H_9 = 10.2,$ $\quad H_{10} = 8.1,$
$P_{11} = 7.94,$ $\quad P_{12} = 8.6,$ $\quad P_{13} = 9.4,$
$H_{11} = 5.9,$ $\quad H_{12} = 3.84,$ $\quad H_{13} = 1.74.$

It will be at once seen that the reaction of P_6 at **A** is the same as of P_6 at **B**, or equal to $10 - P_6$; while the horizontal thrust is the same for both. We need them only to find **P** and **H** for weights 1 to 7, and can then at once write down the others. We are now ready either to calculate or diagram the strains.

Thus, for instance, for P_{10} (see Fig. X.), we lay off the reaction at **A** upwards to scale from **C** to **D**, then the horizontal thrust at **A** from **D** to 1, then the equal thrust at **B** from 1 back to **D**, then the reaction at **B** from **D** to 8, and finally the weight down from 8 back to **C**, thus closing the polygon for the exterior forces. Lines parallel to the pieces then give the strains. Thus the thrust and reaction at **A** are in equilibrium with **E** and 1 2. Then 1 2 is in equilibrium with **B** and 2 3, and so on. Observe that the diagram *checks itself*. Thus the last diagonal 7 8 must be in equilibrium with 6 7 and **G** (flange **H** being removed), and that this is so is shown by the strain in 7 8 passing exactly through 8, thus making the strain in **H** zero. We can also check the work by calculating the strain in the last flange **D** by moments. Thus for P_{10}

$$\mathbf{D} \times 9.43 = 7.2 \times 72.8 - 8.1 \times 19.1 - 10 \times 25 = 119.45,$$
or $\mathbf{D} = +12.6$.

If this agrees with **D** as found by diagram, and if the diagram also checks, we may have confidence in the accuracy of the work, and at once scale off the strains. Observe that diagonals 4 5 and 5 6 are both tension; also that **F** and **G** are tension.

We have given also the diagram for P_{11}, which the reader can easily follow through for himself. **F** and **G** are both tension, 3 4 and 4 5 both compression. The horizontal thrust is 8 b, and the reaction at $\mathbf{A} = b\,1$.

We thus make a diagram for each separate weight, and then taking the dead load at $\frac{1}{2}$ the live, we can form the following table of strains. Since we wish only the maximum strains on one-half the arch, those on the other half being precisely similar, we can diagram the strains due to all the weights upon the right half at *once* by taking the sum of their reactions and thrust at **A**. We have then the following table:

APPENDIX.] THE BRACED ARCH. 389

HINGED AT CROWN AND AT EXTREMITIES OF LOWER FLANGE.

	P_{2-6}	P_{1-7}	P_8	P_9	P_{10}	P_{11}	P_{12}	P_{13}	Total Strains from live load.	Uniform Load $\frac{1}{4}$.	Maximum Strains.
B	− 3.4	− 4.7	+ 2.1	+ 6.0	+ 8.7	+ 13.7	+ 17.0	+ 8.1	+ 55.6 − 8.1	+ 23.7	+ 79.3 ……
C	+ 2.1	+ 2.5	+ 9.0	+ 15.6	+ 22.5	+ 16.8	+ 10.5	+ 4.8	+ 83.8 ……	+ 41.9	+ 125.7 ……
D	+ 15.3	+ 20.3	+ 19.5	+ 15.5	+ 12.0	+ 9.5	+ 6.1	+ 2.6	+ 100.8 ……	+ 50.4	+ 151.2 ……
E	+ 27.3	+ 36.6	+ 12.2	+ 8.5	+ 4.5	+ 0.3	− 3.3	− 7.4	+ 89.4 − 10.7	+ 39.3	+ 128.7 ……
F	+ 25.5	+ 34.2	+ 7.2	+ 0.5	− 7.0	− 14.2	− 6.7	− 4.2	+ 67.4 − 32.1	+ 17.6	+ 85.0 − 14.5
G	+ 16.2	+ 21.7	− 1.3	− 10.3	− 7.8	− 6.5	− 3.6	− 1.4	+ 37.9 − 29.6	+ 4.2	+ 42.1 − 25.4
12	− 2.3	− 3.3	+ 1.4	+ 4.1	+ 6.7	+ 9.4	+ 12.0	+ 14.6	+ 48.2 − 5.6	+ 21.3	+ 69.5 ……
23	+ 2.0	+ 3.1	− 1.3	− 3.6	− 5.9	− 8.3	− 10.3	− 1.1	+ 6.2 − 29.4	− 11.6	…… − 41.0
34	+ 3.8	+ 5.3	+ 4.5	+ 6.1	+ 8.1	+ 9.6	− 6.0	− 3.0	+ 37.5 − 9.0	+ 14.3	+ 51.8 ……
45	− 3.1	− 4.1	+ 4.4	+ 6.5	+ 8.7	+ 3.5	+ 2.5	− 1.1	+ 7.1 − 26.8	− 9.8	…… − 36.6
56	+ 8.5	+ 11.3	+ 6.1	+ 6.7	+ 8.7	+ 6.3	− 4.0	− 2.2	+ 32.6 − 12.5	+ 10.1	+ 42.7 − 2.4
67	− 8.0	− 11.9	+ 7.7	+ 6.1	+ 4.7	+ 3.5	+ 1.8	+ 0.9	+ 17.0 − 28.5	− 5.7	+ 11.3 − 34.2
78	+ 11.2	+ 15.1	− 9.5	− 7.1	− 5.4	− 4.5	− 2.9	− 1.1	+ 26.3 − 30.5	− 2.1	+ 24.2 − 32.6

21. Arch hinged at Apex S and at the Extremities of the centre Line; Flanges A being retained, and only H removed.—The method of solution is precisely the same as before, the only difference being that the span is now 175 ft. instead of 170.6, and the rise 25 ft. instead of 29.5. The reactions and thrust will then be somewhat different. Thus, for the left abutment **A**,

$P_1 = 0.71$, $\quad P_2 = 1.42$, $\quad P_3 = 2.14$, $\quad P_4 = 2.85$, $\quad P_5 = 3.57$,

$H_1 = 2.48$, $\quad H_2 = 4.97$, $\quad H_3 = 7.5$, $\quad H_4 = 9.97$, $\quad H_5 = 12.5$,

$P_6 = 4.28$, $\quad P_7 = 5.00$, $\quad P_8 = 5.72$, $\quad P_9 = 6.43$, $\quad P_{10} = 7.15$, etc.

$H_6 = 14.98$, $\quad H_7 = 17.5$, $\quad H_8 = 14.98$, $\quad H_9 = 12.5$, $\quad H_{10} = 9.97$, etc.

We have therefore the following table:

THE BRACED ARCH.

HINGED AT CROWN AND AT EXTREMITIES OF CENTRE LINE.

	P_{1-7}	P_{2-6}	P_3	P_4	P_{10}	P_{11}	P_{12}	P_{13}	Total Strains for live load.	Uniform Load ↓	Maximum Strains.
A	+ 21.2	+ 15.8	+ 8.1	+ 7.3	+ 6.1	+ 5.2	+ 4.2	+ 3.2	+ 71.1	+ 35.5	+ 106.6 ......
B	+ 10.4	+ 7.7	+ 7.5	+ 10.6	+ 9.3	+ 17.0	+ 20.4	+ 10.5	+ 93.4	+ 46.7	+ 140.1 ......
C	+ 12.0	+ 9.3	+ 12.3	+ 18.6	+ 24.6	+ 19.1	+ 12.6	+ 6.7	+ 116.1	+ 58.1	+ 174.2 ......
D	+ 28.5	+ 21.0	+ 21.8	+ 18.1	+ 21.0	+ 10.8	+ 6.9	+ 4.0	+ 132.1	+ 66.1	+ 108.2 ......
E	+ 27.6	+ 20.7	+ 8.8	+ 5.1	+ 1.8	− 1.6	− 5.1	− 8.6	+ 63.9	+ 24.3	+ 88.2 ......
F	+ 30.4	+ 23.0	+ 6.3	− 0.9	− 7.4	− 6.0	− 10.1	− 5.4	+ 59.7	+ 14.9	+ 74.6 − 14.9
G	+ 21.1	+ 16.4	− 0.9	− 10.4	− 7.6	− 15.0	− 4.0	− 2.8	+ 37.5	− 1.6	+ 37.5 − 42.3
12	− 8.7	− 6.7	− 1.1	+ 2.0	+ 4.7	+ 8.0	+ 11.0	+ 13.8	+ 39.5	+ 11.5	+ 51.0 − 5.0
23	+ 4.8	+ 3.7	− 0.4	− 2.7	− 5.0	− 7.8	− 10.2	+ 1.0	+ 0.5	− 8.3	+ 1.2 − 34.4
34	+ 0.7	+ 0.5	+ 2.6	+ 4.6	+ 6.4	+ 8.6	− 7.0	− 3.4	+ 23.4	+ 6.5	+ 29.9 − 10.4
45	− 2.5	− 1.7	− 3.7	− 6.0	− 8.3	+ 3.9	+ 2.8	+ 1.0	+ 7.7	− 7.3	+ 0.4 − 29.5
56	+ 8.8	+ 6.5	+ 5.1	+ 6.2	+ 9.4	+ 7.1	+ 4.9	− 2.3	+ 26.6	+ 1.4	+ 28.0 − 22.3
67	− 11.4	− 8.7	− 7.3	− 5.7	− 5.1	+ 3.6	+ 2.5	+ 1.5	+ 18.4	+ 4.5	+ 13.9 − 31.9
78	+ 15.1	+ 11.3	− 8.9	− 7.4	− 5.1	− 4.3	− 2.6	− 1.7	+ 26.4	− 1.8	+ 24.6 − 31.8

A comparison of this table with the preceding shows that we have gained nothing by introducing two end flanges at **A** at each end, and pivoting the arch at the extremities of the centre line. We have indeed slightly diminished the strains in the lower flanges **E** and **F**, as also in the bracing, but the other strains are much greater than before—a result which might have been anticipated, since the effect of hinging at the centre, instead of at the extremities of the lower flange, is simply to *reduce the effective height* or rise from 29.5 ft. to 25 ft. In our example, since the depth of arch is half the whole rise of the centre line, this reduction is considerable.

For a much longer span and smaller proportional depth the difference would not be so marked, but it would seem that the strains in the second case must always be greater than in the first. The best construction, then, seems to require the hinge in the upper flange at the crown, and at the extremities of the lower flange at the abutments. By this means, the greatest effective rise is obtained, and both ribs aid in supporting the load. Were the hinges all three in the same rib, then, for uniform load, that rib alone is the sole supporting member, and is unassisted by the other. This should then be avoided.

22. Arch continuous at Crown, and hinged at Ends of Lower Rib.—For this case, referring to Art. 159, we have simply to interpolate from our table there given the values of **A**, **B** and y_0 in the equation,

$$y = \frac{1 + \mathbf{B}\kappa}{1 - \mathbf{A}\kappa} y_0,$$

and thus plot the curve $c\,d\,e\,i\,k$, Fig. 91. The construction of the reactions and horizontal thrust for each weight is then easy. These once known, we proceed as above, in order to find the strains.

Now, in the formulæ above $\kappa = \dfrac{\mathbf{I}}{\mathbf{A}\,r^2}$, and since we can put for $\dfrac{\mathbf{I}}{\mathbf{A}}$ the square of the radius of gyration, and this radius is approximately the half depth of the arch, we have

$$\kappa = \frac{25}{(170.6)^2} = \frac{25}{29104.36} = \frac{1}{1164}.$$

Now **B** and **A** are, as we see from the table, small, and hence in our present example the terms containing κ can be disreguarded, and the value of y can be taken directly from the table for y_0, given in Art. 159. For $a = 25°\ 45'$, then we have at once, since $h = 19.5$,

for $\beta = 0$, $y = 1.295\ h = 25.25$ ft.,
$\beta = 0.2\ a$, $y = 1.304\ h = 25.42$ ft.,
$\beta = 0.3\ a$, $y = 1.335\ h = 26.03$ ft., etc.

The corresponding value of x is $R \cos \beta$.

Having thus plotted the curve, and constructed the reaction and thrust for each weight, the diagram for strains proceeds as before. We thus form the following table:

394   NOTE TO CHAP. XIV.   [APPENDIX.

CONTINUOUS AT CROWN—HINGED AT EXTREMITIES OF LOWER FLANGE.

	P_{2-3}	P_{1-7}	P_8	P_9	P_{10}	P_{11}	P_{12}	P_{13}	Total Strains from live load		Uniform Load $\frac{1}{4}$	Maximum Strains	
B	−17.3	−17.8	−2.3	+0.8	+4.0	+8.1	+15.0	+5.9	+24.4	−37.4	−1.5	+32.9	−38.9
C	−20.9	−16.8	+1.7	+7.6	+14.3	+7.6	+7.0	+1.5	+39.7	−37.7	+1.0	+40.7	−36.7
D	−13.4	−18.5	+11.1	+7.4	+2.6	−2.3	+0.8	−1.3	+21.9	−35.5	−6.8	+15.1	−42.3
E	+47.8	+58.2	+18.8	+16.3	+12.3	+8.1	+1.3	−4.5	+162.8	−4.5	+79.1	+241.9	...
F	+50.0	+65.1	+17.3	+12.1	+4.8	−2.3	−2.7	+0.3	+155.3	+5.0	+75.1	+230.4	...
G	+58.5	+59.0	+10.8	+3.0	+5.7	+8.5	+3.7	+3.2	+146.4	...	+73.2	+219.6	...
H	+40.2	+37.1	+11.4	+11.9	+14.4	+15.5	+8.3	+5.0	+143.8	...	+71.9	+215.7	...
1 2	−12.3	−12.4	−1.4	+0.6	+3.0	+5.7	+10.3	+13.3	+32.9	−26.0	+3.4	+36.3	−22.6
2 3	+10.5	+10.9	+1.3	−0.6	−2.7	−5.0	−9.2	+2.3	+25.0	−17.5	+3.7	+28.7	−13.8
3 4	−0.9	+1.0	+2.8	+4.5	+6.3	+7.8	+6.8	+3.7	+22.4	−11.4	+5.5	+27.9	−5.9
4 5	+4.2	+2.3	−2.3	−4.3	−6.4	−6.3	−3.5	+2.2	+18.5	−13.0	+2.7	+21.2	−10.3
5 6	+6.6	+10.7	+5.8	+7.0	+8.8	−7.1	−4.6	−2.0	+20.1	−22.5	−1.2	+18.9	−23.7
6 7	−2.8	−7.3	+0.3	+6.2	+6.4	+5.7	+3.5	+1.5	+23.3	−16.4	+3.4	+26.7	−13.0
7 8	+13.7	+18.7	−7.8	−5.1	−4.6	−3.3	−2.0	−0.7	+32.4	−23.5	+4.4	+36.8	−18.1

Comparing these results with those in our table above, Art. 19, for the same case *not* continuous at the crown, we see that the strains in the upper flanges are much less, and are, moreover, of opposite character; while the strains in the lower flanges are greatly increased, and nothing is gained. This result might also have been anticipated, since the effect of inserting the flange **H** is to reduce the effective height from 29.5 to 19.5 ft., and, moreover, for total dead and live load, nearly the whole weight comes directly upon the continuous lower rib, and the upper aids but very little.

23. Strains due to Temperature.—We have, in addition, strains due to change of temperature to be taken into account in determining the total maximum strains.

For the present case we have, from Art. 165, for the thrust due to change of temperature,

$$H = \frac{15\,E\,I\,A\,\epsilon\,t}{8\,A\,h^2 + 15\,I},$$

or, substituting in the place of $\frac{I}{A}$ the square of the radius of gyration $= g^2$, we have

$$H = \frac{15\,E\,A\,g^2\,\epsilon\,t}{8\,h^2 + 15\,g^2}.$$

Now g is approximately the half depth of arch; hence $g^2 = 25$ sq. ft. $= 3600$ sq. inches. Taking 5 tons to the square inch as our unit strain, we may take the area of our flanges, as determined from the above table of strains at about 25 square inches. Hence $A = 50$. Taking $E = 14,000$ tons per sq. inch, $h^2 = 19.5^2 = 54,756$ sq. inches, and supposing the temperature to vary 25° (Centigrade) on each side of the mean, we have, assuming ϵ at 0.000012, the thrust $H =$ about 25 tons.

It is easy to find either by moments or diagram, or both, the strains due to this thrust. Since the temperature varies between 25° on *both* sides of the mean, this thrust can be both positive and negative, and the corresponding strains have, therefore, double sign. We find, therefore,

$B = \mp 24.5$	$C = \mp 40.4$	$D = \mp 49.1$	$E = \pm 37.5$
$F = \pm 55.9$	$G = \pm 66.9$	$H = \pm 69.6$	
$1\,2 = \mp 17.1$	$2\,3 = \pm 15.0$	$3\,4 = \mp 8.6$	$4\,5 = \pm 12.5$
$5\,6 = \mp 2.2$	$6\,7 = \pm 9.7$	$7\,8 = \mp 5.7$	

Hence we have, for the *total* maximum strains for the case of the preceding article,

B = + 57.4 − 63.4   C = + 81.1 − 77.1   D = + 64.2 − 91.4
E = + 279.4   F = + 286.3   G = + 286.5
H = + 285.3   1 2 = + 53.4 − 39.7   2 3 = + 43.7 − 28.8
3 4 = + 37.3 − 22.4   4 5 = + 40.4 − 28.4   5 6 = + 23.4 − 12.5
6 7 = + 36.4 − 22.7   7 8 = + 42.5 − 23.8

24. Arch continuous at the Crown, pivoted at the extremities of the Centre Line.—The method of solution is precisely similar to the preceding. We have only to take the rise of the centre line, or $h = 20$, instead of $h = 19.5$, as before, and the radius and span of centre line, instead of the radius and span of the lower rib.

One point only needs to be noticed. Having found the reaction and thrust for any weight, these forces now act at the extremity of the centre line. We can therefore form the strain diagram as follows:

First calculate by moments the strain in **A** and **E**. Then, in diagram (*c*), Fig. X., having laid off the thrust *o* **C** and the reaction **C** *b*, draw from *o* and *b* lines parallel to **A** and **E**, and lay off *o* **A** equal to the strain in **A**, and *b d* to the strain in **E**. Then, if these strains have been correctly found, the line **A** *d* must be parallel to and give the strain in diagonal 1 2. The diagram thus commenced, can then be continued as shown, and the strains in all the pieces determined.

We may also form the strain diagram without calculating **A** and **E**. Thus *o b* is the resultant acting at the end of the centre line. Since it acts then half way between **A** and **E**, bisect it in *a*, and draw *a* **A** perpendicular to flange **A**. Then *o* **A** is the strain in **A**, and drawing **A** *d* parallel to diagonal 1 2, we have at once the strain in **E** and 1 2.

Performing the operations indicated, we obtain the following table of strains:

APPENDIX.] THE BRACED ARCH. 397

CONTINUOUS AT CROWN—HINGED AT EXTREMITIES OF CENTRE LINE.

	$P_{3\cdot6}$	P_{1-7}	P_6	P_9	P_{10}	P_{11}	P_{12}	P_{13}	Total Strains.		Uniform Load $\frac{1}{4}$.	Maximum Strains.	
												+	−
A	+ 18.8	+ 23.8	+ 8.6	+ 8.3	+ 7.7	+ 6.6	+ 5.3	+ 3.9	+ 83.0	.....	+ 41.5	+ 124.5	.....
B	+ 3.5	+ 6.7	+ 6.3	+ 10.1	+ 12.0	+ 14.9	+ 18.7	+ 9.1	+ 81.3	.....	+ 40.7	+ 122.0	.....
C	+ 1.0	+ 4.7	+ 9.8	+ 16.8	+ 20.8	+ 14.5	+ 8.9	+ 4.2	+ 76.6	.....	+ 38.3	+ 114.9	.....
D	+ 10.4	+ 18.2	+ 18.9	+ 15.4	+ 7.9	+ 5.2	+ 2.0	+ 0.6	+ 78.6	.....	+ 30.3	+ 117.9	.....
E	+ 29.8	+ 35.3	+ 10.6	+ 7.3	+ 5.7	+ 1.9	− 2.3	− 7.1	+ 90.6	− 9.4	+ 40.6	+ 131.2	.....
F	+ 37.0	+ 43.4	+ 9.4	+ 3.1	− 1.0	− 8.4	− 5.0	− 1.8	+ 92.9	− 16.2	+ 38.3	+ 131.2	.....
G	+ 32.7	+ 37.3	+ 3.2	− 5.8	+ 0.8	+ 1.8	+ 2.5	+ 2.0	+ 80.3	− 5.8	+ 37.2	+ 117.5	.....
H	+ 17.7	+ 16.5	+ 4.6	+ 5.2	+ 10.3	+ 8.6	+ 7.8	+ 4.7	+ 75.4	.....	+ 37.7	+ 113.1	.....
1 2	− 12.2	− 13.3	− 2.0	+ 0.5	+ 2.4	+ 5.4	+ 9.1	+ 12.6	+ 30.0	− 27.5	+ 1.2	+ 31.2	− 26.3
2 3	+ 7.8	+ 8.9	+ 0.5	− 1.7	− 3.2	+ 5.6	+ 8.7	+ 2.2	+ 19.4	− 19.2	+ 0.1	+ 19.5	− 19.1
3 4	− 1.9	− 2.1	+ 1.8	+ 3.7	+ 5.1	+ 7.3	− 8.2	− 4.2	+ 17.9	− 16.4	+ 0.8	+ 18.7	− 15.6
4 5	+ 1.3	+ 0.7	+ 2.9	+ 5.1	− 6.5	+ 5.5	+ 4.2	+ 2.1	+ 13.8	− 14.5	− 0.3	+ 13.5	− 14.8
5 6	+ 6.0	+ 8.2	+ 4.9	+ 6.1	− 10.1	+ 7.4	+ 5.3	− 2.7	+ 25.2	− 25.5	− 0.2	+ 25.0	− 25.7
6 7	− 6.2	− 9.4	+ 6.7	+ 6.6	+ 6.3	+ 4.9	+ 3.8	+ 2.0	+ 21.6	− 22.3	− 0.3	+ 21.3	− 22.6
7 8	+ 12.7	+ 16.4	− 0.2	− 7.2	− 5.6	− 3.9	− 2.8	− 1.3	+ 29.1	− 30.0	− 0.4	+ 28.7	− 30.4

A comparison of the above with the same case hinged at the ends of the lower rib, shows a decided gain. The effective height is increased, being now 20 ft., in place of 19.5; in addition to which both ribs under total load bear their proper proportion. If, then, we wish the arch continuous at crown, *both* ribs should butt against an end plate, pivoted in the centre. This is preferable to hinging the lower rib at its extremities, and removing the end flanges **A**.

25. Temperature Strains.—For the strains due to temperature, we may take, as before, the thrust $H = 25$ tons, and find thus

$A = \pm 11.2 \quad B = \mp 13.2 \quad C = \mp 29.4 \quad D = \mp 37.6$

$E = \pm 27.0 \quad F = \pm 45.5 \quad G = \pm 56.4 \quad H = \pm 59.2$

$1\,2 = \mp 17.8 \quad 2\,3 = \pm 13.9 \quad 3\,4 = \mp 9.8 \quad 4\,5 = \pm 11.3$

$5\,6 = \mp 2.8 \quad 6\,7 = \pm 8.1 \quad 7\,8 = \pm 4.3$

—all somewhat less than in the previous case, as they should be, since the point of application is at the centre between the flanges.

We have then from the preceding table the *total* maximum strains:

$A = + 135.7 \quad B = + 135.2 \quad C = + 144.3 \quad D = + 155.5$

$E = + 158.2 \quad F = {+ 176.7 \atop - 7.2} \quad G = {+ 173.9 \atop - 19.2} \quad H = {+ 172.3 \atop - 25.5}$

$1\,2 = {+ 49.0 \atop - 44.1} \quad 2\,3 = {+ 33.4 \atop - 33.0} \quad 3\,4 = {+ 28.5 \atop - 25.4} \quad 4\,5 = {+ 24.8 \atop - 26.1}$

$5\,6 = {+ 27.8 \atop - 28.5} \quad 6\,7 = {+ 29.4 \atop - 30.7} \quad 7\,8 = {+ 33.0 \atop - 34.7}$

26. Arch continuous at Crown and fixed at the Ends.
—From our table, Art. 160, we have directly for $a = 25° 45'$, and $h = 20$, y being now measured above the horizontal tangent at crown of centre line,

for $\beta = 0$, $\quad y = 0.209\, h = 4.18$ ft.

$\beta = 0.2a$, $\quad y = 0.208\, h = 4.16$

$\beta = 0.4a$, $\quad y = 0.206\, h = 4.12$

$\beta = 0.6a$, $\quad y = 0.201\, h = 4.02$

$\beta = 0.8a$, $\quad y = 0.198\, h = 3.96$

$\beta = 1.0a$, $\quad y = 0.189\, h = 3.78$

Also from our formulæ of Art. 160, we have

$$c_1 = -\frac{40}{15(87.5+x)}\left[87.5 - 5x - 246.1\right]$$

$$c_2 = -\frac{40}{15(87.5+x)}\left[87.5 + 5x - 246.1\right]$$

Hence for

P_1	$x = 75$	$c_1 = +8.7$	$c_2 = -3.5$
P_2	$x = 62.5$	$c_1 = +8.3$	$c_2 = -2.7$
P_3	$x = 50$	$c_1 = +7.9$	$c_2 = -1.7$
P_4	$x = 37.5$	$c_1 = +7.3$	$c_2 = -0.61$
P_5	$x = 25$	$c_1 = +6.7$	$c_2 = +0.8$
P_6	$x = 12.5$	$c_1 = +5.8$	$c_2 = +2.5$
P_7	$x = 0$	$c_1 = +4.8$	$c_2 = +4.8$
P_8	$x = -12.5$	$c_1 = +2.5$	$c_2 = +5.8$
P_9	$x = -25$	$c_1 = +0.8$	$c_2 = +6.7$, etc.,

negative values of c being laid off above the ends of centre line.

We can therefore easily construct the left reactions, as explained in Art. 160, Fig. 92. We thus obtain

$P_1 = 0.5,$ $\quad P_2 = 0.9,$ $\quad P_3 = 1.2,$
$H_1 = 4.4,$ $\quad H_2 = 8.6,$ $\quad H_3 = 13.1,$
$M_1 = +38.3,$ $\quad M_2 = +71.4,$ $\quad M_3 = +103.5,$

$P_4 = 2.3,$ $\quad P_5 = 3.1,$ $\quad P_6 = 4.4,$
$H_4 = 15.9.$ $\quad H_5 = 18.9,$ $\quad H_6 = 21.1.$
$M_4 = +116.1,$ $\quad M_5 = +126.6,$ $\quad M_6 = +122.4,$

$P_7 = 5.0,$ $\quad P_8 = 5.6,$ $\quad P_9 = 6.9,$
$H_7 = 22.4,$ $\quad H_8 = 21.1,$ $\quad H_9 = 18.9,$
$M_7 = +107.5,$ $\quad M_8 = +52.7,$ $\quad M_9 = +15.1,$

$P_{10} = 7.7,$ $\quad P_{11} = 8.8,$ $\quad P_{12} = 9.1,$ $\quad P_{13} = 9.5,$
$H_{10} = 15.9,$ $\quad H_{11} = 13.1,$ $\quad H_{12} = 8.6,$ $\quad H_{13} = 4.4,$
$M_{10} = -9.7,$ $\quad M_{11} = -22.3,$ $\quad M_{12} = -23.2,$ $\quad M_{13} = -15.3.$

A positive moment indicates tension in the lower flange at abutment, and compression in upper.

We can therefore easily calculate the strains in flanges **A** and **E** for each weight.

Thus, for ₁₁

$$\mathbf{A} \times 9.7 = 8.8 \times 2.2 + 13.1 \times 4.5 - 22.3,$$

or
$$\mathbf{A} = + 5.7.$$

$$\mathbf{E} \times 10.4 = - 8.8 \times 12.5 + 13.1 \times 10.8 + 22.3,$$

or
$$\mathbf{E} = + 5.1.$$

The strain diagram can then be commenced, as shown in diagram (c), Fig. X., and explained above. We can find by calculation the flange **H**, and thus check our diagram.

Proceeding thus, we obtain the following table of strains:

APPENDIX.] THE BRACED ARCH. 401

	P_{1-7}	P_{1-1}	P_8	P_9	P_{10}	P_{11}	P_{12}	P_{13}	Total Strains.	Uniform Load $\frac{1}{4}$.	Maximum Strains. +	Maximum Strains. -
A	+67.4	+54.7	+16.4	+11.8	+8.1	+5.7	+3.6	+2.6	+170.3	+85.1	+255.4	.....
B	+36.2	+30.0	+9.8	+10.6	+11.4	+14.7	+16.8	+7.7	+137.2	+68.6	+205.8	.....
C	+21.4	+14.4	+9.9	+15.3	+20.2	+15.7	+8.4	+3.2	+108.5	+54.2	+162.7	.....
D	+25.9	+17.0	+17.1	+12.1	+8.5	+8.3	+2.8	+0.0	+92.6	+46.3	+138.0	.....
E	+12.2	+8.3	+10.1	+9.8	+8.2	+5.1	+0.2	−5.3	+53.9	+24.3	+78.2	.....
F	+34.2	+27.4	+12.7	+7.5	+1.7	−6.7	−2.7	−0.3	+83.5	+36.9	+120.4	.....
G	+37.4	+31.1	+8.8	−0.3	−2.9	−2.7	−3.4	−3.2	+82.5	+41.1	+123.6	.....
H	+23.4	+20.3	+11.1	+13.1	+10.8	+7.2	+7.0	+5.2	+98.1	+49.1	+147.2	.....
1 2	−26.9	−20.9	−5.8	+1.9	+1.8	+5.9	+9.7	+8.1	+23.7	−15.9	+7.8	−71.4
2 3	+13.9	+10.6	+2.6	+0.2	+2.9	+6.1	+8.9	+2.5	+20.6	+5.7	+35.3	−12.4
3 4	−13.0	−13.2	+0.8	+2.5	+5.3	+8.4	+7.4	−3.7	+16.2	−10.9	+5.3	−40.0
4 5	+5.5	+6.7	+1.1	+3.0	+6.8	+4.8	+3.4	+2.5	+22.9	+5.5	+28.4	−6.3
5 6	+0.9	+1.0	+3.6	+5.0	+9.3	+6.2	+4.3	−2.3	+11.4	+5.3	+16.7	−16.8
6 7	−5.0	−3.8	−5.0	−8.8	−5.8	+3.5	+2.8	+1.6	+22.5	+4.1	+26.6	−10.3
7 8	+12.4	+9.9	+8.8	+8.1	−4.3	−2.7	−1.8	−1.1	+22.3	−2.7	+19.6	−29.5

26

The strains in the present case are, we see, much greater than for any of the others. Unless the maximum of stiffness is essential, it would appear, then, undesirable to fix the arch at the ends *for an arch of the above dimensions*.

27. Temperature Strains.—The strains due to temperature are also very great. Thus, from Art. 165, we have

$$H = \frac{45 \, E I A \, \epsilon \, t}{4 \, A \, h^2 + 45 \, I},$$

and for the distance of the point of action of this thrust, below the crown of the centre line,

$$e_0 = \frac{(A \, a^2 + 6 \, I) \, h}{3 \, A \, a^2},$$

or since $\frac{I}{A} = g^2 = 25$ ft.,

$$H = \frac{45 \, E A \, g^2 \, \epsilon \, t}{4 \, h^2 + 45 \, g^2}, \quad e_0 = \frac{h \, (a^2 + 6 \, g^2)}{3 \, a^2}.$$

For $A = 60$ square in., $a = 87.5$ ft., $h = 20$ ft. $= 240$ in., $g = 60$ in. $= 5$ ft., $\epsilon = 0.000012$, $E = 14,000$ tons, $t = 30°$, we have

$$H = 125 \text{ tons} \quad \text{and} \quad e_0 = 6.7 \text{ ft.}$$

Hence we have the strains

$A = \pm 228$, $C = \pm 29.0$, $E = \mp 30.0$, $G = \pm 118.5$,
$B = \pm 112.5$, $D = \mp 13.0$, $F = \mp 65.0$, $H = \pm 137.5$,
$1\,2 = \mp 96.0$, $3\,4 = \mp 67.0$, $5\,6 = \mp 29.5$, $7\,8 = \pm 2.5$.
$2\,3 = \pm 51.0$, $4\,5 = \pm 39.5$, $6\,7 = \pm 25.0$,

Therefore the *total* strains are

$A = \begin{matrix} + 483.4 \\ - 142.9 \end{matrix}$ $B = \begin{matrix} + 318.3 \\ - 43.9 \end{matrix}$ $C = + 191.7$ $D = + 151.9$

$E = \begin{matrix} + 108.2 \\ - 5.7 \end{matrix}$ $F = \begin{matrix} + 185.4 \\ - 28.1 \end{matrix}$ $G = \begin{matrix} + 242.1 \\ - 77.4 \end{matrix}$ $H = \begin{matrix} + 234.7 \\ - 88.4 \end{matrix}$

$1\,2 = \begin{matrix} + 103.8 \\ - 167.4 \end{matrix}$ $2\,3 = \begin{matrix} + 86.3 \\ - 63.4 \end{matrix}$ $3\,4 = \begin{matrix} + 72.3 \\ - 116.0 \end{matrix}$ $4\,5 = \begin{matrix} + 67.9 \\ - 45.3 \end{matrix}$

$5\,6 = \begin{matrix} + 46.2 \\ - 46.3 \end{matrix}$ $6\,7 = \begin{matrix} + 51.6 \\ - 35.3 \end{matrix}$ $7\,8 = \begin{matrix} + 22.1 \\ - 31.7 \end{matrix}$

With the above we close our discussion of the braced arch. Our design has been to illustrate the application of the formulæ and methods of Chapter XIV., and to show that by their aid such a structure can be calculated with ease and certainty.

In short, the difficulty is but little if any greater than for a simple girder, only for a long span and many panels the work becomes tedious and wearisome.

In such a case, perhaps the method of moments will be found preferable to diagrams. Thus, for any condition of loading, we can easily find the strains at certain given intervals or portions of the span, as $\frac{1}{10}$th, $\frac{2}{10}$ths, etc. These strains being plotted to scale along the span, we have a *curve* from which we can readily determine the strain at other points.

The strains in the flanges being thus known, we can readily determine the transverse force, or force at right angles to the rib, at any point. This force causes strain in the diagonals, and has simply to be multiplied by the secant of the angle made with it by any diagonal.

As to the effects of temperature, the remarks of Art. 166 do not seem to be substantiated by our results. It would seem that, according to the received formulæ, the strains due to temperature are very great, and that by far the best form of construction for short spans is that in which the arch is hinged at both abutments and crown.

28. Advantage of Arch with fixed Ends for long Spans. We cannot conclude from our results above anything as to the comparative advantages or disadvantages of the arch with fixed ends. Different proportions will give altogether different results. We can only say that for small spans the arch with three hinges is undoubtedly the best construction. The advantages of continuity will be apparent only for long spans where the point of inflexion is distant from the ends by a greater proportion of the span. We have already seen the same to be true of the continuous girder. If we were to judge from comparisons of short spans only, we should be inclined to discredit any great advantage for continuity. If, however, we take longer spans, so as to bring the points of inflection well out, we find a marked saving.*

We had intended to give here a comparison of the strains in a hinged arch with those in the central span of the St. Louis bridge, as given in the *Report* of Capt. Eads *to the Illinois and St. Louis Bridge Co. for May*, 1868.

As this goes to press, however, our attention has been called

---

* Art. 17 of this Appendix.

to an article in the *Trans. of the Am. Soc. of Civil Eng.* for May, 1875, by Mr. S. H. Shreve, which, although written with precisely the opposite intention, seems to prove so clearly the superiority for long spans of the arch without hinges, that it is unnecessary to give a comparison here. We have only to take Mr. Shreve's results and properly interpret them.

Thus, while ostensibly investigating the strains in the centre arch of the St. Louis bridge—an arch which is continuous at crown and fixed at the end—Mr. Shreve uses the formula given in Art. 27 of the Supplement to Chap. XIV., viz., $H = \dfrac{p\,a^2}{2\,h}$.

That is, he considers the arch as *having hinges at both crown and ends*.

Then, supposing the arch to be affected by temperature, he applies the above formula to an arch hinged at crown in lower chord and at ends in upper chord of the same dimensions as the St. Louis bridge. It is hardly necessary to point out here that if the arch *is* really thus hinged, or can be supposed thus hinged, there can be *no* temperature strains. If, however, it is *not* hinged, then the above formula does not apply. The one assumption contradicts the other. The formula $H = \dfrac{p\,a^2}{2\,h}$ can be applied to no arch which is strained by temperature. Such a treatment would seem justified on Mr. Shreve's part in view of the statement of Capt. Eads, that for the greatest rise of temperature above the mean, the lower arch does *all* the duty at crown, and the upper at the ends. If this were accurately so, then Mr. Shreve's results would give the true strains. All that Capt. Eads evidently intended to imply was, that a rise of temperature relieved the upper chord at crown of a great part of its compression and increased that of the lower. It does not by any means follow that the upper chord is *entirely* relieved, under which supposition only can the lower chord be supposed hinged. On the contrary, for an equal *fall* of temperature below the mean, the lower chord is relieved and extra strain brought in the upper chord at crown. If the adjustment were just such that the previous compression in the lower chord should be *exactly* neutralized, then the arch might be considered as hinged at the upper flange and lower ends, and thus Mr. Shreve should *increase* the rise of his arch by the depth, which would decrease greatly his strains. The one supposition is as much justified by the remarks of Capt. Eads, which he quotes, as the other—and neither are correct. Apart, however, from the merits of the controversy, with which we have nothing to do, Mr. Shreve's results are undoubtedly correct for an arch of the same dimensions as the St. Louis—uniformly loaded and hinged at the ends in upper flange and at the crown in lower. If, then, a comparison of these results with those given by Capt. Eads shows them all too large, then, since Capt. Eads' formulæ are, as we have seen, undoubtedly correct, it clearly shows the superiority of the arch without hinges. This is the only legitimate deduction which can be made.

Mr. Shreve's formulæ are undoubtedly as "true as the principles of the lever," and apply, beyond question, to an arch hinged as he supposes. Our formulæ in Art. 27 of the Supplement to Chap. XIV. are also as true as these principles; but to apply correctly even so simple a principle as that of the lever, demands a knowledge of *all* the forces and their points of application. From our formulæ, as we have shown in Art. 34 of the above Supplement, we may easily deduce Capt. Eads', thus proving the accuracy of both. Though the "calculus will not determine the *strains* affecting a truss, whether arched or horizontal," it may nevertheless be exceedingly serviceable in determining the *forces* which act upon the truss—without an accurate knowledge of which the "principle of the lever" can only mislead. This principle, upon which Mr. Shreve lays such stress, is precisely that which we have employed so often in this work, and shown to be of universal application. In Art. 36 of this Appendix we have made use of it, just as Mr. Shreve does, in the calculation of an arch similar to the St. Louis. Our results differ from those he would obtain, simply because we take into account a force and lever arm whose existence he ignores. Mr. Shreve assumes that V and H and the load are all the forces which act, and these are all of which his formula takes account. In common with Capt. Eads, we take in *addition* a moment due to the continuity of the ends, while V and H themselves, by reason of this continuity, have very different values.

Thus, for full load, we have from eq. (81), Art. 34 of Supplement to Chap. XIV.,

$$H = \frac{p\, a^2}{2\, h} \frac{4\, h^2}{45\, g^2 + 4\, h^2},$$

and from eq. (84)

$$M_0 = -\frac{3}{70} p h^2 \frac{4 h^4 + 175 g^2 a^2}{4 h^4 + 45 g^2 h^2}.$$

Thus, instead of $H = \dfrac{p a^2}{2 h}$, as given by Mr. Shreve, we have this into a certain coefficient which is less than unity.

Taking $p a = 936{,}000$ lbs., $g = 6.025$ ft., $a = 257.88$ ft., and $h = 46.65$ ft. for centre line, we have $H = 2{,}178{,}317$ for thrust at crown, instead of $2{,}586{,}184.9$ lbs., as given by Mr. Shreve. This thrust alone would cause, then, $1{,}089{,}158$ lbs. compression in each flange. But due to continuity of ends and crown, we have *also* a moment at crown $M_0 = -6{,}587{,}335$, which being negative causes *tension* in lower flange at crown. Dividing by 12.05, the depth of arch, we have $546{,}666$ lbs. tension, and therefore only $1{,}089{,}158 - 546{,}666$ or $542{,}492$ lbs. resulting compression. This at 27,500 lbs. per square inch, requires 19.72 square inches area, while Mr. Shreve requires in his arch 126.42 square inches area. It is, however, but just to notice, that while this loading (uniform) causes the maximum compression in lower flange at crown for Mr. Shreve's arch, it does not for the arch fixed at ends and continuous at crown.

In this latter case, as we may see at once from the table for M_s of Art. 18, Supplement to Chap. XIV., a load within the centre half anywhere causes tension in the lower flange, and the maximum compression is when the flanks are loaded and this portion is empty. It is with the maxima that the comparison must be made, and as Capt. Eads has, very properly, taken the rolling load into account, it is with these maxima that the comparison *has* been made. From such comparison Mr. Shreve finds that "every member of the two tubes is deficient in area, many containing much less than half the material that is necessary." As his results are correctly calculated for a hinged arch, and Capt. Eads' results are also correct for an arch *without* hinges, we can only conclude—*not* that "the great importance of immediately strengthening the ribs of the St. Louis bridge can no longer be ignored," but rather that, for long spans of small relative rise, the arch without hinges is much preferable and more economical. The case is, indeed, perfectly analogous to that of the continuous girder. Here also we have end moments, and here also for long spans the advantage over the simple girder is marked.

In Mr. Shreve's arch it is, indeed, perfectly true that, "when one segment is loaded, any weight whatever in any other position on the other segment will lessen the tension on the lower arc of the loaded segment." In the arch without hinges the case is altogether different, owing to the influence of the end moments, which Mr. Shreve so persistently ignores.

The two cases have, indeed, nothing whatever in common, and from the strains in one no conclusion whatever can be drawn as to what should be the strains in the other. With the same propriety might one comparing the strains in the same girder fixed at ends and free at ends, as given in Art. 17 of this Appendix, infer that the strains in the first were unduly small. The only legitimate conclusion from such comparison is the one there drawn, viz., that the one in which the strains are least is the one most economical of material. In this respect, and in this only, Mr. Shreve's results are valuable, and we can only thank him for having saved us the labor of making the comparison for ourselves.

As a case in point bearing out our conclusion above, we may instance the *Coblentz bridge*, which, as originally constructed, was continuous at the crown, but pivoted at the ends of the centre line, as in our example, Art. 20. But unlike that example, owing to the length of span being much greater, and the rise and depth much less in proportion, it was found advantageous to *block up* the ends after erection, and thus fix it at the ends.

If Mr. Shreve's deductions are to be believed, this was a very dangerous thing to do; but, as experience has proved, greater rigidity has thereby been secured, and no evil effects have as yet been perceptible. It is, however, quite possible that before thus blocking the ends, the effect of the end moments thus brought into play was duly considered; and in view of the result, it would appear as if they really *had* some influence upon the character and distribution of the strains.

It would seem, therefore, that, for the present at least, the "strengthening" of the arches of the St. Louis bridge by *hinging them* (!) at crown and ends may be safely postponed until it can be satisfactorily shown in what manner, for rise of temperature, the end moments mysteriously disappear, and the previously existing compression, due to load in upper flange at crown and lower at ends, is exactly and entirely neutralized.

Meanwhile it would seem that the St. Louis arch, as constructed, is far superior to the same arch hinged, more economical of material and more rigid, and sanctioned alike by theory and precedent.

www.ingramcontent.com/pod-product-compliance
Lightning Source LLC
Chambersburg PA
CBHW022110300426
44117CB00007B/656